Lecture Notes in Computer Science 9954

Commenced Publication in 1973
Founding and Former Series Editors:
Gerhard Goos, Juris Hartmanis, and Jan van Leeuwen

More information about this series at http://www.springer.com/series/7407

Shunsuke Inenaga · Kunihiko Sadakane
Tetsuya Sakai (Eds.)

String Processing and Information Retrieval

23rd International Symposium, SPIRE 2016
Beppu, Japan, October 18–20, 2016
Proceedings

 Springer

Editors
Shunsuke Inenaga
Informatics
Kyushu University
Fukuoka
Japan

Tetsuya Sakai
Computer Science and Engineering
Waseda University
Tokyo
Japan

Kunihiko Sadakane
Mathematical Informatics
University of Tokyo
Tokyo
Japan

ISSN 0302-9743   ISSN 1611-3349 (electronic)
Lecture Notes in Computer Science
ISBN 978-3-319-46048-2   ISBN 978-3-319-46049-9 (eBook)
DOI 10.1007/978-3-319-46049-9

Library of Congress Control Number: 2016950414

LNCS Sublibrary: SL1 – Theoretical Computer Science and General Issues

Printed on acid-free paper

This Springer imprint is published by Springer Nature
The registered company is Springer International Publishing AG
The registered company address is: Gewerbestrasse 11, 6330 Cham, Switzerland

Preface

This volume contains the papers presented at SPIRE 2016, the 23rd International Symposium on String Processing and Information Retrieval, held October 18–20, 2016 in Beppu, Japan. Following the tradition from previous years, the focus of SPIRE this year was on fundamental studies on string processing and information retrieval, as well as application areas such as bioinformatics, Web mining, and so on.

The call for papers resulted in 46 submissions. Each submitted paper was reviewed by at least three Program Committee members. Based on the thorough reviews and discussions by the Program Committee members and additional subreviewers, the Program Committee decided to accept 25 papers.

The main conference featured three keynote speeches by Kunsoo Park (Seoul National University), Koji Tsuda (University of Tokyo), and David Hawking (Microsoft & Australian National University), together with presentations by authors of the 25 accepted papers. Prior to the main conference, two satellite workshops were held: String Masters in Fukuoka, held October 12–14, 2016 in Fukuoka, and the 11th Workshop on Compression, Text, and Algorithms (WCTA 2016), held on October 17, 2016 in Beppu. String Masters was coordinated by Hideo Bannai, and WCTA was coordinated by Simon J. Puglisi and Yasuo Tabei. WCTA this year featured two keynote speeches by Juha Kärkkäinen (University of Helsinki) and Yoshitaka Yamamoto (University of Yamanashi).

We would like to thank the SPIRE Steering Committee for giving us the opportunity to host this wonderful event. Also, many thanks go to the Program Committee members and the additional subreviewers, for their valuable contribution ensuring the high quality of this conference. We appreciate Springer for their professional publishing work and for sponsoring the Best Paper Award for SPIRE 2016. We finally thank the Local Organizing Team (led by Hideo Bannai) for their effort to run the event smoothly.

October 2016

Shunsuke Inenaga
Kunihiko Sadakane
Tetsuya Sakai

Organization

Program Committee

Leif Azzopardi	University of Glasgow, UK
Philip Bille	Technical University of Denmark, Denmark
Praveen Chandar	University of Delware, USA
Raphael Clifford	University of Bristol, UK
Shane Culpepper	RMIT University, Australia
Zhicheng Dou	Renmin University of China, China
Hui Fang	University of Delaware, USA
Simone Faro	University of Catania, Italy
Johannes Fischer	TU Dortmund, Germany
Sumio Fujita	Yahoo! Japan Research, Japan
Travis Gagie	University of Helsinki, Finland
Pawel Gawrychowski	University of Wroclaw, Poland and University of Haifa, Israel
Simon Gog	Karslruhe Institute of Technology, Germany
Roberto Grossi	Università di Pisa, Italy
Ankur Gupta	Butler University, USA
Wing-Kai Hon	National Tsing Hua University, Taiwan
Shunsuke Inenaga	Kyushu University, Japan
Makoto P. Kato	Kyoto University, Japan
Gregory Kucherov	CNRS/LIGM, France
Moshe Lewenstein	Bar Ilan University, Israel
Yiqun Liu	Tsinghua University, China
Mihai Lupu	Vienna University of Technology, Austria
Florin Manea	Christian-Albrechts-Universität zu Kiel, Germany
Gonzalo Navarro	University of Chile, Chile
Yakov Nekrich	University of Waterloo, Canada
Tadashi Nomoto	National Institute of Japanese Literature, Japan
Iadh Ounis	University of Glasgow, UK
Simon Puglisi	University of Helsinki, Finland
Kunihiko Sadakane	University of Tokyo, Japan
Tetsuya Sakai	Waseda University, Japan
Hiroshi Sakamoto	Kyushu Institute of Technology, Japan
Leena Salmela	University of Helsinki, Finland
Srinivasa Rao Satti	Seoul National University, South Korea
Ruihua Song	Microsoft Research Asia, China
Young-In Song	Wider Planet, South Korea
Kazunari Sugiyama	National University of Singapore, Singapore

Aixin Sun	Nanyang Technological University, Singapore
Wing-Kin Sung	National University of Singapore, Singapore
Julián Urbano	University Carlos III of Madrid, Spain
Sebastiano Vigna	Università degli Studi di Milano, Italy
Takehiro Yamamoto	Kyoto University, Japan

Additional Reviewers

Bingmann, Timo
Bouvel, Mathilde
Chikhi, Rayan
Cicalese, Ferdinando
Conte, Alessio
Farach-Colton, Martin
Fici, Gabriele
Fontaine, Allyx
Frith, Martin
Ganguly, Arnab
I, Tomohiro
Jo, Seungbum

Kempa, Dominik
Kosolobov, Dmitry
Lee, Joo-Young
Liu, Xitong
Mercas, Robert
Ordóñez Pereira, Alberto
Pisanti, Nadia
Rosone, Giovanna
Schmid, Markus L.
Starikovskaya, Tatiana
Thankachan, Sharma V.
Välimäki, Niko

Keynote Speeches

Indexes for Highly Similar Sequences

Kunsoo Park

Department of Computer Science and Engineering, Seoul National University,
Seoul, South Korea
kpark@theory.snu.ac.kr

The 1000 Genomes Project aims at building a database of a thousand individual human genome sequences using a cheap and fast sequencing, called next generation sequencing, and the sequencing of 1092 genomes was announced in 2012. To sequence an individual genome using the next generation sequencing, the individual genome is divided into short segments called reads and they are aligned to the human reference genome. This is possible because an individual genome is more than 99 % identical to the reference genome. This similarity also enables us to store individual genome sequences efficiently.

Recently many indexes have been developed which not only store highly similar sequences efficiently but also support efficient pattern search. To exploit the similarity of the given sequences, most of these indexes use classical compression schemes such as run-length encoding and Lempel-Ziv compression.

We introduce a new index for highly similar sequences, called FM index of alignment. We start by finding common regions and non-common regions of highly similar sequences. We need not find a multiple alignment of non-common regions. Finding common and non-common regions is much easier and simpler than finding a multiple alignment, especially in the next generation sequencing. Then we make a transformed alignment of the given sequences, where gaps in a non-common region are put together into one gap. We define a suffix array of alignment on the transformed alignment, and the FM index of alignment is an FM index of this suffix array of alignment. The FM index of alignment supports the LF mapping and backward search, the key functionalities of the FM index. The FM index of alignment takes less space than other indexes and its pattern search is also fast.

This research was supported by the Bio & Medical Technology Development Program of the NRF funded by the Korean government, MSIP (NRF-2014M3C9A3063541).

Simulation in Information Retrieval: With Particular Reference to Simulation of Test Collections

David Hawking

Microsoft, Canberra, Australia
david.hawking@acm.org

Keywords: Information retrieval · Simulation · Modeling

Simulation has a long history in the field of Information Retrieval. More than 50 years ago, contractors for the US Office of Naval Research (ONR) were working on simulating information storage and retrieval systems.[1]

The purpose of simulation is to predict the behaviour of a system over time, or under conditions in which a real system can't easily be observed. My talk will review four general areas of simulation activity. First is the simulation of entire information retrieval systems, as for example exemplified by Blunt (1965):

> A general time-flow model has been developed that enables a systems engineer to simulate the interactions among personnel, equipment and data at each step in an information processing effort.

and later by Cahoon and McKinley (1996).

A second area is the simulation of behaviour when a person interacts with an information retrieval service, with particular interest in multi-turn interactions. For example user simulation has been used to study implicit feedback systems (White et al., 2004), PubMed browsing strategies (Lin and Smucker, 2007), and query suggestion algorithms (Jiang and He, 2013).

A third area has been little studied – simulating an information retrieval service (in the manner of Kemelen's 1770 Automaton Chess Player) in order to study the behaviour of real users when confronted with a retrieval service which hasn't yet been built.

The final area is that of simulation of test collections. It is an area in which I have been working recently, with my colleagues Bodo Billerbeck, Paul Thomas and Nick Craswell. My talk will include some preliminary results.

As early as 1973, Michael Cooper published a method for generating artificial documents and queries in order to, "evaluate the effect of changes in characteristics of the query and document files on the quantity of material retrieved." More recently, Azzopardi and de Rijke (2006) have studied the automated creation of known-item test collections.

---

[1] "System" used in the Systems Theory sense.

Organizations like Microsoft have a need to develop, tune and experiment with information retrieval services using simulated versions of private or confidential data. Furthermore, there may be a need to predict the performance of a retrieval service when an existing data set is scaled up or altered in some way.

We have been studying how to simulate text corpora and query sets for such purposes. We have studied many different corpora with a wide range of different characteristics. Some of the corpora are readily available to other researchers; others we are unable to share. With accurate simulation models we may be able to share sufficient characteristics of those data sets to enable others to reproduce our results.

The models underpinning our simulations include:

1. Models of the distribution of document lengths.
2. Models of the distribution of word frequencies. (Revisiting Zipf's law.)
3. Models of term dependence.
4. Models of the representation of indexable words.
5. Models of how these change as the corpus grows. (e.g. revisiting the models due to Herdan and Heaps.)

We have implemented a document generator based on these models and software for estimating model parameters from a real corpus. We test the models by running the generator with extracted parameters and comparing various properties of the resulting corpus with those of the original. In addition, we test the growth model by extracting parameters from 1 % samples and simulating a corpus 100 times larger. In early experimentation we have found reasonable agreement between the properties of the real corpus and its scaled-up emulation.

The value gained from a simulation approach depends heavily on the accuracy of the system model, but a highly accurate model may be very complex and may be over-fitted to the extent that it doesn't generalise. We study what is required to achieve high fidelity but also discuss simpler forms of model which may be sufficiently accurate for less demanding requirements.

References

1. Blunt, C.R.: An information retrieval system model. Report of Contract Nonr. 3818(00), ONR (1965). http://www.dtic.mil/dtic/tr/fulltext/u2/623590.pdf
2. Cooper, M.D.: A simulation model of a retrieval system. Inf. Storage Retrieval **9**, 13–32 (1973)

Significant Pattern Mining: Efficient Algorithms and Biomedical Applications

Koji Tsuda

Department of Computational Biology and Medical Sciences, Graduate School
of Frontier Sciences, The University of Tokyo, Kashiwa, Japan

Pattern mining techniques such as itemset mining, sequence mining and graph mining have been applied to a wide range of datasets. To convince biomedical researchers, however, it is necessary to show statistical significance of obtained patterns to prove that the patterns are not likely to emerge from random data. The key concept of significance testing is family-wise error rate, i.e., the probability of at least one pattern is falsely discovered under null hypotheses. In the worst case, FWER grows linearly to the number of all possible patterns. We show that, in reality, FWER grows much slower than the worst case, and it is possible to find significant patterns in biomedical data. The following two properties are exploited to accurately bound FWER and compute small p-value correction factors. (1) Only closed patterns need to be counted. (2) Patterns of low support can be ignored, where the support threshold depends on the Tarone bound. We introduce efficient depth-first search algorithms for discovering all significant patterns and discuss about parallel implementations.

Contents

RLZAP: Relative Lempel-Ziv with Adaptive Pointers

Anthony J. Cox[1], Andrea Farruggia[2], Travis Gagie[3,4(✉)], Simon J. Puglisi[3,4], and Jouni Sirén[5]

[1] Illumina Cambridge Ltd., Cambridge, UK
[2] University of Pisa, Pisa, Italy
a.farruggia@di.unipi.it
[3] Helsinki Institute for Information Technology, Espoo, Finland
[4] University of Helsinki, Helsinki, Finland
travis.gagie@gmail.com, simon.j.puglisi@gmail.com
[5] Wellcome Trust Sanger Institute, Hinxton, UK
jouni.siren@iki.fi

Abstract. Relative Lempel-Ziv (RLZ) is a popular algorithm for compressing databases of genomes from individuals of the same species when fast random access is desired. With Kuruppu et al.'s (SPIRE 2010) original implementation, a reference genome is selected and then the other genomes are greedily parsed into phrases exactly matching substrings of the reference. Deorowicz and Grabowski (*Bioinformatics*, 2011) pointed out that letting each phrase end with a mismatch character usually gives better compression because many of the differences between individuals' genomes are single-nucleotide substitutions. Ferrada et al. (SPIRE 2014) then pointed out that also using relative pointers and run-length compressing them usually gives even better compression. In this paper we generalize Ferrada et al.'s idea to handle well also short insertions, deletions and multi-character substitutions. We show experimentally that our generalization achieves better compression than Ferrada et al.'s implementation with comparable random-access times.

1 Introduction

Next-generation sequencing technologies can quickly and cheaply yield far more genetic data than can fit into an everyday computer's memory, so it is important to find ways to compress it while still supporting fast random access. Often the data is highly repetitive and can thus be compressed very well with LZ77 [1], but then random access is slow. For many applications, however, we need store only a database of genomes from individuals of the same species, which are not only highly repetitive collectively but also but also all very similar to each other.

Supported by the Academy of Finland through grants 258308, 268324, 284598 and 285221 and by the Wellcome Trust grant 098051. Parts of this work were done during the second author's visit to the University of Helsinki and during the third author's visits to Illumina Cambridge Ltd. and the University of A Coruña, Spain.

© Springer International Publishing AG 2016
S. Inenaga et al. (Eds.): SPIRE 2016, LNCS 9954, pp. 1–14, 2016.
DOI: 10.1007/978-3-319-46049-9_1

Kuruppu, Puglisi and Zobel [2] proposed choosing one of the genomes as a reference and then greedily parsing each of the others into phrases exactly matching substrings of that reference. They called their algorithm Relative Lempel-Ziv (RLZ) because it can be viewed as a version of LZ77 that looks for phrase sources only in the reference, which greatly speeds up random access later. (Ziv and Merhav [3] introduced a similar algorithm for estimating the relative entropy of the sources of two sequences.) RLZ is now is popular for compressing not only such genomic databases but also other kinds of repetitive datasets; see, e.g., [4,5]. Deorowicz and Grabowski [6] pointed out that letting each phrase end with a mismatch character usually gives better compression on genomic databases because many of the differences between individuals' genomes are single-nucleotide substitutions, and gave a new implementation with this optimization. Ferrada, Gagie, Gog and Puglisi [7] then pointed out that often the current phrase's source ends two characters before the next phrase's source starts, so the distances between the phrases' starting positions and their sources' starting positions are the same. They showed that using relative pointers and run-length compressing them usually gives even better compression on genomic databases.

In this paper we generalize Ferrada et al.'s idea to handle well also short insertions, deletions and substitutions. In the Sect. 2 we review in detail RLZ and Deorowicz and Grabowski's and Ferrada et al.'s optimizations. We also discuss how RLZ can be used to build relative data structures and why the optimizations that work to better compress genomic databases fail for this application. In Sect. 3 we explain the design and implementation of RLZ with adaptive pointers (RLZAP): in short, after parsing each phrase, we look ahead several characters to see if we can start a new phrase with a similar relative pointer; if so, we store the intervening characters as mismatch characters and store the new relative pointer encoded as its difference from the previous one. We present our experimental results in Sect. 4, showing that RLZAP achieves better compression than Ferrada et al.'s implementation with comparable random-access times. Our implementation and datasets are available for download from http://github.com/farruggia/rlzap.

2   Preliminaries

In this section we discuss the previous work that is the basis and motivation for this paper. We first review in greater detail Kuruppu et al.'s implementation of RLZ and Deorowicz and Grabowski's and Ferrada et al.'s optimizations. We then quickly summarize the new field of *relative data structures* — which concerns when and how we can use compress a new instance of a data structure, using an instance we already have for a similar dataset — and explain how it uses RLZ and why it needs a generalization of Deorowicz and Grabowski's and Ferrada et al.'s optimizations.

2.1 RLZ

To compute the RLZ parse of a string $S[1; n]$ with respect to a reference string R using Kuruppu et al.'s implementation, we greedily parse S from left to right into phrases

$$S[p_1 = 1; p_1 + \ell_1 - 1]$$
$$S[p_2 = p_1 + \ell_1; p_2 + \ell_2 - 1]$$
$$\vdots$$
$$S[p_t = p_{t-1} + \ell_{t-1}; p_t + \ell_t - 1 = n]$$

such that each $S[p_i; p_i + \ell_i - 1]$ exactly matches some substring $R[q_i; q_i + \ell_i - 1]$ of R — called the ith phrase's *source* — for $1 \leq i \leq t$, but $S[p_i; p_i + \ell_i]$ does not exactly match any substring in R for $1 \leq i \leq t - 1$. For simplicity, we assume R contains every distinct character in S, so the parse is well-defined.

Suppose we have constant-time random access to R. To support constant-time random access to S, we store an array $Q[1; t]$ containing the starting positions of the phrases' sources, and a compressed bitvector $B[1; n]$ with constant query time (see, e.g., [8] for a discussion) and 1 s marking the first character of each phrase. Given a position j between 1 and n, we can compute in constant time

$$S[j] = R\left[Q[B.\mathsf{rank}(j)] + j - B.\mathsf{select}(B.\mathsf{rank}(j))\right].$$

If there are few phrases then Q is small and B is sparse, so we use little space.

For example, if

$$R = \mathsf{ACATCATTCGAGGACAGGTATAGCTACAGTTAGAA}$$
$$S = \mathsf{ACATGATTCGACGACAGGTACTAGCTACAGTAGAA}$$

then we parse S into

$$\mathsf{ACAT, GA, TTCGA, CGA, CAGGTA, CTA, GCTACAGT, AGAA,}$$

and store

$$Q = 1, 10, 7, 9, 15, 24, 23, 32$$
$$B = 10001010000100100000100100000001000.$$

To compute $S[25]$, we compute $B.\mathsf{rank}(25) = 7$ and $B.\mathsf{select}(7) = 24$, which tell us that $S[25]$ is $25 - 24 = 1$ character after the initial character in the 7th phrase. Since $Q[7] = 23$, we look up $S[25] = R[24] = \mathsf{C}$.

2.2 GDC

Deorowicz and Grabowski [6] pointed out that with Kuruppu et al.'s implementation of RLZ, single-character substitutions usually cause two phrase breaks:

e.g., in our example $S[1;11] = \mathsf{ACATGATTCGA}$ is split into three phrases, even though the only difference between it and $R[1;11]$ is that $S[5] = \mathsf{G}$ and $R[5] = \mathsf{C}$. They proposed another implementation, called the Genome Differential Compressor (GDC), that lets each phrase end with a mismatch character — as the original version of LZ77 does — so single-character substitutions usually cause only one phrase break. Since many of the differences between individuals' DNA are single-nucleotide substitutions, GDC usually compresses genomic databases better than Kuruppu et al.'s implementation.

Specifically, with GDC we parse S from left to right into phrases $S[p_1; p_1 + \ell_1], S[p_2 = p_1 + \ell_1 + 1; p_2 + \ell_2], \ldots, S[p_t = p_{t-1} + \ell_{t-1} + 1; p_t + \ell_t = n]$ such that each $S[p_i; p_i + \ell_i - 1]$ exactly matches some substring $R[q_i; q_i + \ell_i - 1]$ of R — again called the ith phrase's source — for $1 \leq i \leq t$, but $S[p_i; p_i + \ell_i]$ does not exactly match any substring in R, for $1 \leq i \leq t - 1$.

Suppose again that we have constant-time random access to R. To support constant-time random access to S, we store an array $Q[1;t]$ containing the starting positions of the phrases' sources, an array $M[1;t]$ containing the last character of each phrase, and a compressed bitvector $B[1;n]$ with constant query time and 1 s marking the last character of each phrase. Given a position j between 1 and n, we can compute in constant time

$$S[j] = \begin{cases} M[B.\mathsf{rank}(j)] & \text{if } B[j] = 1, \\ R[Q[B.\mathsf{rank}(j) + 1] + j - B.\mathsf{select}(B.\mathsf{rank}(j)) - 1] & \text{otherwise,} \end{cases}$$

assuming $B.\mathsf{select}(0) = 0$.

In our example, we parse S into

$$\mathsf{ACATG}, \mathsf{ATTCGAC}, \mathsf{GACAGGTAC}, \mathsf{TAGCTACAGT}, \mathsf{AGAA},$$

and store

$$Q = 1, 6, 13, 21, 32$$
$$M = \mathsf{GCCTA}$$
$$B = 00001000000100000000100000000010001.$$

To compute $S[25]$, we compute $B[25] = 0$, $B.\mathsf{rank}(25) = 3$ and $B.\mathsf{select}(3) = 21$, which tell us that $S[25]$ is $25 - 21 - 1 = 3$ characters after the initial character in the 4th phrase. Since $Q[4] = 21$, we look up $S[25] = R[24] = \mathsf{C}$.

2.3 Relative Pointers

Ferrada, Gagie, Gog and Puglisi [7] pointed out that after a single-character substitution, the source of the next phrase in GDC's parse often starts two characters after the end of the source of the current phrase: e.g., in our example the source for $S[1;5] = \mathsf{ACATG}$ is $R[1;4] = \mathsf{ACAT}$ and the source for $S[6;12] = \mathsf{ATTCGAC}$ is $R[6;11] = \mathsf{ATTCGA}$. This means the distances between the phrases' starting positions and their sources' starting positions are the same. They proposed an implementation of RLZ that parses S like GDC does but keeps a relative pointer, instead of the explicit pointer, and stores the list of those relative pointers run-length compressed. Since the relative pointers usually do not change after single-nucleotide substitutions, RLZ with relative pointers usually gives even better compression than GDC on genomic databases. (We note that Deorowicz, Danek and Niemiec [9] recently proposed a new version of GDC, called GDC2, that has improved compression but does not support fast random access.)

Suppose again that we have constant-time random access to R. To support constant-time random access to S, we store the array M of mismatch characters and the bitvector B as with GDC. Instead of storing Q, we build an array $D[1;t]$ containing, for each phrase, the difference $q_i - p_i$ between its source's starting position and its own starting position. We store D run-length compressed: i.e., we partition it into maximal consecutive subsequences of equal values, store an array V containing one copy of the value in each subsequence, and a bitvector $L[1;t]$ with constant query time and 1 s marking the first value of each subsequence. Given k between 1 and t, we can compute in constant time

$$D[k] = V[L.\mathsf{rank}(k)].$$

Given a position j between 1 and n, we can compute in constant time

$$S[j] = \begin{cases} M[B.\mathsf{rank}(j)] & \text{if } B[j] = 1, \\ R[D[B.\mathsf{rank}(j)+1]+j] & \text{otherwise.} \end{cases}$$

In our example, we again parse S into

$$\mathsf{ACATG, ATTCGAC, GACAGGTAC, TAGCTACAGT, AGAA,}$$

and store

$$M = \mathsf{GCCTA}$$
$$B = 0000100000010000000100000000010001,$$

but now we store $D = 0, 0, 0, -1, 0$ as $V = 0, -1, 0$ and $L = 10011$ instead of storing Q. To compute $S[25]$, we again compute $B[25] = 0$ and $B.\mathsf{rank}(25) = 3$, which tell us that $S[25]$ is in the 4th phrase. We add 25 to the 4th relative pointer $D[4] = V[L.\mathsf{rank}(4)] = -1$ and obtain 24, so $S[25] = R[24]$.

A single-character insertion or deletion usually causes only a single phrase break in the parse but a new run in D, with the values in the run being one less or one more than the values in the previous run. In our example, the insertion of $S[21] = \mathsf{C}$ causes the value to decrement to -1, and the deletion of $R[26] = \mathsf{T}$ (or, equivalently, of $R[27] = \mathsf{T}$) causes the value to increment to 0 again. In larger examples, where the values of the relative pointers are often a significant fraction of n, it seems wasteful to store a new value uncompressed when it differs only by 1 from the previous value.

For example, suppose R and S are thousands of characters long,

$$R[1783; 1817] = \ldots \mathsf{ACATCATTCGAGGACAGGTATAGCTACAGTTAGAA} \ldots$$
$$S[2009; 2043] = \ldots \mathsf{ACATGATTCGACGACAGGTACTAGCTACAGTAGAA} \ldots$$

and GDC still parses $S[2009; 2043]$ into the same phrases as before, with their sources in $R[1783; 1817]$. The relative pointers for those phrases are $-136, -136, -136, -137, -136$, so we store $-136, -137, -136$ for them in V, which takes at least a couple of dozen bits without further compression.

2.4   Relative Data Structures

As mentioned in Sect. 1, the new field of relative data structures concerns when and how we can use compress a new instance of a data structure, using an instance we already have for a similar dataset. Suppose we have a basic FM-index [10] for R — i.e., a rank data structure over the Burrows-Wheeler Transform (BWT) [11] of R, without a suffix-array sample — and we want to use it to build a very compact basic FM-index for S. Since R and S are very similar, it is not surprising that their BWTs are also fairly similar:

$$\mathsf{BWT}(R) = \mathsf{AAGGT\$TTGCCTCCAAATTGAGCAAAGACTAGATGA}$$

$$\mathsf{BWT}(S) = \mathsf{AAGGT\$GTTTCCCGAAAATGAACCTAAGACGGCTAA}.$$

Belazzougui, Gog, Gagie, Manzini and Sirén [12] (see also [13]) showed how we can implement such a relative FM-index for S by choosing a common subsequence of the two BWTs and then storing bitvectors marking the characters not in that common subsequence, and rank data structures over those characters. They also showed how to build a relative suffix-array sample to obtain a fully-functional relative FM-index for S, but reviewing that is beyond the scope of this paper.

An alternative to Belazzougui et al.'s basic approach is to compute the RLZ parse of $\mathsf{BWT}(S)$ with respect to $\mathsf{BWT}(R)$ and then store the rank for each character just before the beginning of each phrase. We can then answer a rank query $\mathsf{BWT}(S).\mathsf{rank}_X(j)$ by finding the beginning $\mathsf{BWT}(S)[p]$ of the phrase containing $\mathsf{BWT}(S)[j]$ and the beginning $\mathsf{BWT}(R)[q]$ of that phrase's source, then computing

$$\mathsf{BWT}(S).\mathsf{rank}_X(p-1) + \mathsf{BWT}(R).\mathsf{rank}_X(q+j-p) - \mathsf{BWT}(R).\mathsf{rank}_X(q-1).$$

Unfortunately, single-character substitutions between R and S usually cause insertions, deletions and multi-character substitutions between $\mathsf{BWT}(R)$ and $\mathsf{BWT}(S)$, so Deorowicz and Grabowski's and Ferrada et al.'s optimizations no longer help us, even when the underlying strings are individuals' genomes. On the other hand, on average those insertions, deletions and multi-character substitutions are fairly few and short [14], so there is still hope that those optimized parsing algorithms can be generalized and applied to make this alternative practical.

Our immediate concern is with a recent implementation of relative suffix trees [15], which uses relative FM-indexes and relatively-compressed longest-common-prefix (LCP) arrays. Deorowicz and Grabowski's and Ferrada et al.'s optimizations also fail when we try to compress the LCP arrays, and when we use

Kuruppu et al.'s implementation of RLZ the arrays take a substantial fraction of the total space. In our example, however,

$$\mathsf{LCP}(R) = 0,1,1,4,3,1,2,2,3,2,1,2,2,0,3,2,3,1,1,0,2,2,1,1,2,1,2,0,2,3,2,1,2,1,2$$
$$\mathsf{LCP}(S) = 0,1,1,4,3,2,2,1,2,2,2,1,2,0,3,2,1,4,1,3,0,2,3,2,1,1,1,3,0,3,2,3,1,1,1$$

are quite similar: e.g., they have a common subsequence of length 26, almost three quarters of their individual lengths. LCP values tend to grow at least logarithmically with the size of the strings, so good compression becomes more important.

3   Adaptive Pointers

We generalize Ferrada et al.'s optimization to handle short insertions, deletions and substitutions by introducing *adaptive pointers* and by allowing more than one mismatch character at the end of each phrase. An adaptive pointer is represented as the difference from the previous non-adaptive pointer. Henceforth we say a phrase is *adaptive* if its pointer is adaptive, and *explicit* otherwise. In this section we first describe our parsing strategy and then describe how we can support fast random access.

3.1   Parsing

The parsing strategy is a generalization of the Greedy approach for adaptive phrases. The parser first compute the *matching statistics* between input S and reference R: for each suffix $S[i; n]$ of S, a suffix of R with the longest LCP with $S[i]$ is found; let $R[k; m]$ be that suffix. Let MatchPtr(i) be the relative pointer $k - i$ and MatchLen(i) be the length of the LCP between the two suffixes $S[i; n]$ and $R[k; m]$.

Parsing scans S from left to right, in one pass. Let us assume S has already been parsed up to a position i, and let us assume the most recent explicit phrase starts at position h. The parser first tries to find an adaptive phrase (*adaptive step*); if it fails, looks for an explicit phrase (*explicit step*). Specifically:

1. *adaptive step*: the parser checks, for the current position i if (i) the relative pointer MatchPtr(i) can be represented as an adaptive pointer, that is, if the differential MatchPtr(i) -MatchPtr(j) can be represented as a signed binary integer of at most DeltaBits bits, and (ii) if it is convenient to start a new adaptive phrase instead of representing literals as they are, that is, whether MatchLen(i) · $\log \sigma$ > DeltaBits, where σ is the alphabet size. The parser outputs the adaptive phrase and advances MatchLen(i) positions if both conditions are satisfied; otherwise, it looks for the leftmost position k in range $i + 1$ up to $i +$ LookAhead where both conditions are satisfied. If it finds such position k, the parser outputs literals $S[i; k - 1]$ and an adaptive phrase; otherwise, it goes to step 2.

2. *explicit step*: in this step the parser goes back to position i and scans forward until it has found a match starting at position $k \geq i$ where at least one of these two conditions is satisfied: (i) match length MatchLen(k) is greater than a parameter Explicit$_{Len}$; (ii) the match, if selected as explicit phrase, is followed by an adaptive phrase. It then outputs a literal range $S[i; k-1]$ and the explicit phrase found.

The purpose of the two conditions on the explicit phrase is to avoid having spurious explicit phrases which are not associated to a meaningfully aligned substrings.

It is important to notice that our data structure logically represents an adaptive/explicit phrase followed by a literal run as a single phrase: for example, an adaptive phrase of length 5 followed by a literal sequence GAT is represented as an adaptive phrase of length 8 with the last 3 symbols represented as literals.

3.2   Representation

In order to support fast random access to S, we deploy several data structures, which can be grouped into two sets with different purposes:

1. **Storing the parsing:** a set of data structures mapping any position i to some useful information about the phrase P_i containing $S[i]$, that is: (i) the position Start(i) of the first symbol in P_i; (ii) P_i's length Len(i); (iii) its relative pointer Rel(i); (iv) the number of phrases Prev(i) preceding P_i in the parsing, and (v) the number of *explicit* phrases Abs(i) $\leq$ Prev(i) preceding P_i.
2. **Storing the literals:** a set of data structures which, given a position i and the information about phrase P_i, tells whether $S[i]$ is a literal in the parsing and, if this is the case, returns $S[i]$.

Here we provide a detailed illustration of these data structures.

Storing the Parsing. The parsing is represented by storing two bitvectors. The first bitvector P has $|S|$ entries, marking with a 1 characters in S at the beginning of a new phrase in the parsing. The second bitvector E has m entries, one for every phrases in the parsing, and marks every explicit phrase in the parsing with a 1, otherwise 0. A rank/select data structure is built on top of P, and a rank data structure on top of E. In this way, given i we can efficiently compute the phrase index Prev(i) as P.rank(i), the explicit phrase index Abs(i) as E.rank(p_i) and the phrase beginning Start(i) as P.select(p_i).

Experimentally, bitvector P is sparse, while E is usually dense. Bitvector P can be represented with any efficient implementation for sparse bitvectors; our implementation, detailed in Sect. 4, employs the Elias-Fano based SDarrays data structure of Okanohara and Sadakane [16], which requires $m \log \frac{|S|}{m} + O(m)$ bits and supports rank in $O(\log \frac{|S|}{m})$ time and select in constant time. Bitvector E is represented plainly, taking m bits, with any $o(m)$-space $O(1)$-time rank implementation on top of it [16,17]. In particular, it is interesting to notice that only one rank query is needed for extracting an unbounded number of consecutive symbols from E, since each starting position of consecutive phrases can be accessed with a single select query, which has very efficient implementations on sparse bitvectors.

Both explicit and relative pointers are stored in tables A and R, respectively. These integers are stored in binary, and so not compressed using statistical encoding, because this would prevent efficient random access to the sequence. Each explicit and relative pointer takes thus $\lceil \log n \rceil$ and DeltaBits bits of space, respectively. To compute Rel(i), we first check if the phrase is explicit by checking if $E[\mathsf{Prev}(i)]$ is set to one; if it is, then $\mathsf{Rel}(i) = A[\mathsf{Abs}(i)]$, otherwise it is $\mathsf{Rel}(i) = A[\mathsf{Abs}(i)] + R[\mathsf{Prev}(i) - \mathsf{Abs}(i)]$.

Storing Literals. Literals are extracted as follows. Let us assume we are interested in accessing $S[i]$, which is contained in phrase P_j. First, it is determined whether $S[i]$ is a literal or not. Since literals in a phrase are grouped at the end of the phrase itself, it is sufficient to store, for every phrase P_k in the parsing, the number of literals Lits(k) at its end. Thus, knowing the starting position Start(j) and length Len(j) of phrase P_j, symbol $S[i]$ is a literal if and only if $i \geq \mathsf{Start}(j) + \mathsf{Len}(j) - \mathsf{Lits}(j)$.

All literals are stored in a table L, where $L[k]$ is the k-th literal found by scanning the parsing from left to right. How we represent L depends on the kind of data we are dealing with. In our experiments, described in Sect. 4, we consider differentially-encoded LCP arrays and DNA. For DLCP values, L simply stores all values using minimal binary codes. For DNA values, a more refined implementation (which we describe in a later paragraph) is needed to use less than 3 bits on average for each symbol. So, in order to display the literal $S[i]$, we need a way to compute its index in L, which is equal to Start(j) – Len(j) – Lits(k) plus the prefix sum $\sum_{k=1}^{j-1} \mathsf{Lits}(k)$. In the following paragraph we detail two solutions for efficiently storing Lits(k) values and computing prefix sums.

Storing Literal Counts. Here we detail a simple and fast data structure for storing Lits(−) values and for computing prefix sums on them. The basic idea is to store Lits(−) values explicitly, and accelerate prefix sums by storing the prefix sum of some regularly sampled positions. To provide fast random access, the maximum number of literals in a phrase is limited to $2^{\mathsf{MaxLit}} - 1$, where MaxLit is a parameter chosen at construction time. Every value Lits(−) is thus collected in a table L, stored using MaxLit bits each. Since each phrase cannot have more than $2^{\mathsf{MaxLit}} - 1$ literals, we split each run of more than $2^{\mathsf{MaxLit}} - 1$ literals into the minimal number of phrases which do meet the limit. In order to speed-up the prefix sum computation on L, we sample one every SampleInt positions and store prefix sums of sampled positions into a table Prefix. To accelerate further prefix sum computation, we employ a 256-entries table Byte_Σ which maps any sequence of 8/MaxLit elements into their sum. Here, we constrain MaxLit as a power of two not greater than 8 (that is, either 1, 2, 4 or 8) and SampleInt as a multiple of 8/MaxLit. In this way we can compute the prefix sum by just one look-up into Prefix and at most $\frac{\mathsf{SampleInt}}{8/\mathsf{MaxLit}}$ queries into Byte_Σ. Using Byte_Σ is faster than summing elements in L because it replaces costly bitshift operations with efficient byte-accesses to L. This is because 8/MaxLit elements of L fit into one byte; moreover, those bytes are aligned to byte-boundaries because SampleInt is a multiple of 8/MaxLit, which in turn implies that the sampling interval spans entire bytes of L.

Storing DNA Literals. Every literal is collected into a table J, where each element is represented using a fixed number of bits. For the DNA sequences we consider in our experiments, this would imply using 3 bits, since the alphabet is $\{A, C, G, T, N\}$. However, since symbols N occur less often than the others, it is more convenient to handle those as exceptions, so other literals can be stored in just 2 bits. In particular, every N in table J is stored as one of the other four symbols in the alphabet (say, A) and a bit-vector Exc marks every position in J which corresponds to an N. Experimentally, bitvector Exc is sparse and the 1 are usually clustered together into a few regions. In order to reduce the space needed to store Exc, we designed a simple bit-vector implementation to exploit this fact. In our design, Exc is divided into equal-sized chunks of length C. A bitvector Chunk marks those chunks which contain at least one bit set to 1. Marked chunks of Exc are collected into a vector V. Because of the clustering property we just mentioned, most of the chunks are not marked, but marked chunks are locally dense. Because of this, bitvector Chunk is implemented using a sparse representation, while each chunk employs a dense representation. Good experimental values for C are around $16 - 32$ bits, so each chunk is represented with a fixed-width integer. In order to check whether a position i is marked in Exc, we first check if chunk $c = \lfloor i/C \rfloor$ is marked in Chunk. If it is marked, we compute Chunk.rank(c) to get the index of the marked chunk in V.

4   Experiments

We implemented RLZAP in C++11 with bitvectors from Gog et al.'s sdsl library (https://github.com/simongog/sdsl-lite), and compiled it with gcc version 4.8.4 with flags -O3, -march=native, -ffast-math, -funroll-loops and -DNDEBUG. We performed our experiments on a computer with a 6-core Intel Xeon X5670 clocked at 2.93 GHz, 40 GiB of DDR3 ram clocked at 1333 MHz and running Ubuntu 14.04. As noted in Sect. 1, our code is available at http://github.com/farruggia/rlzap.

We performed our experiments on the following four datasets:

- Cere: the genomes of 39 strains of the *Saccharomyces cerevisiae* yeast;
- E. Coli: the genomes of 33 strains of the *Escherichia coli* bacteria;
- Para: the genomes of 36 strains of the *Saccharomyces paradoxus* yeast;
- DLCP: differentially-encoded LCP arrays for three human genomes, with 32-bit entries.

These files are available from http://acube.di.unipi.it/rlzap-dataset.

For each dataset we chose the file (i.e., the single genome or DLCP array) with the lexicographically largest name to be the *reference*, and made the concatenation of the other files the *target*. We then compressed the target against the reference with Ferrada et al.'s optimization of RLZ — which reflects the current state of the art, as explained in Sect. 1 — and with RLZAP. For the DNA files (i.e., Cere, E. Coli and Para) we used LookAhead = 32, $\text{Explicit}_{\text{Len}}$ = 32, DeltaBits = 2 MaxLit = 4 and SampleInt = 64, while for DLCP we used LookAhead = 8, $\text{Explicit}_{\text{Len}}$ = 4, DeltaBits = 4, MaxLit = 2 and SampleInt = 64. We chose these parameters during a calibration step performed on a different dataset, which we will describe in the full version of this paper.

Table 1 shows the compression achieved by RLZ and RLZAP. (We note that, since the DNA datasets are each over an alphabet of $\{A, C, G, T, N\}$ and Ns are rare, the targets for those datasets can be compressed to about a quarter of their size even with only, e.g., Huffman coding.) Notice RLZAP consistently achieves better compression than RLZ, with its space usage ranging from about 17 % less for Cere to about 32 % less for DLCP.

Table 1. Compression achieved by RLZ and RLZAP. For each dataset we report in MiB (2^{20} bytes) the size of the reference and the size of the target uncompressed and compressed with each method

Dataset	Reference	Target	Compressed target size (MiB)	
	Size (MiB)	Size (MiB)	RLZ	RLZAP
Cere	12.0	451	9.16	7.61
E. Coli	4.8	152	30.47	21.51
Para	11.3	398	15.57	10.49
DLCP	11,582	23,392	1,745.33	1,173.81

Table 2 shows extraction times for RLZ- and RLZAP-compressed targets. RLZAP is noticeably slower than RLZ for DNA, while it is slightly faster for the DLCP dataset when at least four characters are extracted. We believe RLZAP outperforms RLZ on the DLCP because its parsing is generally more cache-friendly: our measurements indicate that on this dataset RLZAP causes about 36 % fewer L2 and L3 cache misses than RLZ. Even for DNA, RLZAP is still fast in absolute terms, taking just tens of nanoseconds per character when extracting at least four characters.

On DNA files, RLZAP achieves better compression at the cost of slightly longer extraction times. On differentially-encoded LCP arrays, RLZAP outperforms RLZ in all regards, except for a slight slowdown when extraction substrings of length less than 4. That is, RLZAP is competitive with the state of the art even for compressing DNA and, as we hoped, advances it for relative data structures. Our next step will be to integrate it into the implementation of relative suffix trees mentioned in Subsect. 2.4.

Table 2. Extraction times per character from RLZ- and RLZAP-compressed targets. For each file in each target, we compute the mean extraction time for $2^{24}/\ell$ pseudorandomly chosen substrings; take the mean of these means

Dataset	Algorithm	Mean extraction time per character (ns)					
		1	4	16	64	256	1024
Cere	RLZ	234	59	16.4	4.4	1.47	0.55
	RLZAP	274	70	19.5	5.7	2.34	1.26
E. Coli	RLZ	225	62	20.1	7.7	4.34	3.34
	RLZAP	322	91	31.3	15.3	10.78	9.47
Para	RLZ	235	59	17.2	5.2	2.23	1.03
	RLZAP	284	74	21.2	6.9	3.09	2.26
DLCP	RLZ	756	238	61.5	20.5	9.00	6.00
	RLZAP	826	212	57.5	19.0	8.00	4.50

5   Future Work

In the near future we plan to perform more experiments to tune RLZAP and discover its limitations. For example, we will test it on the balanced-parentheses representations of suffix trees' shapes, which are an alternative to LCP arrays, and on the BWTs in relative FM-indexes. We also plan to investigate how to minimize the bit-complexity of our parsing — i.e., how to choose the phrases and sources so as to minimize the number of bits in our representation — building on the results by Farruggia, Ferragina and Venturini [18,19] about minimizing the bit-complexity of LZ77.

RLZAP can be viewed as a bounded-lookahead greedy heuristic for computing a glocal alignment [20] or S against R. Such an alignment allows for genetic recombination events, in which potentially large sections of DNA are rearranged. We note that standard heuristics for speeding up edit-distance computation and global alignment do not work here, because even a low-cost path through the dynamic programming matrix can occasionally jump arbitrarily far from the diagonal. RLZAP runs in linear time, which is attractive, but it may produce a suboptimal alignment — i.e., it is not an admissible heuristic. In the longer term, we are interested in finding practical admissible heuristics.

Apart from the direct biological interest of computing optimal or nearly optimal glocal alignments, they can also help us design more data structures. For example, consider the problem of representing the mapping between orthologous genes in several species' genomes; see, e.g., [21]. Given two genomes' indices and the position of a base-pair in one of those genomes, we would like to return quickly the positions of all corresponding base-pairs in the other genome. Only a few base-pairs correspond to two base-pairs in another genome and, ignoring those, this problem reduces to representing compressed permutations. A feature of these permutations is that base-pairs tend to be mapped in blocks, possibly with some slight reordering within each block. We can extract this block structure by computing a glocal alignment, either between the genomes or between the permutation and its inverse.

References

1. Ziv, J., Lempel, A.: A universal algorithm for sequential data compression. IEEE Trans. Inf. Theor. **23**, 337–343 (1977)
2. Kuruppu, S., Puglisi, S.J., Zobel, J.: Relative Lempel-Ziv compression of genomes for large-scale storage and retrieval. In: Chavez, E., Lonardi, S. (eds.) SPIRE 2010. LNCS, vol. 6393, pp. 201–206. Springer, Heidelberg (2010)
3. Ziv, J., Merhav, N.: A measure of relative entropy between individual sequences with application to universal classification. IEEE Trans. Inf. Theor. **39**, 1270–1279 (1993)
4. Hoobin, C., Puglisi, S.J., Zobel, J.: Sample selection for dictionary-based corpus compression. In: Proceedings of SIGIR, pp. 1137–1138 (2011)
5. Hoobin, C., Puglisi, S.J., Zobel, J.: Relative Lempel-Ziv factorization for efficient storage and retrieval of web collections. Proc. VLDB **5**, 265–273 (2011)
6. Deorowicz, S., Grabowski, S.: Robust relative compression of genomes with random access. Bioinformatics **27**, 2979–2986 (2011)
7. Ferrada, H., Gagie, T., Gog, S., Puglisi, S.J.: Relative Lempel-Ziv with constant-time random access. In: Moura, E., Crochemore, M. (eds.) SPIRE 2014. LNCS, vol. 8799, pp. 13–17. Springer, Heidelberg (2014)
8. Kärkkäinen, J., Kempa, D., Puglisi, S.J.: Hybrid compression of bitvectors for the FM-index. In: Proceedings of DCC, pp. 302–311 (2014)
9. Deorowicz, S., Danek, A., Niemiec, M.: GDC2: compression of large collections of genomes. Sci. Rep. **5**, 1–12 (2015)
10. Ferragina, P., Manzini, G.: Indexing compressed text. J. ACM **52**, 552–581 (2005)
11. Burrows, M., Wheeler, D.J.: A block sorting lossless data compression algorithm. Technical report 124, Digital Equipment Corporation (1994)
12. Belazzougui, D., Gagie, T., Gog, S., Manzini, G., Sirén, J.: Relative FM-indexes. In: Moura, E., Crochemore, M. (eds.) SPIRE 2014. LNCS, vol. 8799, pp. 52–64. Springer, Heidelberg (2014)
13. Boucher, C., Bowe, A., Gagie, T., Manzini, G., Sirén, J.: Relative select. In: Iliopoulos, C., Puglisi, S., Yilmaz, E. (eds.) SPIRE 2015. LNCS, vol. 9309, pp. 149–155. Springer, Heidelberg (2015)
14. Léonard, M., Mouchard, L., Salson, M.: On the number of elements to reorder when updating a suffix array. J. Discrete Algorithms **11**, 87–99 (2012)
15. Gagie, T., Navarro, G., Puglisi, S.J., Sirén, J.: Relative compressed suffix trees. Technical report 1508.02550 (2015). arxiv.org
16. Okanohara, D., Sadakane, K.: Practical entropy-compressed rank/select dictionary. In: Proceedings of ALENEX (2007)
17. Raman, R., Raman, V., Satti, S.R.: Succinct indexable dictionaries with applications to encoding k-ary trees, prefix sums and multisets. ACM Trans. Algorithms **3**, 43 (2007)
18. Farruggia, A., Ferragina, P., Venturini, R.: Bicriteria data compression. In: Proceedings of SODA, pp. 1582–1595 (2014)
19. Farruggia, A., Ferragina, P., Venturini, R.: Bicriteria data compression: efficient and usable. In: Schulz, A.S., Wagner, D. (eds.) ESA 2014. LNCS, vol. 8737, pp. 406–417. Springer, Heidelberg (2014)

20. Brudno, M., Malde, S., Poliakov, A., Do, C.B., Couronne, O., Dubchak, I., Batzoglou, S.: Glocal alignment: finding rearrangements during alignment. In: Proceedings of ISMB, pp. 54–62 (2003)
21. Kubincová, P.: Mapping between genomes. Bachelor thesis, Comenius University, Slovakia Supervised by Broňa Brejová (2014)

A Linear-Space Algorithm for the Substring Constrained Alignment Problem

Yoshifumi Sakai[⊠]

Graduate School of Agricultural Science, Tohoku University,
1-1, Amamiyamachi, Tsutsumidori, Aobaku, Sendai 981-8555, Japan
sakai@biochem.tohoku.ac.jp

Abstract. In a string similarity metric adopting affine gap penalties, we propose a quadratic-time, linear-space algorithm for the following constrained string alignment problem. The input of the problem is a pair of strings to be aligned and a pattern given as a string. Let an occurrence of the pattern in a string be a minimal substring of the string that is most similar to the pattern. Then, the output of the problem is a highest-scoring alignment of the pair of strings that matches an occurrence of the pattern in one string and an occurrence of the pattern in the other, where the score of the alignment excludes the similarity between the matched occurrences of the pattern. This problem may arise when we know that each of the strings has exactly one meaningful occurrence of the pattern and want to determine a putative pair of such occurrences based on homology of the strings.

1 Introduction

Constructing a highest-scoring alignment is a common way to analyze how two strings are similar to each other [7], because it is well known that, using the dynamic programming technique, we can obtain such an alignment of an arbitrary m-length string A and an arbitrary n-length string B in $O(mn)$ time [10]. As a more appropriate analysis of the similarity in the case where we know that a common pattern string P occurs both in A and B and that these occurrences should be matched in the alignment, Tsai [12] proposed the constrained longest common subsequence (LCS) problem. This problem consists of finding an arbitrary LCS containing P as a subsequence, where an LCS can be thought of as a highest-scoring alignment in a certain simple similarity metric. Chin et al. [4] showed that this problem is solvable in $O(mnr)$ time and $O(nr)$ space, where r is the length of P and $m \geq n \geq r$. Recently, as one of the generalized constrained LCS problems, Chen and Chao [2] proposed the STR-IC-LCS problem, which consists of finding an arbitrary LCS of A and B that contains P as a substring, instead of as a subsequence. Deorowicz [5] showed that this problem is solvable in $O(mn)$ time and $O(mn)$ space. The difference between the alignments found in these problems is whether the score of the alignment takes the similarity between the matched occurrences of P in X and Y into account or not. The STR-IC-LCS problem may arise when we know that each of the strings

© Springer International Publishing AG 2016
S. Inenaga et al. (Eds.): SPIRE 2016, LNCS 9954, pp. 15–21, 2016.
DOI: 10.1007/978-3-319-46049-9_2

has exactly one meaningful occurrence of the pattern and want to determine a putative pair of such occurrences based on homology of the strings.

In comparing strings over an alphabet set with various levels of symbol similarity, such as amino acid sequences of proteins, however, the LCS metric is sometimes too naive to adopt as a similarity metric. In the present article, we consider generalized similarity metrics, including the metric based on an amino acid substitution matrix with affine gap penalties, which is widely used to estimate the similarity between amino acid sequences [7]. This similarity metric is also adopted by another generalized constrained LCS problem, the regular expression constrained alignment problem [1,3,9], in which the pattern P is given as a regular expression.

The present article propose an $O(mn)$-time, $O(n)$-space algorithm for the problem consisting of finding a highest-scoring alignment of A and B that matches an occurrence of P in A and an occurrence of P in B. In this problem, we treat an arbitrary minimal substring of a string most similar to P as an occurrence of P in the string and ignore the similarity between the matched occurrences of P when estimating the score of the alignment. The proposed algorithm achieves the same asymptotic execution time and required space as the algorithm for the (non-constrained) alignment problem based on the divide-and-conquer technique of Hirschberg [8]. Furthermore, since the problem we consider is identical to the STR-IC-LCS problem if we adopt the LCS metric, the proposed algorithm improves space complexity of the STR-IC-LCS problem achieved by the algorithm of Deorowicz [5] from quadratic to linear.

2   Preliminaries

A string is a sequence of symbols. For any string X, $|X|$ denotes the length of X, $X[i]$ denotes the symbol in X at position i, and $X(i', i]$ denotes the substring of X at position between $i' + 1$ and i. The concatenation of string X' followed by string X'' is denoted by $X'X''$.

Let Σ be an alphabet set of a constant number of symbols. Let – denote a gap symbol that does not belong to Σ. A gap is a string consisting only of more than zero gap symbols. We use + and / to represent the first and last gap symbols in a gap of length more than one, respectively, and * to represent the only gap symbol in a gap of length one. In what follows, we use – to represent a gap symbol in a gap of length more than two other than the first and last gap symbols. Let $\Gamma = \{+, -, /, *\}$ and let $\tilde{\Sigma} = \Sigma \cup \Gamma$. Let a gapped string of a string X over Σ be a string over $\tilde{\Sigma}$ obtained from X by inserting a concatenation of zero or more gaps at position between i and $i + 1$ for each index i with $0 \leq i \leq |X|$. Although concatenations of two or more gaps inserted in a string may look uncommon, we adopt this definition of a gapped string for a technical reason mentioned later. We sometimes use the index representation, denoted $I_{\tilde{X}}$, of a gapped string $\tilde{X}$ of a substring of X, in which $X[i]$ is represented as index i and any gap symbol γ in Γ that appears in the concatenation of gaps inserted in X at position between i and $i + 1$ is represented as γ with subscript i.

For any strings X and Y over Σ, an alignment of X and Y is a pair of a gapped string $\tilde{X}$ of X and a gapped string $\tilde{Y}$ of Y with $|\tilde{X}| = |\tilde{Y}|$ such that $\tilde{X}[q]$ or $\tilde{Y}[q]$ is not a gap symbol in Γ for any index q with $1 \leq q \leq |\tilde{X}|$ $(= |\tilde{Y}|)$. Let a symbol similarity score table s consist of values $s(a,b)$ indicating how much a is similar to b for all ordered pair (a,b) of symbols in $\tilde{\Sigma}$ other than pairs of gap symbols in Γ. A typical setting, adopted in affine gap penalty metrics, is $s(a,+) = s(a,*) = s(+,a) = s(*,a) = gip + gep$ and $s(a,-) = s(a,/) = s(-,a) = s(/,a) = gep$ for any symbol a in Σ, where gip is a gap insertion penalty representing the penalty for each insertion of a gap and gep is a gap extension penalty representing the penalty for each one-symbol extension of a gap. How well an alignment $(\tilde{X}, \tilde{Y})$ makes a connection between symbols in X and symbols in Y is estimated by the score $s(\tilde{X}, \tilde{Y}) = \sum_{1 \leq q \leq |\tilde{X}|} s(\tilde{X}[q], \tilde{Y}[q])$ of the alignment. For any strings X and Y over Σ, let how much X is similar to Y be defined as $Sim(X,Y) = \max_{(\tilde{X},\tilde{Y})} s(\tilde{X}, \tilde{Y})$, where $(\tilde{X}, \tilde{Y})$ ranges over all alignments of X and Y. We define an occurrence of a pattern in a string as a minimal substring of the string that is most similar to the patter in the sense of the following definition.

Definition 1. For any strings X and Y over Σ, let a substring X' of X be an occurrence of Y in X if $Sim(X',Y) \geq Sim(X'',Y)$ for any substring X'' of X and $Sim(X',Y) > Sim(X'',Y)$ for any substring X'' of X' with $|X''| < |X'|$.

The present article considers the following problem.

Definition 2. Given strings, A of length m, B of length n, and P of length r, over Σ with $m \geq n \geq r$, let the substring constrained alignment (StrCA) problem consist of finding an arbitrary pair of an occurrence A_{occ} of P in A and an occurrence B_{occ} of P in B such that

$$Sim(A_{pref}, B_{pref}) + Sim(A_{suff}, B_{suff})$$

is maximum, where $A = A_{pref} A_{occ} A_{suff}$ and $B = B_{pref} B_{occ} B_{suff}$. (If arbitrary highest-scoring alignments of A_{pref} and B_{pref} and of A_{suff} and B_{suff} are necessary after the StrCA problem is solved, we can obtain such alignments in $O(mn)$ time and $O(n)$ space based on the divide-and-conquer technique of Hirschberg [8].)

3   Algorithm

This section proposes an $O(mn)$-time, $O(n)$-space algorithm for the StrCA problem. In order to design the proposed algorithm, we introduce several lemmas each with no proof, due to limitation of space. However, they can be proven easily in a straightforward manner.

The algorithm we propose is based on the dynamic programming technique. We use edge-weighted directed acyclic graphs (DAGs) to represent dynamic programming (DP) tables as follows.

Definition 3. Let G be an arbitrary edge-weighted DAG. For any edge e in G, let $w(e)$ denote the weight of e. We also use $w(u, v)$ to denote $w(e)$ if e is from vertex u to vertex v. For any path π in G, let the weight $w(\pi)$ of π be the sum of $w(e)$ over all edges e in π. For any vertex v in G, let $to(v)$ denote the set of all vertices u such that G has an edge from u to v. If no such vertices u exist, then v is a source vertex. Any vertex u not appearing in $to(v)$ for any vertex v in G is a sink vertex. We focus only on edge-weighted DAGs having exactly one source vertex and one sink vertex. For any vertex v in G, we use $dp(v)$ to denote the value of v in the DP table with respect to G. This value is defined recursively as $dp(v) = 0$, if v is the source vertex, or $dp(v) = \max_{u \in to(v)}(dp(u) + w(u, v))$, otherwise. Hence, $dp(v)$ represents the weight of any heaviest path from the source vertex to v.

To solve the StrCA problem, we utilize an edge-weighted DAG, called the StrCA DAG, that reduces the StrCA problem to the problem of finding an arbitrary one of certain edges through which a heaviest path from the source vertex to the sink vertex passes. Applying the same idea as the algorithm of Deorowicz [5] for the STR-IC-LCS problem to this DAG, we can immediately obtain an algorithm for the SrtCA problem. However, as mentioned later, the algorithm proposed in the present article uses this DAG in a different way in order to save a great deal of space required.

The StrCA DAG is defined as a certain variant of the following edge-weighted DAG, called the alignment DAG, which is based on an idea similar to the algorithm of Gotoh [6] for the alignment problem with affine gap penalties. This DAG is designed such that any two-edge path corresponds to a pair of consecutive positions in some alignment of two strings and vice versa. The reason for the uncommon definition of a gapped string is because of a close relationship between paths in the DAG and alignments of substrings of the strings.

Definition 4. For any strings X and Y over Σ, let the alignment DAG, denoted $G(X, Y)$, for X and Y be the edge-weighted DAG consisting of vertices

- $d(i, j)$ for all index pairs (i, j) with $0 \le i \le |X|$ and $0 \le j \le |Y|$,
- $h(i, j)$ for all index pairs (i, j) with $0 \le i \le |X|$ and $0 < j < |Y|$, and
- $v(i, j)$ for all index pairs (i, j) with $0 < i < |X|$ and $0 \le j \le |Y|$

and edges

- $e(i, j)$ of weight $s(X[i], Y[j])$ from $d(i - 1, j - 1)$ to $d(i, j)$,
- $e(+_i, j)$ of weight $s(+, Y[j])$ from $d(i, j - 1)$ to $h(i, j)$,
- $e(-_i, j)$ of weight $s(-, Y[j])$ from $h(i, j - 1)$ to $h(i, j)$,
- $e(/_i, j)$ of weight $s(/, Y[j])$ from $h(i, j - 1)$ to $d(i, j)$,
- $e(*_i, j)$ of weight $s(*, Y[j])$ from $d(i, j - 1)$ to $d(i, j)$,
- $e(i, +_j)$ of weight $s(X[i], +)$ from $d(i - 1, j)$ to $v(i, j)$,
- $e(i, -_j)$ of weight $s(X[i], -)$ from $v(i - 1, j)$ to $v(i, j)$,
- $e(i, /_j)$ of weight $s(X[i], /)$ from $v(i - 1, j)$ to $d(i, j)$, and
- $e(i, *_j)$ of weight $s(X[i], *)$ from $d(i - 1, j)$ to $d(i, j)$

for all possible index pairs (i, j). Let the ith row of $G(X, Y)$ consist of all vertices $d(i, j)$ with $0 \le j \le |Y|$, $h(i, j)$ with $0 < j < |Y|$, and $v(i, j)$ with $0 \le j \le |Y|$.

Lemma 1. *Any path $\pi = e(\tilde{\imath}_1, \tilde{\jmath}_1)e(\tilde{\imath}_2, \tilde{\jmath}_2) \cdots e(\tilde{\imath}_p, \tilde{\jmath}_p)$ in $G(X, Y)$ from $d(i', j')$ to $d(i, j)$ bijectively corresponds to the alignment $(\tilde{X}, \tilde{Y})$ of $X[i'+1..i]$ and $Y[j'+1..j]$ with $I_{\tilde{X}} = \tilde{\imath}_1\tilde{\imath}_2 \cdots \tilde{\imath}_p$ and $I_{\tilde{Y}} = \tilde{\jmath}_1\tilde{\jmath}_2 \cdots \tilde{\jmath}_p$. Furthermore, for any such pair of a path π and an alignment $(\tilde{X}, \tilde{Y})$, $w(\pi) = s(\tilde{X}, \tilde{Y})$ holds.*

Before presenting the StrCA DAG, we show that all occurrences of a pattern in a string can be found in quadratic time and linear space, if we use the following variant of the alignment DAG. This DAG is based on an idea similar to the algorithm of Smith and Waterman [11] for the local alignment problem.

Definition 5. For any strings X and Y over Σ, let the occurrence DAG, denoted $G_{occ}(X, Y)$, of Y in X be the edge-weighted DAG obtained from $G(X, Y)$ by adding two vertices src and snk, bypass edges $in(i')$ of weight zero from src to $d(i', 0)$ for all indices i' with $0 \le i' \le |X|$, and bypass edges $out(i)$ of weight zero from $d(i, |Y|)$ to snk for all indices i with $0 \le i \le |X|$. For any vertex v in $G_{occ}(X, Y)$ other than src, let $i'(v)$ be the greatest index i' such that some heaviest path from src to v passes through bypass edge $in(i')$.

Lemma 2. *Substring $X(i', i]$ is an occurrence of Y in X if and only if some heaviest path in $G_{occ}(X, Y)$ from src to snk passes through $out(i)$, $i'(d(i, |Y|)) = i'$, and no substrings $X(i', i'']$ with $i' < i'' < i$ are occurrences of Y in X.*

Lemma 3. *For any vertex v in $G_{occ}(X, Y)$ other than src, $i'(v)$ is equal to the maximum of $i'(u)$ over all vertices u in $to(v)$ with $dp(v) = dp(u) + w(u, v)$, where we treat $i'(u) = i'$ if $u = src$ and $v = d(i', 0)$.*

Let $DP_{occ}(i)$ and $I'(i)$ denote the array of DP table values $dp(v)$ and the array of indices $i'(v)$ for all vertices v in the ith row of $G_{occ}(X, Y)$, respectively. It then follows from the recurrence relation of DP table value $dp(v)$ given in Definition 3 that $DP_{occ}(i)$ can be constructed in $O(|Y|)$ time from scratch, if $i = 0$, or from $DP_{occ}(i-1)$, otherwise. Similarly, we can obtain $I'(i)$ in $O(|Y|)$ time from scratch, if $i = 0$, or from $DP_{occ}(i-1)$, $I'(i-1)$, and $DP(i)_{occ}$, otherwise, based on Lemma 3. Thus, we obtain Algorithm findOcc(X, Y) presented in Fig. 1 as an $O(|X||Y|)$-time, $O(|Y|)$-space algorithm that enumerates all occurrences of Y in X. In this algorithm, lines 1 through 4 prepare $dp(snk)$, the weight of any heaviest path from src to snk, as the value of variable dp_{snk}. Using this value, each iteration of lines 7 through 9 applies Lemma 2, where index variable i' in line 8 is maintained so as to indicate that, if $i' \ge 0$, then some substring $X(i', i'']$ with $i' < i'' < i$ is an occurrence of Y in X.

Lemma 4. *For any strings X and Y over Σ, Algorithm findOcc(X, Y) enumerates all occurrences $X(i', i]$ of Y in X in ascending order with respect to i and, hence, with respect to i' in $O(|X||Y|)$ time and $O(|Y|)$ space.*

Now we present the StrCA DAG, together with the properties crucial to designing the proposed algorithm.

1: Let $dp_{snk} = 0$;
2: for each index i from 0 to $|X|$,
3:      construct $DP_{occ}(i)$; delete $DP_{occ}(i-1)$ if $i \geq 1$;
4:      let $dp_{snk} = \max(dp_{snk}, dp(d(i, |Y|)) + w(out(i)))$;
5: let $i' = -1$;
6: for each index i from 0 to $|X|$,
7:      construct $DP_{occ}(i)$ and $I'(i)$; delete $DP_{occ}(i-1)$ and $I'(i-1)$ if $i \geq 1$;
8:      if $dp(d(i, |Y|)) = dp_{snk}$ and $i' < i'(d(i, |Y|))$, then
9:          let $i' = i'(d(i, |Y|))$; report that $X(i', i]$ is an occurrence of Y in X.

Fig. 1. Algorithm findOcc(X, Y)

1: Obtain all occurrences of P in B by executing Algorithm findOcc(B, P);
2: let $i' = 0$; let $i = 0$;
3: for each occurrence $A(i'_P, i_P]$ of P in A reported by Algorithm findOcc(A, P),
            which is executed along with iterations of this sentence,
4:      while $i' \leq i'_P$,
5:          compute $DP_{pref}(i')$; delete $DP_{pref}(i'-1)$ if $i' \geq 1$;
6:          increase i' by one;
7:      while $i \leq i_P$, or $i \leq m$ if $A(i'_P, i_P]$ is the last occurrence of P in A,
8:          compute $DP_{suff}(i)$ and $TR(i)$; delete $DP_{suff}(i-1)$ and $TR(i-1)$ if $i \geq 1$;
9:          increase i by one;
10: output $(A(i', i], B(j', j])$, where the edge in $tr(d_{suff}(m, n))$ obtained as an element of $TR(m)$ is from $d_{pref}(i', j')$ to $d_{suff}(i, j)$.

Fig. 2. Algorithm solveSrtCA(A, B, P)

Definition 6. Let G_{pref} and G_{suff} be copies of $G(A, B)$ and let vertices in them be indicated by subscripts pref and suff, respectively. Let the StrCA DAG, denoted G_{StrCA}, be the edge-weighted DAG obtained from G_{pref} and G_{suff} by adding a transition edge of weight zero from $d_{pref}(i', j')$ to $d_{suff}(i, j)$ for any pair of an occurrence $A(i', i]$ of P in A and an occurrence $B(j', j]$ of P in B and adding a dummy transition edge of weight $-\infty$ from $d_{pref}(0, 0)$ to $d_{suff}(0, 0)$. For any vertex v in G_{suff}, let $tr(v)$ represent an arbitrary transition edge through which some heaviest path from $d_{pref}(0, 0)$ to v passes.

Lemma 5. *Substring pair $(A(i', i], B(j', j])$ is a solution of the StrCA problem if and only if the transition edge from $d_{pref}(i', j')$ to $d_{suff}(i, j)$ is passed through by some heaviest path in G_{StrCA} from $d_{pref}(0, 0)$ to $d_{suff}(m, n)$. Hence, $tr(d_{suff}(m, n))$ gives a solution of the StrCA problem.*

Lemma 6. *For any vertex v in G_{suff} and any vertex u in $to(v)$ with $dp(v) = dp(u) + w(u, v)$, $tr(u)$ is an instance of $tr(v)$, where we treat the transition edge from u to v as $tr(u)$ if u is a vertex in G_{pref}.*

The proposed algorithm solves the StrCA problem based on Lemma 5. The key idea to achieve linear-space computation of $tr(d_{suff}(m, n))$ is to successively focus on which transition edge some heaviest path in G_{StrCA} from $d_{pref}(0, 0)$

to each vertex v in G_{suff} passes through. According to the recurrence relation of $tr(v)$ given in Lemma 6, the algorithm determines $tr(v)$ for each vertex v in G_{suff} and forget previously determined $tr(u)$ no longer in use successively. This is unlike in the case of the algorithm adopting an approach similar to the quadratic-space algorithm of Deorowicz [5] for the STR-IC-LCS problem, which simultaneously determines how much any heaviest path from $d_{\text{pref}}(0,0)$ to $d_{\text{suff}}(m,n)$ passing through each of all transition edges weighs.

Let $DP_{\text{pref}}(i')$ denote the array of DP table values $dp(v)$ for all vertices v in the i'th row of G_{pref} and let $DP_{\text{suff}}(i)$ and $TR(i)$ denote the array of DP table values $dp(v)$ and the array of transition edges $tr(v)$ for all vertices v in the ith row of G_{suff}, respectively. Then, $DP_{\text{pref}}(i')$ can be constructed in $O(n)$ time from scratch, if $i' = 0$, or from $DP_{\text{pref}}(i'-1)$, otherwise. Furthermore, $DP_{\text{suff}}(i)$ and $TR(i)$ can be constructed in $O(n)$ time from scratch, if $i = 0$, or otherwise from $DP_{\text{suff}}(i-1)$ and $TR(i-1)$, together with $DP_{\text{pref}}(i')$ if A has an occurrence $A(i',i]$ of P for some index i'. Thus, we eventually obtain Algorithm solveStrCA(A,B,P) presented in Fig. 2 as the proposed algorithm for the StrCA problem, which satisfies the following theorem.

Theorem 1. *The StrCA problem is solvable in $O(mn)$ time and $O(n)$ space by executing Algorithm* solveStrCA(A,B,P).

References

1. Arslan, A.N.: Regular expression constrained sequence alignment. J. Discrete Algorithms **5**, 647–661 (2007)
2. Chen, Y.-C., Chao, K.-M.: On the generalized constrained longest common subsequence problems. J. Comb. Optim. **21**, 383–392 (2011)
3. Chung, Y.-S., Lu, C.L., Tang, C.Y.: Efficient algorithms for regular expression constrained sequence alignment. Inf. Process. Lett. **103**, 240–246 (2007)
4. Chin, F.Y.L., De Santis, A., Ferrara, A.L., Ho, N.L., Kim, S.K.: A simple algorithm for the constrained sequence problems. Inf. Process. Lett. **90**, 175–179 (2004)
5. Deorowicz, S.: Quadratic-time algorithm for a string constrained LCS problem. Inf. Process. Lett. **112**, 423–426 (2012)
6. Gotoh, O.: An improved algorithm for matching biological sequences. J. Mol. Biol. **162**, 705–708 (1982)
7. Gusfield, D.: Algorithms on Strings, Trees, and Sequences. Cambridge University Press, Cambridge (1997)
8. Hirschberg, D.S.: Algorithms for the longest common subsequence problem. J. ACM **24**, 664–675 (1977)
9. Kucherov, G., Pinhas, T., Ziv-Ukelson, M.: Regular language constrained sequence alignment revisited. J. Comput. Biol. **18**, 771–781 (2011)
10. Needleman, S.B., Wunsch, C.D.: A general method applicable to the search for similarities in the amino acid sequence of two proteins. J. Mol. Biol. **48**, 443–453 (1970)
11. Smith, T.F., Waterman, M.S.: Identification of common molecular subsequences. J. Mol. Biol. **147**, 195–197 (1981)
12. Tsai, Y.-T.: The constrained longest common subsequence problem. Inf. Process. Lett. **88**, 173–176 (2003)

Near-Optimal Computation of Runs over General Alphabet via Non-Crossing LCE Queries

Maxime Crochemore[1(✉)], Costas S. Iliopoulos[1], Tomasz Kociumaka[2],
Ritu Kundu[1], Solon P. Pissis[1], Jakub Radoszewski[1,2], Wojciech Rytter[2],
and Tomasz Waleń[2]

[1] Department of Informatics, King's College London, London, UK
{maxime.crochemore,costas.iliopoulos,ritu.kundu,solon.pissis}@kcl.ac.uk
[2] Faculty of Mathematics, Informatics and Mechanics,
University of Warsaw, Warsaw, Poland
{kociumaka,jrad,rytter,walen}@mimuw.edu.pl

Abstract. Longest common extension queries (LCE queries) and runs
are ubiquitous in algorithmic stringology. Linear-time algorithms computing runs and preprocessing for constant-time LCE queries have been
known for over a decade. However, these algorithms assume a linearly-
sortable integer alphabet. A recent breakthrough paper by Bannai et al.
(SODA 2015) showed a link between the two notions: all the runs in a
string can be computed via a linear number of LCE queries. The first to
consider these problems over a general ordered alphabet was Kosolobov
(*Inf. Process. Lett.*, 2016), who presented an $\mathcal{O}(n(\log n)^{2/3})$-time algorithm for answering $\mathcal{O}(n)$ LCE queries. This result was improved by
Gawrychowski et al. (CPM 2016) to $\mathcal{O}(n \log \log n)$ time. In this work
we note a special *non-crossing* property of LCE queries asked in the
runs computation. We show that any n such non-crossing queries can be
answered on-line in $\mathcal{O}(n\alpha(n))$ time, where $\alpha(n)$ is the inverse Ackermann
function, which yields an $\mathcal{O}(n\alpha(n))$-time algorithm for computing runs.

1 Introduction

Runs (also called *maximal repetitions*) are a fundamental type of repetitions
in a string as they represent the structure of all repetitions in a string in a
succinct way. A run is an inclusion-maximal periodic factor of a string in which
the shortest period repeats at least twice. A crucial property of runs is that
their maximal number in a string of length n is $\mathcal{O}(n)$. This fact was already
observed by Kolpakov and Kucherov [15,16] who conjectured that this number

T. Kociumaka—Supported by Polish budget funds for science in 2013–2017 as a
research project under the 'Diamond Grant' program.

J. Radoszewski—Newton International Fellow.

W. Rytter and T. Waleń—Supported by the Polish National Science Center, grant
no. 2014/13/B/ST6/00770.

S. Inenaga et al. (Eds.): SPIRE 2016, LNCS 9954, pp. 22–34, 2016.
DOI: 10.1007/978-3-319-46049-9_3

is actually smaller than n, which was known as the runs conjecture. Due to the works of several authors [6–8, 12, 19–21] more precise bounds on the number of runs have been obtained, and finally in a recent breakthrough paper [2] Bannai et al. proved the runs conjecture, which has since then become the runs theorem (even more recently in [10] the upper bound of $0.957n$ was shown for binary strings).

Perhaps more important than the combinatorial bounds is the fact that the set of all runs in a string can be computed efficiently. Namely, in the case of a linearly-sortable alphabet Σ (e.g., $\Sigma = \{1, \ldots, \sigma\}$ with $\sigma = n^{\mathcal{O}(1)}$) a linear-time algorithm based on Lempel-Ziv factorization [15, 16] was known for a long time. In the recent papers of Bannai et al. [1, 2] it is shown that to compute the set of all runs in a string, it suffices to answer $\mathcal{O}(n)$ longest common extension (LCE) queries. An LCE query asks, for a pair of suffixes of a string, for the length of their longest common prefix. In the case of $\sigma = n^{\mathcal{O}(1)}$ such queries can be answered on-line in $\mathcal{O}(1)$ time after $\mathcal{O}(n)$-time preprocessing that consists of computing the suffix array with its inverse, the LCP table and a data structure for range minimum queries on the LCP table; see e.g. [5]. The algorithms from [1, 2] use (explicitly and implicitly, respectively) an intermediate notion of Lyndon tree (see [3, 13]) which can, however, also be computed using LCE queries.

Let $T_{\mathrm{LCE}}(n)$ denote the time required to answer on-line n LCE queries in a string. In a very recent line of research, Kosolobov [17] showed that, for a general ordered alphabet, $T_{\mathrm{LCE}}(n) = \mathcal{O}(n(\log n)^{2/3})$, which immediately leads to $\mathcal{O}(n(\log n)^{2/3})$-time computation of the set of runs in a string. In [11] a faster, $\mathcal{O}(n \log \log n)$-time algorithm for answering n LCE queries has been presented which automatically leads to $\mathcal{O}(n \log \log n)$-time computation of runs.

Runs have found a number of algorithmic applications. Knowing the set of runs in a string of length n one can compute in $\mathcal{O}(n)$ time all the local periods and the number of all squares, and also in $\mathcal{O}(n + T_{\mathrm{LCE}}(n))$ time all distinct squares provided that the suffix array of the string is known [9]. Runs were also used in a recent contribution on efficient answering of internal pattern matching queries and their applications [14].

Our Results. We observe that the computation of a Lyndon tree of a string and furthermore the computation of all the runs in a string can be reduced to answering $\mathcal{O}(n)$ LCE queries that are *non-crossing*, i.e., no two queries $\mathrm{LCE}(i, j)$ and $\mathrm{LCE}(i', j')$ are asked with $i < i' < j < j'$ or $i' < i < j' < j$. Let $T_{\mathrm{ncLCE}}(n)$ denote the time required to answer n such queries on-line in a string of length n over a general ordered alphabet. We show that $T_{\mathrm{ncLCE}}(n) = \mathcal{O}(n\alpha(n))$, where $\alpha(n)$ is the inverse Ackermann function. As a consequence, we obtain $\mathcal{O}(n\alpha(n))$-time algorithms for computing the Lyndon tree, the set of all runs, the local periods and the number of all squares in a string over a general ordered alphabet.

Our solution relies on a trade-off between two approaches. The results of [11] let us efficiently compute the LCEs if they are short, while LCE queries with similar arguments and a large answer yield structural properties of the string, which we discover and exploit to answer further such queries.

Our approach for answering non-crossing LCE queries is described in three sections: in Sect. 3 we give an overview of the data structure, in Sect. 4 we present the details of the implementation, and in Sect. 5 we analyse the complexity of answering the queries. The applications including runs computation are detailed in Sect. 6.

2   Preliminaries

Strings. Let Σ be a finite ordered alphabet of size σ. A string w of length $|w| = n$ is a sequence of letters $w[1] \ldots w[n]$ from Σ. By $w[i, j]$ we denote the *factor* of w being a string of the form $w[i] \ldots w[j]$. A factor $w[i, j]$ is called proper if $w[i, j] \neq w$. A factor is called a *prefix* if $i = 1$ and a *suffix* if $j = n$. We say that p is a *period* of w if $w[i] = w[i + p]$ for all $i = 1, \ldots, n - p$. If p is a period of w, the prefix $w[1, p]$ is called a *string period* of w.

By an interval $[\ell, r]$ we mean the set of integers $\{\ell, \ldots, r\}$. If w is a string of length n, then an interval $[a, b]$ is called a *run* in w if $1 \leq a < b \leq n$, the shortest period p of $w[a, b]$ satisfies $2p \leq b - a + 1$ and none of the factors $w[a - 1, b]$ and $w[a, b + 1]$ (if it exists) has the period p. An example of a run is shown in Fig. 1.

$$w = \quad \text{a b a b a a b a a b b b a a}$$

$$\phantom{w = \quad \text{a b }} 3 \phantom{\text{ a b a a b a a }} 10$$

Fig. 1. Example of a run $[3, 10]$ with period 3 in the string $w = $ ababaabaabbbaa. This string contains also other runs, e.g. $[10, 12]$ with period 1 and $[1, 5]$ with period 2.

Lyndon Words and Trees. By $\prec = \prec_0$ we denote the order on Σ and by $\prec_1$ we denote the reverse order on Σ. We extend each of the orders $\prec_r$ for $r \in \{0, 1\}$ to a lexicographical order on strings over Σ. A string w is called an *r-Lyndon word* if $w \prec_r u$ for every non-empty proper suffix u of w. The *standard factorization* of an r-Lyndon word w is a pair (u, v) of r-Lyndon words such that $w = uv$ and v is the longest proper suffix of w that is an r-Lyndon word.

The *r-Lyndon tree* of an r-Lyndon word w, denoted as $LTree_r(w)$, is a rooted full binary tree defined recursively on $w[1, n]$ as follows:

- $LTree_r(w[i, i])$ consists of a single node labeled with $[i, i]$
- if $j - i > 1$ and (u, v) is the standard factorization of $w[i, j]$, then the root of $LTree_r(w)$ is labeled by $[i, j]$, has left child $LTree_r(u)$ and right child $LTree_r(v)$.

See Fig. 2 for an example. We can also define the r-Lyndon tree of an arbitrary string. Let $\$_0, \$_1$ be special characters smaller than and greater than all the letters from Σ, respectively. We then define $LTree_r(w)$ as $LTree_r(\$_r w)$; note that $\$_r w$ is an r-Lyndon word.

LCE Queries. For two strings u and v, by $\mathrm{lcp}(u, v)$ we denote the length of their longest common prefix. Let w be a string of length n. An LCE query $LCE(i, j)$

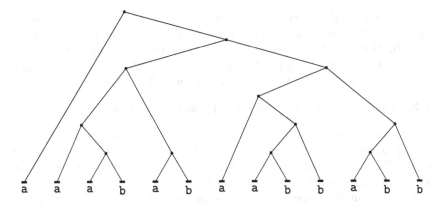

Fig. 2. The Lyndon tree $LTree_0(w)$ of a Lyndon word $w = $ aaababaabbabb.

computes $\mathrm{lcp}(w[i, n], w[j, n])$. An ℓ-limited LCE query Limited-LCE$_{\leq \ell}(i, j)$ computes $\min(\mathrm{LCE}(i, j), \ell)$. Such queries can be answered efficiently as follows; see Lemma 14 in [11].

Lemma 1 ([11]). *A sequence of q queries* Limited-LCE$_{\leq \ell_p}(i_p, j_p)$ *can be answered on-line in* $\mathcal{O}((n + \sum_{p=1}^{q} \log \ell_p)\alpha(n))$ *time over a general ordered alphabet.*

The following observation shows a relation between LCE queries and periods in a string that we use in our data structure; for an illustration see Fig. 3.

Observation 2. *Assume that the factors $w[a, d_A - 1]$ and $w[b, d_B - 1]$ have the same string period, but neither $w[a, d_A]$ nor $w[b, d_B]$ has this string period. Then*

$$\mathrm{LCE}(a, b) = \begin{cases} \min(d_A - a, d_B - b) & \text{if } d_A - a \neq d_B - b, \\ d_A - a + \mathrm{LCE}(d_A, d_B) & \text{otherwise.} \end{cases}$$

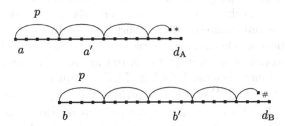

Fig. 3. In this example figure $d_A - a = 14$, $d_B - b = 18$, and $p = 4$. We have $\mathrm{LCE}(a, b) = 14$ and $\mathrm{LCE}(a', b') = 8 + \mathrm{LCE}(d_A, d_B)$.

Non-Crossing Pairs. For a positive integer n, we define the set of pairs

$$P_n = \{(a, b) \in \mathbb{Z}^2 : 1 \leq a \leq b \leq n\}.$$

Pairs (a, b) and (a', b') are called *crossing* if $a < a' < b < b'$ or $a' < a < b' < b$. A subset $S \subseteq P_n$ is called *non-crossing* if it does not contain crossing pairs.

A graph G is called *outerplanar* if it can be drawn on a plane without crossings in such a way that all vertices belong to the unbounded face. An outerplanar graph on n vertices has less than $2n$ edges (at most $2n - 3$ for $n \geq 2$).

Fact 3. *A non-crossing set of pairs $S \subseteq P_n$ has less than $3n$ elements.*

Proof. We associate $S \setminus \{(a, a) : 1 \leq a \leq n\}$ with a plane graph on vertices $\{1, \ldots, n\}$ drawn on a circle in this order, and edges represented as straight-line segments. The non-crossing property of pairs implies that these segments do not intersect. Thus, the graph drawing is outerplanar, and therefore the number of edges is less than $2n$. Accounting for the pairs of the form (a, a), we get the claimed upper bound. □

For a set of pairs $S = \{(a_i, b_i) : 1 \leq i \leq k\}$ and a positive integer t, by $\lceil S/t \rceil$ we denote the set $\{(\lceil \frac{a_i}{t} \rceil, \lceil \frac{b_i}{t} \rceil) : 1 \leq i \leq k\}$.

Observation 4. *If S is non-crossing, then $\lceil S/t \rceil$ is also non-crossing.*

3 High-Level Description of the Data Structure

We say that a sequence of $\mathrm{LCE}(a, b)$ queries, for $a \leq b$, is *non-crossing* if the underlying collection of pairs (a, b) is non-crossing. In this section, we give an overview of our data structure, which answers a sequence of q non-crossing LCE queries on-line in $\mathcal{O}(q + n \cdot \alpha(n))$ total time.

The data structure is composed of $\lceil \log n \rceil$ levels. Function $\mathrm{LCE}^{(i)}(a, b)$ corresponds to the level i and returns $\mathrm{LCE}(a, b)$. In the computation it may make calls to $\mathrm{LCE}^{(i+1)}(a, b)$. However, we make sure that the total number of such calls is bounded. Each original $\mathrm{LCE}(a, b)$ query is first asked at the level 0.

The implementation of $\mathrm{LCE}^{(i)}(a, b)$ consists of two phases. If $\mathrm{LCE}(a, b) \geq 3 \cdot 2^i$, then this $\mathrm{LCE}^{(i)}$ query is called *relevant*; otherwise it is called *short*. In the first phase, we check the type of the query via a Limited-LCE$_{\leq 3 \cdot 2^i}(a, b)$ query. This lets us immediately answer short queries. In the second phase, we know that the query is relevant, and we try to deduce the answer based on data gathered while processing *similar* queries or to learn some information useful for answering future *similar* queries by asking $\mathrm{LCE}^{(i+1)}$ queries.

We shall say that $\mathrm{LCE}^{(i)}$ queries for (a, b) and (a', b') are *similar* if $\lceil \frac{a}{2^i} \rceil = \lceil \frac{a'}{2^i} \rceil$ and $\lceil \frac{b}{2^i} \rceil = \lceil \frac{b'}{2^i} \rceil$. Each equivalence class of this relation is processed by an independent component, called a *block-pair*. A *block* at level i is an interval of the form $[x \cdot 2^i + 1, (x + 1) \cdot 2^i]$, and a *block-pair* is a data structure identified by a pair (A, B) of blocks. If a relevant $\mathrm{LCE}^{(i)}(a, b)$ query satisfies $a \in A$ and $b \in B$ for some block-pair (A, B), we say that the block-pair is *responsible* for the query or that the query *concerns* the block-pair. As we show in Sect. 5, the pairs of interval right endpoints of block-pairs at each level are non-crossing (whereas $\mathrm{LCE}^{(i)}$ queries that will be asked for $i \geq 1$ are non necessarily non-crossing).

The implementation of a block-pair, summarized in the lemma below, is given in Sect. 4.

Lemma 5. *Consider a sequence of relevant* $\mathrm{LCE}^{(i)}$ *queries concerning a block-pair* (A, B). *The block-pair can answer these queries on-line in worst-case constant time plus the time to answer at most four* $\mathrm{LCE}^{(i+1)}(a, b)$ *queries, such that each either corresponds to the currently processed* $\mathrm{LCE}^{(i)}$ *query or satisfies* $a < b \leq a + 2^{i+1}$.

Structural conditions stated in Lemma 5 let us characterize the set of queries passed to the next level. The complexity analysis in Sect. 5 relies on this characterization.

4   Block-Pair Implementation

Our aim in this section is to prove Lemma 5. Information stored by a block-pair changes through the course of the algorithm, and the implementation of the query algorithm depends on what is currently stored. We distinguish four states of a block pair (A, B) at level i. Figure 4 illustrates two of the states.

state(A, B)	description
initial	No auxiliary data is stored
visited(a_0, b_0, L)	$a_0 \in A$ and $b_0 \in B$ are the arguments of the first query that concerns this block pair, $L = \mathrm{LCE}(a_0, b_0) \geq 3 \cdot 2^i$
full(d_A, d_B)	$\exists_{p \in [1, 2^{i+1}]} : w[\max A, d_A - 1]$ and $w[\max B, d_B - 1]$ have common period p and length at least $p + 2^i$, but neither $w[\max A, d_A]$ nor $w[\max B, d_B]$ has period p
full$^+$ (d_A, d_B, L')	As in **full**(d_A, d_B) plus $L' = \mathrm{LCE}(d_A, d_B)$.

4.1   Initial State

In this state, we simply forward the query to the level $i + 1$, return the obtained $\mathrm{LCE}(a, b)$ value, and change the state to **visited**$(a, b, \mathrm{LCE}(a, b))$.

Algorithm 1. Initial-LCE$^{(i)}_{(A,B)}(a, b)$

Require: $\mathrm{LCE}^{(i)}(a, b)$ concerns (A, B), whose state is **initial**
$L \leftarrow \mathrm{LCE}^{(i+1)}(a, b)$;   ▷ **higher level call**
transform (A, B) to state **visited**(a, b, L);
return L;

---

visited(a_0, b_0, L)

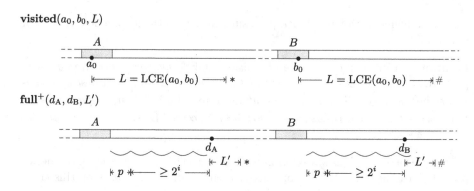

full$^+(d_A, d_B, L')$

Fig. 4. Block-pair (A, B) in states **visited**(a_0, b_0, L) and **full**$^+(d_A, d_B, L')$.

4.2   Visited State

In state **visited**(a_0, b_0, L), we can immediately determine LCE(a, b) if (a, b) is a shift of (a_0, b_0). Otherwise, we apply Lemma 6 to move to state **full**.

Lemma 6. *Let* LCE$^{(i)}(a, b)$, LCE$^{(i)}(a', b')$ *be similar and relevant queries and let* $p = |(b - b') - (a - a')|$. *If* $p \neq 0$ *and* $b' \leq b$, *then* LCE$(a, a + p) \geq 2^{i+1}$, *i.e.,* p *is a (not necessarily shortest) period of the factor* $w[a, a + 2^{i+1} + p - 1]$.

Proof. We shall first prove that LCE$(a, a + q) \geq 3 \cdot 2^i - (b - b')$ where $q = (b - b') - (a - a')$. First, observe that $a + q = a' + (b - b')$, and thus LCE$(a + q, b) =$ LCE$(a' + (b - b'), b' + (b - b')) \geq 3 \cdot 2^i - (b - b')$ because LCE$^{(i)}(a', b')$ is relevant. Since LCE$^{(i)}(a, b)$ is also relevant, we have LCE$(a, b) \geq 3 \cdot 2^i \geq 3 \cdot 2^i - (b - b')$. Combining these two inequalities, we immediately get LCE$(a, a + q) \geq \min($LCE$(a, b),$ LCE$(a + q, b)) \geq 3 \cdot 2^i - (b - b')$, as claimed.

If $q > 0$, we have $q = p$, and thus LCE$(a, a + p) \geq 3 \cdot 2^i - (b - b')$. Since the two LCE$^{(i)}$ queries are similar, we have $3 \cdot 2^i - (b - b') \geq 2^{i+1}$, so LCE$(a, a + p) \geq 2^{i+1}$. See Fig. 5 for an illustration of this case.

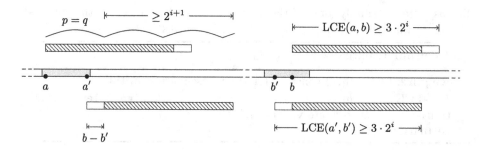

Fig. 5. Illustration of Lemma 6: case $q > 0$. We assume that LCE$(a + q, b) \leq$ LCE(a, b). The marked fragments correspond to LCE$(a, a + q) =$ LCE$(a + q, b)$.

Otherwise, $q = -p$, and we have $\mathrm{LCE}(a, a - p) \geq 3 \cdot 2^i - (b - b')$, which implies $\mathrm{LCE}(a + p, a) \geq 3 \cdot 2^i - (b - b') + q = 3 \cdot 2^i - (a - a')$. Again, the fact that the queries are similar yields $3 \cdot 2^i - (a - a') \geq 2^{i+1}$, and consequently $\mathrm{LCE}(a, a + p) \geq 2^{i+1}$. $\qquad\square$

In the query algorithm, we first check if $a - a_0 = b - b_0$. If so, let us denote the common value by Δ. Note that $|\Delta| \leq 2^i$, $\mathrm{LCE}(a, b) \geq 3 \cdot 2^i$, and $\mathrm{LCE}(a_0, b_0) \geq 3 \cdot 2^i$. This clearly yields $\mathrm{LCE}(a, b) = \mathrm{LCE}(a_0, b_0) + \Delta$, which lets us compute the result in constant time.

Algorithm 2. Visited-$\mathrm{LCE}_{(A,B)}^{(i)}(a, b)$

Require: $\mathrm{LCE}^{(i)}(a, b)$ concerns (A, B), whose state is **visited**(a_0, b_0, L)
if $a - a_0 = b - b_0$ **then**
    return $L + a - a_0$;
else
    $p \leftarrow |(a - a_0) - (b - b_0)|$;
    $a' \leftarrow \max A$; $b' \leftarrow \max B$;
    $d_A \leftarrow a' + p + \mathrm{LCE}^{(i+1)}(a', a' + p)$;   ▷ **higher level call**
    $d_B \leftarrow b' + p + \mathrm{LCE}^{(i+1)}(b', b' + p)$;   ▷ **higher level call**
    transform (A, B) to state **full**(d_A, d_B);
    return Full-$\mathrm{LCE}_{(A,B)}^{(i)}(a, b)$;   ▷ **recursive call on state full**

Otherwise, our aim is to change the state of the block-pair to **full**. Lemma 6 lets us deduce that $\mathrm{LCE}(\bar{a}, \bar{a} + p) \geq 2^{i+1}$ for some $\bar{a} \in \{a, a_0\}$ and (by symmetry) $\mathrm{LCE}(\bar{b}, \bar{b} + p) \geq 2^{i+1}$ for some $\bar{b} \in \{b, b_0\}$, where $p = |(a - a_0) - (b - b_0)|$ ($\bar{a}$ and $\bar{b}$ depend on the relative order of b, b_0 and a, a_0, respectively). Let $a' = \max A$ and $b' = \max B$. We have $\mathrm{LCE}(a', a' + p) \geq 2^i$ and $\mathrm{LCE}(b', b' + p) \geq 2^i$ because $a' - 2^i < \bar{a} \leq a'$ and $b' - 2^i < \bar{b} \leq b'$. Such a situation allows for a move to state **full**. The exact values of d_A and d_B are computed using a higher level call, which lets us determine $\mathrm{LCE}(a', a' + p)$ and $\mathrm{LCE}(b', b' + p)$. Note that $p \leq 2^{i+1}$ implies that these queries satisfy the condition of Lemma 5. The answer to the initial $\mathrm{LCE}^{(i)}(a, b)$ query is computed by the routine for state **full**, which we give below.

4.3   Full States

In state **full**$^+$ we can answer every relevant query in constant time. In state **full** we can either answer the query in constant time or make the final query at level $i + 1$ to transform the state to **full**$^+$; see the following lemma.

Lemma 7. *Consider a relevant $\mathrm{LCE}^{(i)}(a, b)$ query concerning a block-pair (A, B) in state* **full**(d_A, d_B) *or* **full**$^+(d_A, d_B, L')$. *Then*

$$\mathrm{LCE}(a, b) = \begin{cases} \min(d_A - a, d_B - b) & \textit{if } d_A - a \neq d_B - b, \\ d_A - a + \mathrm{LCE}(d_A, d_B) & \textit{otherwise.} \end{cases}$$

Proof. Let $a_0 = \max A$, $b_0 = \max B$ and let p be the witness period of the state of (A, B). Let us define $\Delta = \max(a_0 - a, b_0 - b)$, $a' = a + \Delta$, and $b' = b + \Delta$. Observe that $\Delta \leq 2^i$, $a_0 \leq a' \leq a_0 + 2^i$, and $b_0 \leq b' \leq b_0 + 2^i$. The fact that the query is relevant yields $\mathrm{LCE}(a, b) \geq 3 \cdot 2^i \geq p + \Delta$, so $\mathrm{LCE}(a, b) = \Delta + \mathrm{LCE}(a', b')$ and $\mathrm{LCE}(a', b') \geq p$. Moreover, $d_A \geq p + 2^i + a_0$ and $d_B \geq p + 2^i + b_0$ implies that fragments $w[a', d_A - 1]$ and $w[b', d_B - 1]$ have length at least p, and thus they are right-maximal with period p. Consequently, the fragments $w[a', d_A - 1]$ and $w[b', d_B - 1]$ have the same string period of length p. This lets us apply Observation 2, which gives

$$\mathrm{LCE}(a', b') = \begin{cases} \min(d_A - a', d_B - b') & \text{if } d_A - a' \neq d_B - b', \\ d_A - a' + \mathrm{LCE}(d_A, d_B) & \text{otherwise.} \end{cases}$$

Since $a' = a + \Delta$, $b' = b + \Delta$, and $\mathrm{LCE}(a, b) = \Delta + \mathrm{LCE}(a', b')$, this is clearly equivalent to the claimed formula for $\mathrm{LCE}(a, b)$. $\qquad\square$

Algorithm 3. Full-LCE$^{(i)}_{(A,B)}(a, b)$

Require: $\mathrm{LCE}^{(i)}(a, b)$ concerns (A, B), whose state is **full**(d_A, d_B) or **full**$^+(d_A, d_B, L')$

if $d_A - a \neq d_B - b$ **then**
    return $\min(d_A - a, d_B - b)$;
else
    if (A, B) *is in state* **full**(d_A, d_B) **then**
        $L' \leftarrow \mathrm{LCE}^{(i+1)}(a, b) - (d_A - a)$;   ▷ higher level call
        transform (A, B) to state **full**$^+(d_A, d_B, L')$;
    return $d_A - a + L'$;

4.4 Proof of Lemma 5

Lemma 5. *Consider a sequence of relevant* $\mathrm{LCE}^{(i)}$ *queries concerning a block-pair (A,B). The block-pair can answer these queries on-line in worst-case constant time plus the time to answer at most four* $\mathrm{LCE}^{(i+1)}(a, b)$ *queries, such that each either corresponds to the currently processed* $\mathrm{LCE}^{(i)}$ *query or satisfies* $a < b \leq a + 2^{i+1}$.

Proof. Algorithms 1, 2 and 3 answer queries concerning the block-pair (A, B), and use constant time. The level $i + 1$ call is only made when the state changes. The original query is forwarded during a shift from state **initial** to **visited** and from state **full** to **full**$^+$, while during a shift from **visited** to **full** two LCE queries are asked, both with arguments at distance $p \leq 2^{i+1}$, as claimed. $\qquad\square$

5   Complexity Analysis

Algorithm 4 summarizes the implementation of the $\mathrm{LCE}^{(i)}(a,b)$ function. As mentioned in Sect. 3, we first compute Limited-LCE$_{\leq 3 \cdot 2^i}(a,b)$, which might immediately give us the sought value $\mathrm{LCE}(a,b)$. Otherwise the query is relevant, and we refer to the block-pair (A,B) which is responsible for the query.

Algorithm 4. $\mathrm{LCE}^{(i)}(a,b)$

$\ell \leftarrow$ Limited-LCE$_{\leq 3 \cdot 2^i}(a,b)$;
if $\ell < 3 \cdot 2^i$ **then**   ▷ **short query**
    return ℓ;
else   ▷ **relevant query**
    $(A,B) \leftarrow$ block-pair responsible for the query (a,b) at level i;
    return:
        Initial-LCE$^{(i)}_{(A,B)}(a,b)$ **if** (A,B) is in state **initial**
        Visited-LCE$^{(i)}_{(A,B)}(a,b)$ **if** (A,B) is in state **visited**
        Full-LCE$^{(i)}_{(A,B)}(a,b)$ **if** (A,B) is in state **full** or **full$^+$**

Let $S_i = \{(a,b) : \mathrm{LCE}^{(i)}(a,b) \text{ is called}\}$. Then $\lceil S_i/2^i \rceil$ corresponds to the set of pairs of interval right endpoints of block-pairs at level i.

Fact 8. *The set* $\lceil S_i/2^i \rceil$ *is non-crossing.*

Proof. We proceed by induction on i. The base case is trivial from the assumption on the input sequence. Lemma 5 proves that $S_{i+1} \subseteq S_i \cup \{(a,b) : a < b \leq a + 2^{i+1}\}$. Hence, $\lceil S_{i+1}/2^{i+1} \rceil \subseteq \lceil \lceil S_i/2^i \rceil /2 \rceil \cup \{(a,b) : a \leq b \leq a+1\}$. The first component is non-crossing by the inductive hypothesis combined with Observation 4. Pairs of the form (a,a) and $(a, a+1)$ do not cross any other pair, so adding them to a non-crossing family preserves this property.   □

Consequently, Fact 3 proves that the number of block-pairs responsible for a query at level $i-1$ is bounded by $\frac{3n}{2^{i-1}}$. Each of them yields at most 4 queries at level i. This leads straight to the following bound.

Observation 9. $|S_i| \leq \frac{24n}{2^i}$ *for* $i \geq 1$.

If we stored the block-pairs using a hash table, we could retrieve the internal data of the block-pair responsible for (a,b) in randomised constant time. However, in the case of non-crossing LCE queries we can make this time worst-case.

Recall from Fact 3 that for a set $S \subseteq P_n$ of non-crossing pairs we can identify $S \setminus \{(a,a) : 1 \leq a \leq n\}$ with an outerplanar graph on vertices $\{1, \ldots, n\}$. We say that a simple undirected graph has *arboricity* at most c if it can be partitioned into c forests. Outerplanar graphs have arboricity at most 2 (see [18]) which lets us use the following theorem to store $S \setminus \{(a,a) : 1 \leq a \leq n\}$. Membership queries for pairs (a,a) are trivial to support using an array.

Theorem 10. ([4]). *Consider a graph of arboricity c with vertices given in advance and edges revealed on-line. One can support adjacency queries, asking to return the edge between two given vertices or **nil** if it does not exist, in worst-case $\mathcal{O}(c)$ time, with edge insertions processed in amortized constant time.*

The following corollary shows, by Fact 8, that indeed the block-pairs at each level can be retrieved in worst-case constant time.

Corollary 11. *Consider a set $S \subseteq P_n$ of non-crossing pairs arriving on-line. One can support membership queries (asking if $(a, b) \in S$ and, if so, to return data associated with this pair) in worst-case constant time with insertions processed in amortized constant time.*

Theorem 12. *In a string of length n, a sequence of q non-crossing LCE queries can be answered in total time $\mathcal{O}(q + n \cdot \alpha(n))$.*

Proof. For $i > 0$, an $\text{LCE}^{(i)}$ query, excluding the $\text{LCE}^{(i+1)}$ queries called, requires $\mathcal{O}(i \cdot \alpha(n))$ time for answering a Limited-LCE$_{\leq 3 \cdot 2^i}$ query by Lemma 1 plus $O(1)$ additional time by Lemma 5. For $i = 0$ we may compute Limited-LCE$_{\leq 3}$ naïvely in constant time, so the running time is constant.

The number of $\text{LCE}^{(0)}$ queries is q, while the number of $\text{LCE}^{(i)}$ queries for $i \geq 1$ is $\mathcal{O}(\frac{n}{2^i})$ by Observation 9. The total running time is therefore

$$\mathcal{O}\left(q + n \cdot \alpha(n) \cdot \sum_{i=1}^{\infty} \frac{i}{2^i}\right) = \mathcal{O}(q + n \cdot \alpha(n)). \qquad \square$$

6   Computing Runs

Bannai et al. [1, 2] presented an algorithm for computing all the runs in a string of length n that works in time proportional to answering $\mathcal{O}(n)$ LCE queries on the string or on its reverse. As main tool they used Lyndon trees. We note here that the LCE queries asked by their algorithm can be divided into a constant number of groups, each consisting of non-crossing LCE queries. Roughly speaking, this is based on the obvious fact that intervals in a Lyndon tree form a *laminar family*, i.e., for every two they are either disjoint or one of them contains the other.

In the first phase, given a string w, the algorithm of [1, 2] constructs $LTree_0(w)$ and $LTree_1(w)$. For each $r \in \{0, 1\}$, the construction of $LTree_r(w)$ goes from right to left. Before the k-th step (for $k = n, \ldots, 1$), we store on a stack the roots of subtrees of $LTree_r(w)$ that correspond to $w[k+1, n]$. Hence, the intervals corresponding to the roots on the stack are disjoint and cover the interval $[k + 1, n]$. In the k-th step we push on the stack a single node corresponding to $[k, k]$. Afterwards, as long as the stack contains at least two elements and the top element $[k, l]$ and the second to top element $[a, b]$ satisfy $w[k, l] \prec_r w[a, b]$, we pop the two subtrees from the stack and push one subtree with the root $[k, b]$. The lexicographical comparison is performed via an $\text{LCE}(k, a)$ query.

Observation 13. *The* LCE *queries asked in the construction of LTree$_r$(w) are non-crossing.*

Proof. In the k-th step of the algorithm we only ask LCE(i, j) queries for $i = k$. Suppose towards contradiction that in the course of the algorithm we ask two LCE queries with (i, j) and (i', j') such that $i < i' < j < j'$. The latter is asked at step i', and at that moment $[i', j' - 1]$ is a root of a subtree of *LTree$_r$(w)*. Then the former is asked at step i, and then $[i, j - 1]$ is a root of a subtree of *LTree$_r$(w)*. This contradicts the fact that the intervals in *LTree$_r$(w)* form a laminar family. □

In the second phase, for each node $[a, b]$ of each Lyndon tree *LTree$_r$(w)* we check if there is a run with period $p = b - a + 1$ that contains $w[a, b]$. To this end we check how long does the periodicity with period p extend to the right and to the left of $w[a, b]$. The former obviously reduces to an LCE$(a, b + 1)$ query and the latter to an LCE query in the reverse of w, which is totally symmetric. As the intervals in *LTree$_r$(w)* form a laminar family, we arrive at the following.

Observation 14. *The* LCE *queries asked when right-extending the periodicity of the intervals from LTree$_r$(w) are non-crossing.*

By Observations 13 and 14, Theorem 12 yields the following result and its immediate corollary.

Theorem 15. *The Lyndon tree and the set of all runs in a string of length n over a general ordered alphabet can be computed in $\mathcal{O}(n\alpha(n))$ time.*

Corollary 16. *All the local periods and the number of all squares in a string of length n over a general ordered alphabet can be computed in $\mathcal{O}(n\alpha(n))$ time.*

References

1. Bannai, H., I, T., Inenaga, S., Nakashima, Y., Takeda, M., Tsuruta, K.: A new characterization of maximal repetitions by Lyndon trees. In: Indyk, P. (ed.) 26th Annual ACM-SIAM Symposium on Discrete Algorithms, SODA 2015, pp. 562–571. SIAM (2015)
2. Bannai, H., I, T., Inenaga, S., Nakashima, Y., Takeda, M., Tsuruta, K.: The "runs" theorem (2015). arXiv:1406.0263v7
3. Barcelo, H.: On the action of the symmetric group on the free Lie algebra and the partition lattice. J. Comb. Theory, Ser. A **55**(1), 93–129 (1990)
4. Brodal, G.S., Fagerberg, R.: Dynamic representations of sparse graphs. In: Dehne, F., Gupta, A., Sack, J.-R., Tamassia, R. (eds.) WADS 1999. LNCS, vol. 1663, pp. 342–351. Springer, Heidelberg (1999)
5. Crochemore, M., Hancart, C., Lecroq, T.: Algorithms on Strings. Cambridge University Press, New York (2007)
6. Crochemore, M., Ilie, L.: Analysis of maximal repetitions in strings. In: Kučera, L., Kučera, A. (eds.) MFCS 2007. LNCS, vol. 4708, pp. 465–476. Springer, Heidelberg (2007)

7. Crochemore, M., Ilie, L.: Maximal repetitions in strings. J. Comput. Syst. Sci. **74**(5), 796–807 (2008)
8. Crochemore, M., Ilie, L., Tinta, L.: Towards a solution to the "runs" conjecture. In: Ferragina, P., Landau, G.M. (eds.) CPM 2008. LNCS, vol. 5029, pp. 290–302. Springer, Heidelberg (2008)
9. Crochemore, M., Iliopoulos, C.S., Kubica, M., Radoszewski, J., Rytter, W., Waleń, T.: Extracting powers and periods in a word from its runs structure. Theor. Comput. Sci. **521**, 29–41 (2014)
10. Fischer, J., Holub, Š., I, T., Lewenstein, M.: Beyond the runs theorem. In: Iliopoulos, C., Puglisi, S., Yilmaz, E. (eds.) SPIRE 2015. LNCS, vol. 9309, pp. 277–286. Springer, Heidelberg (2015)
11. Gawrychowski, P., Kociumaka, T., Rytter, W., Waleń, T.: Faster longest common extension queries in strings over general alphabets. In: Grossi, R., Lewenstein, M. (eds.) 27th Annual Symposium on Combinatorial Pattern Matching, CPM 2016. LIPIcs, vol. 54, pp. 5:1–5:13. Schloss Dagstuhl (2016)
12. Giraud, M.: Not so many runs in strings. In: Martín-Vide, C., Otto, F., Fernau, H. (eds.) LATA 2008. LNCS, vol. 5196, pp. 232–239. Springer, Heidelberg (2008)
13. Hohlweg, C., Reutenauer, C.: Lyndon words, permutations and trees. Theor. Comput. Sci. **307**(1), 173–178 (2003)
14. Kociumaka, T., Radoszewski, J., Rytter, W., Waleń, T.: Internal pattern matching queries in a text and applications. In: Indyk, P. (ed.) 26th Annual ACM-SIAM Symposium on Discrete Algorithms, SODA 2015, pp. 532–551. SIAM (2015)
15. Kolpakov, R.M., Kucherov, G.: Finding maximal repetitions in a word in linear time. In: 40th Annual Symposium on Foundations of Computer Science, FOCS 1999, pp. 596–604. IEEE Computer Society (1999)
16. Kolpakov, R.M., Kucherov, G.: On maximal repetitions in words. J. Discrete Algorithms, 159–186. Special Issue of Matching Patterns, Hermes Science Publishing (2000). https://www.amazon.com/Matching-Patterns-Crochemore/dp/190339807X
17. Kosolobov, D.: Computing runs on a general alphabet. Inf. Process. Lett. **116**(3), 241–244 (2016)
18. Nash-Williams, C.S.J.A.: Decompositions of finite graphs into forests. J. London Math. Soc. **39**, 12 (1964)
19. Puglisi, S.J., Simpson, J., Smyth, W.F.: How many runs can a string contain? Theor. Comput. Sci. **401**(1–3), 165–171 (2008)
20. Rytter, W.: The number of runs in a string: improved analysis of the linear upper bound. In: Durand, B., Thomas, W. (eds.) STACS 2006. LNCS, vol. 3884, pp. 184–195. Springer, Heidelberg (2006)
21. Rytter, W.: The number of runs in a string. Inf. Comput. **205**(9), 1459–1469 (2007)

The Smallest Grammar Problem Revisited

Danny Hucke[(⊠)], Markus Lohrey, and Carl Philipp Reh

University of Siegen, Siegen, Germany
{hucke,lohrey,reh}@eti.uni-siegen.de

Abstract. In a seminal paper of Charikar et al. on the smallest grammar problem, the authors derive upper and lower bounds on the approximation ratios for several grammar-based compressors, but in all cases there is a gap between the lower and upper bound. Here we close the gaps for LZ78 and BISECTION by showing that the approximation ratio of LZ78 is $\Theta((n/\log n)^{2/3})$, whereas the approximation ratio of BISECTION is $\Theta((n/\log n)^{1/2})$. We also derive a lower bound for a smallest grammar for a word in terms of its number of LZ77-factors, which refines existing bounds of Rytter. Finally, we improve results of Arpe and Reischuk relating grammar-based compression for arbitrary alphabets and binary alphabets.

1 Introduction

The idea of grammar-based compression is based on the fact that in many cases a word w can be succinctly represented by a context-free grammar that produces exactly w. Such a grammar is called a *straight-line program* (SLP) for w. In the best case, one gets an SLP of size $O(\log n)$ for a word of length n, where the size of an SLP is the total length of all right-hand sides of the rules of the grammar. A grammar-based compressor is an algorithm that produces for a given word w an SLP $\mathbb{A}$ for w, where, of course, $\mathbb{A}$ should be smaller than w. Grammar-based compressors can be found at many places in the literature. Probably the best known example is the classical LZ78-compressor of Lempel and Ziv [17]. Indeed, it is straightforward to transform the LZ78-representation of a word w into an SLP for w. Other well-known grammar-based compressors are BISECTION [9], SEQUITUR [13], and RePair [10], just to mention a few.

One of the first appearances of straight-line programs in the literature are [2,5], where they are called *word chains* (since they generalize addition chains from numbers to words). In [2], Berstel and Brlek prove that the function $g(k, n) = \max\{g(w) \mid w \in \{1, \ldots, k\}^n\}$, where $g(w)$ is the size of a smallest SLP for the word w, is in $\Theta(n/\log_k n)$. Note that $g(k, n)$ measures the worst case SLP-compression over all words of length n over a k-letter alphabet. The first systematic investigations of grammar-based compressors are [4,8]. Whereas in [8], grammar-based compressors are used for universal lossless compression (in the information-theoretic sense), Charikar et al. study in [4] the worst case approximation ratio of grammar-based compressors. For a given grammar-based compressor $\mathcal{C}$ that computes from a given word w an SLP $\mathcal{C}(w)$ for w one defines

© Springer International Publishing AG 2016
S. Inenaga et al. (Eds.): SPIRE 2016, LNCS 9954, pp. 35–49, 2016.
DOI: 10.1007/978-3-319-46049-9_4

the approximation ratio of $\mathcal{C}$ on w as the quotient of the size of $\mathcal{C}(w)$ and the size $g(w)$ of a smallest SLP for w. The approximation ratio $\alpha_{\mathcal{C}}(n)$ is the maximal approximation ratio of $\mathcal{C}$ among all words of length n over any alphabet. In [4] the authors compute upper and lower bounds for the approximation ratios of several grammar-based compressors (among them are the compressors mentioned above), but for none of the compressors the lower and upper bounds match. Our first main contribution (Sect. 3) closes the gaps for LZ78 and BISECTION. For this we improve the corresponding lower bounds from [4] and obtain the approximation ratios $\Theta((n/\log n)^{1/2})$ for BISECTION and $\Theta((n/\log n)^{2/3})$ for LZ78. For BISECTION (resp., LZ78), we prove this lower bound for a binary (resp., ternary) alphabet.

In Sect. 4 we compare the size of a smallest SLP for a word w with the number of factors of the LZ77-factorization of w (we denote the latter with $g_{\mathsf{LZ77}}(w)$). Rytter [14] proved for every word w of length n the following bounds on the size $g(w)$ of a smallest SLP for w: $g(w) \geq g_{\mathsf{LZ77}}(w)$ and $g(w) \in O(g_{\mathsf{LZ77}}(w) \cdot \log n)$. This leads to the question whether the upper bound $g(w) \in O(g_{\mathsf{LZ77}}(w) \cdot \log n)$ on $g(w)$ can be improved. This would have immediate consequences for grammar-based compression: If one could construct in polynomial time an SLP of size $o(g_{\mathsf{LZ77}}(w) \cdot \log n)$ for a given word w, then one would obtain a grammar-based compressor with an approximation ratio of $o(\log n)$. Currently, the theoretically best grammar-based compressors (which all work in linear time) achieve an approximation ratio in $O(\log(n/g(w)))$ [4,7,14], and a polynomial time grammar-based compressor with an approximation ratio in $o(\log n / \log \log n)$ would imply a spectacular breakthrough on a long standing open problem on approximating addition chains [4]. Here, we partially answer the above question whether the bound $g(w) \in O(g_{\mathsf{LZ77}}(w) \cdot \log n)$ is sharp. Using a Kolmogorov complexity argument we construct a sequence of words w_n for which $g(w_n) \in \Omega(g_{\mathsf{LZ77}}(w_n) \cdot \log |w_n| / \log \log |w_n|)$.

Our last contribution deals with the hardness of the smallest grammar problem for words over a binary alphabet. The smallest grammar problem is the problem of computing a smallest grammar for a given input word. Storer and Szymanski [15] and Charikar et al. [4] proved that the smallest grammar problem cannot be solved in polynomial time unless $\mathsf{P} = \mathsf{NP}$. Even worse, unless $\mathsf{P} = \mathsf{NP}$ one cannot compute in polynomial time for a given word w an SLP of size $< 8569/8568 \cdot g(w)$ [4]. The construction in [4] uses an alphabet of unbounded size, and it was open whether this complexity lower bound also holds for words over a fixed alphabet. In [4] it is remarked that the construction in [15] shows that the smallest grammar problem for words over a ternary alphabet cannot be solved in polynomial time unless $\mathsf{P} = \mathsf{NP}$. But this is not clear at all, see the recent paper [3] for a detailed explanation. In the same paper [3] it was shown that the smallest grammar problem for an alphabet of size 24 cannot be solved in polynomial time unless $\mathsf{P} = \mathsf{NP}$ using a rather complicated construction [3]. It is far from clear whether this construction can be adapted so that it works also for a binary alphabet. Another idea is to reduce the smallest grammar problem for unbounded alphabets to the smallest grammar problem for a binary alphabet.

This route was investigated in [1], where the following result was shown: If there is a polynomial time grammar-based compressor with approximation ratio c (a constant) on binary words, then there is a polynomial time grammar-based compressor with approximation ratio $24c + \varepsilon$ for every $\varepsilon > 0$ on arbitrary words. The construction in [1] uses a quite technical block encoding of arbitrary alphabets into a binary alphabet. Here, we present a very simple construction that encodes the i-th alphabet symbol by $a^i b$, which yields the same result as [1] but with $24c + \varepsilon$ replaced by 6.

2   Straight-Line Programs

Let $w = a_1 \cdots a_n$ $(a_1, \ldots, a_n \in \Sigma)$ be a *word* over an *alphabet* Σ. The length $|w|$ of w is n and we denote by ε the word of length 0. Let $\Sigma^+ = \Sigma^* \setminus \{\varepsilon\}$ be the set of nonempty words. For $w \in \Sigma^+$, we call $v \in \Sigma^+$ a *factor* of w if there exist $x, y \in \Sigma^*$ such that $w = xvy$. If $x = \varepsilon$ (respectively $y = \varepsilon$) then we call v a *prefix* (respectively *suffix*) of w. A factorization of w is a decomposition $w = f_1 \cdots f_\ell$ into factors $f_1, \ldots, f_\ell$. For words $w_1, \ldots, w_n \in \Sigma^*$, we further denote by $\prod_{i=j}^{n} w_i$ the word $w_j w_{j+1} \cdots w_n$ if $j \leq n$ and ε otherwise.

A *straight-line program*, briefly SLP, is a context-free grammar that produces a single word $w \in \Sigma^+$. Formally, it is a tuple $\mathbb{A} = (N, \Sigma, P, S)$, where N is a finite set of nonterminals with $N \cap \Sigma = \emptyset$, $S \in N$ is the start nonterminal, and P is a finite set of productions (or rules) of the form $A \to w$ for $A \in N$, $w \in (N \cup \Sigma)^+$ such that: (i) For every $A \in N$, there exists exactly one production of the form $A \to w$, and (ii) the binary relation $\{(A, B) \in N \times N \mid (A \to w) \in P, B \text{ occurs in } w\}$ is acyclic. Every nonterminal $A \in N$ produces a unique string $\text{val}_{\mathbb{A}}(A) \in \Sigma^+$. The string defined by $\mathbb{A}$ is $\text{val}(\mathbb{A}) = \text{val}_{\mathbb{A}}(S)$. We omit the subscript $\mathbb{A}$ when it is clear from the context. The *size* of the SLP $\mathbb{A}$ is $|\mathbb{A}| = \sum_{(A \to w) \in P} |w|$. We will use the following lemma which summarizes known results about SLPs.

Lemma 1. *Let Σ be a finite alphabet.*

1. *For every word $w \in \Sigma^+$ of length n, there exists an SLP $\mathbb{A}$ of size $O(n/\log n)$ such that $\text{val}(\mathbb{A}) = w$.*
2. *For an SLP $\mathbb{A}$ and a number $n > 0$, there exists an SLP $\mathbb{B}$ of size $|\mathbb{A}| + O(\log n)$ such that $\text{val}(\mathbb{B}) = \text{val}(\mathbb{A})^n$.*
3. *For SLPs $\mathbb{A}_1$ and $\mathbb{A}_2$ there exists an SLP $\mathbb{B}$ of size $|\mathbb{A}_1| + |\mathbb{A}_2|$ such that $\text{val}(\mathbb{B}) = \text{val}(\mathbb{A}_1)\text{val}(\mathbb{A}_2)$.*
4. *For given words $w_1, \ldots, w_n \in \Sigma^*$, $u \in \Sigma^+$ and SLPs $\mathbb{A}_1, \mathbb{A}_2$ with $\text{val}(\mathbb{A}_1) = u$ and $\text{val}(\mathbb{A}_2) = w_1 x w_2 x \cdots w_{n-1} x w_n$ for a symbol $x \notin \Sigma$, there exists an SLP $\mathbb{B}$ of size $|\mathbb{A}_1| + |\mathbb{A}_2|$ such that $\text{val}(\mathbb{B}) = w_1 u w_2 u \cdots w_{n-1} u w_n$.*

Statement 1 can be found for instance in [2]. Statements 2 and 3 are shown in [4]. The proof of 4 is straightforward: Simply replace in the SLP $\mathbb{A}_2$ every occurrence of the terminal x by the start nonterminal of $\mathbb{A}_1$ and add all rules of $\mathbb{A}_1$ to $\mathbb{A}_2$.

We denote by $g(w)$ the size of a smallest SLP producing the word $w \in \Sigma^+$. The maximal size of a smallest SLP for all words of length n over an alphabet of size k is

$$g(k,n) = \max\{g(w) \mid w \in [1,k]^n\},$$

where $[1,k] = \{1,\ldots,k\}$. By point 1 of Lemma 1 we have $g(k,n) \in O(n/\log_k n)$. In fact, Berstel and Brlek proved in [2] that $g(k,n) \in \Theta(n/\log_k n)$. The following result provides further information about the function $g(k,n)$:

Proposition 2. *Let* $n_k = 2k^2 + 2k + 1$ *for* $k > 0$. *Then (i)* $g(k,n) < n$ *for* $n > n_k$ *and (ii)* $g(k,n) = n$ *for* $n \leq n_k$.

Proof. Let $\Sigma_k = \{a_1,\ldots,a_k\}$ and let $M_{n,\ell} \subseteq \Sigma_k^*$ be the set of all words w where a factor v of length ℓ occurs at least n times without overlap. It is easy to see that $g(w) < |w|$ if and only if $w \in M_{3,2} \cup M_{2,3}$. Hence, we have to show that every word $w \notin M_{3,2} \cup M_{2,3}$ has length at most $2k^2 + 2k + 1$. Moreover, we present words $w_k \in \Sigma_k^*$ of length $2k^2 + 2k + 1$ such that $w_k \notin M_{3,2} \cup M_{2,3}$.

Let $w \notin M_{3,2} \cup M_{2,3}$. Consider a factor $a_i a_j$ of length two. If $i \neq j$ then this factor does not overlap itself, and thus $a_i a_j$ occurs at most twice in w. Now consider $a_i a_i$. Then w contains at most four (possibly overlapping) occurrence of $a_i a_i$, because five occurrences of $a_i a_i$ would yield at least three non-overlapping occurrences of $a_i a_i$. It follows that w has at most $2(k^2 - k) + 4k$ positions where a factor of length 2 starts, which implies $|w| \leq 2k^2 + 2k + 1$.

Now we create a word $w_k \notin M_{3,2} \cup M_{2,3}$ which realizes the above maximal occurrences of factors of length 2:

$$w_k = \left(\prod_{i=1}^{k} a_{k-i+1}^5\right) \prod_{i=1}^{k-1} \left(\prod_{i+2}^{j=k} (a_j a_i)^2\right) a_{i+1} a_i a_{i+1}$$

For example we have $w_3 = a_3^5 a_2^5 a_1^5 (a_3 a_1)^2 a_2 a_1 a_2 a_3 a_2 a_3$. One can check that $|w_k| = 2k^2 + 2k + 1$ and $w_k \notin M_{3,2} \cup M_{2,3}$. □

3   Approximation Ratio

As mentioned in the introduction, there is no polynomial time algorithm that computes a smallest SLP for a given word, unless $P = NP$ [4,15]. This result motivates approximation algorithms which are called *grammar-based compressors*. A grammar-based compressor $\mathcal{C}$ computes for a word w an SLP $\mathcal{C}(w)$ such that $\mathrm{val}(\mathcal{C}(w)) = w$. The *approximation ratio* $\alpha_\mathcal{C}(w)$ of $\mathcal{C}$ for an input w is defined as $|\mathcal{C}(w)|/g(w)$. The worst-case approximation ratio $\alpha_\mathcal{C}(k,n)$ of $\mathcal{C}$ is the maximal approximation ratio over all words of length n over an alphabet of size k:

$$\alpha_\mathcal{C}(k,n) = \max\{\alpha_\mathcal{C}(w) \mid w \in [1,k]^n\} = \max\{|\mathcal{C}(w)|/g(w) \mid w \in [1,k]^n\}$$

If the alphabet size is unbounded, i.e. we allow alphabets of size $|w|$, then we write $\alpha_\mathcal{C}(n)$ instead of $\alpha_\mathcal{C}(n,n)$. This is the definition of the worst-case approximation ratio in [4]. The grammar-based compressors studied in our work are

BISECTION [9] and LZ78 [17]. We will abbreviate the approximation ratio of BISECTION (respectively LZ78) by α_{BI} (respectively α_{LZ78}). The families of words which we will use to prove new lower bounds for $\alpha_{\mathsf{BI}}(n)$ and $\alpha_{\mathsf{LZ78}}(n)$ are inspired by the constructions in [4].

3.1   BISECTION

The BISECTION algorithm [9] first splits an input word w with $|w| \geq 2$ as $w = w_1 w_2$ such that $|w_1| = 2^j$ for the unique number $j \geq 0$ with $2^j < |w| \leq 2^{j+1}$. This process is recursively repeated with w_1 and w_2 until we obtain words of length 1. During the process, we introduce a nonterminal for each distinct factor of length at least two and create a rule with two symbols on the right-hand side corresponding to the split. Note that if $w = u_1 u_2 \cdots u_k$ with $|u_i| = 2^n$ for all $i, 1 \leq i \leq k$, then the SLP produced by BISECTION contains a nonterminal for each distinct word u_i $(1 \leq i \leq k)$.

Example 3. BISECTION constructs an SLP for $w = ababbbaabbaaab$ as follows:

- $w = w_1 w_2$ with $w_1 = abababbaa$, $w_2 = bbaaab$
  Introduced rule: $S \rightarrow W_1 W_2$
- $w_1 = x_1 x_2$ with $x_1 = abab$, $x_2 = bbaa$, and $w_2 = x_2 x_3$ with $x_3 = ab$
  Introduced rules: $W_1 \rightarrow X_1 X_2$, $W_2 \rightarrow X_2 X_3$, $X_3 \rightarrow ab$
- $x_1 = x_3 x_3$, $x_2 = y_1 y_2$ with $y_1 = bb$ and $y_2 = aa$
  Introduced rules: $X_1 \rightarrow X_3 X_3$, $X_2 \rightarrow Y_1 Y_2$, $Y_1 \rightarrow bb$, $Y_2 \rightarrow aa$

BISECTION performs asymptotically optimal on unary words a^n since it produces an SLP of size $O(\log n)$. Therefore $\alpha_{\mathsf{BI}}(1, n) \in \Theta(1)$. The following bounds on the approximation ratio for alphabets of size at least two are proven in [4, Theorems 5 and 6]:

$$\alpha_{\mathsf{BI}}(2, n) \in \Omega(\sqrt{n}/\log n) \tag{1}$$

$$\alpha_{\mathsf{BI}}(n) \in O(\sqrt{n/\log n}) \tag{2}$$

We improve the lower bound (1) so that it matches the upper bound (2):

Theorem 4. *For every $k, 2 \leq k \leq n$ we have $\alpha_{\mathsf{BI}}(k, n) \in \Theta(\sqrt{n/\log n})$.*

Proof. The upper bound (2) implies that $\alpha_{\mathsf{BI}}(k, n) \in O(\sqrt{n/\log n})$ for all $k, 2 \leq k \leq n$. So it suffices to show $\alpha_{\mathsf{BI}}(2, n) \in \Omega(\sqrt{n/\log n})$. We first show that $\alpha_{\mathsf{BI}}(3, n) \in \Omega(\sqrt{n/\log n})$. In a second step, we encode a ternary alphabet into a binary alphabet while preserving the approximation ratio.

For every $k \geq 2$ let $\mathrm{bin}_k : \{0, 1, \ldots, k-1\} \rightarrow \{0, 1\}^{\lceil \log_2 k \rceil}$ be the function where $\mathrm{bin}_k(j)$ $(0 \leq j \leq k-1)$ is the binary representation of j filled with leading zeros (e.g. $\mathrm{bin}_9(3) = 0011$). We further define for every $k \geq 2$ the word

$$u_k = \left(\prod_{j=0}^{k-2} \mathrm{bin}_k(j) a^{m_k} \right) \mathrm{bin}_k(k-1),$$

where $m_k = 2^{k-\lceil \log_2 k \rceil} - \lceil \log_2 k \rceil$. For instance $k = 4$ leads to $m_k = 2$ and $u_4 = 00aa01aa10aa11$. We analyse the approximation ratio $\alpha_{\mathsf{BI}}(s_k)$ for the word

$$s_k = \left(u_k a^{m_k+1}\right)^{m_k} u_k.$$

Claim 1. The SLP produced by BISECTION on input s_k has size $\Omega(2^k)$.

If s_k is split into non-overlapping factors of length $m_k + \lceil \log_2 k \rceil = 2^{k-\lceil \log_2 k \rceil}$, then the resulting set F_k of factors is

$$F_k = \{a^i \mathrm{bin}_k(j) a^{m_k-i} \mid 0 \le j \le k-1,\ 0 \le i \le m_k\}.$$

For example s_4 consecutively consists of the factors $00aa$, $01aa$, $10aa$, $11aa$, $a00a$, $a01a$, $a10a$, $a11a$, $aa00$, $aa01$, $aa10$ and $aa11$. The size of F_k is $(m_k + 1) \cdot k \in \Theta(2^k)$, because all factors are pairwise different and $m_k \in \Theta(2^k/k)$. It follows that the SLP produced by BISECTION on input s_k has size $\Omega(2^k)$, because the length of each factor in F_k is a power of two and thus BISECTION creates a nonterminal for each distinct factor in F_k.

Claim 2. A smallest SLP producing s_k has size $O(k)$.

There is an SLP of size $O(\log m_k) = O(k)$ for the word a^{m_k} by Lemma 1 (point 2). This yields an SLP for u_k of size $O(k) + g(u'_k)$ by Lemma 1 (point 4), where $u'_k = (\prod_{i=0}^{k-2} \mathrm{bin}_k(i)x)\mathrm{bin}_k(k-1)$ is obtained from u_k by replacing all occurrences of a^{m_k} by a fresh symbol x. The word u'_k has length $\Theta(k \log k)$. Applying point 1 of Lemma 1 (note that u'_k is a word over a ternary alphabet) it follows that

$$g(u'_k) \in O\left(\frac{k \log k}{\log(k \log k)}\right) = O\left(\frac{k \log k}{\log k + \log \log k}\right) = O(k).$$

Hence $g(u_k) \in O(k)$. Finally, the SLP of size $O(k)$ for u_k yields an SLP of size $O(k)$ for s_k again using Lemma 1 (points 2 and 3).

In conclusion: We showed that a smallest SLP for s_k has size $O(k)$, while BISECTION produces an SLP of size $\Omega(2^k)$. This implies $\alpha_{\mathsf{BI}}(s_k) \in \Omega(2^k/k)$. Let $n = |s_k|$. Since s_k is the concatenation of $\Theta(2^k)$ factors of length $\Theta(2^k/k)$, we have $n \in \Theta(2^{2k}/k)$ and thus $\sqrt{n} \in \Theta(2^k/\sqrt{k})$. This yields $\alpha_{\mathsf{BI}}(s_k) \in \Omega(\sqrt{n/k})$. Together with $k \in \Theta(\log n)$ we obtain $\alpha_{\mathsf{BI}}(3, n) \in \Omega(\sqrt{n/\log n})$.

Let us now encode words over $\{0, 1, a\}$ into words over $\{0, 1\}$. Consider the homomorphism $f : \{0, 1, a\}^* \to \{0, 1\}^*$ with $f(0) = 00$, $f(1) = 01$ and $f(a) = 10$. Then we can prove the same approximation ratio of BISECTION for the input $f(s_k) \in \{0, 1\}^*$ that we proved for s_k above: The size of a smallest SLP for $f(s_k)$ is at most twice as large as the size of a smallest SLP for s_k, because an SLP for s_k can be transformed into an SLP for $f(s_k)$ by replacing every occurrence of a symbol $x \in \{0, 1, a\}$ by $f(x)$. Moreover, if we split $f(s_k)$ into non-overlapping factors of twice the length as we considered for s_k, then we obtain the factors from $f(F_k)$, whose length is again a power of two. Since f is injective, we have $|f(F_k)| = |F_k| \in \Theta(2^k)$. $\qquad \square$

3.2   LZ78

The LZ78 algorithm on input $w \in \Sigma^+$ implicitly creates a list of words $f_1, \ldots, f_\ell$ (which we call the LZ78-*factorization*) with $w = f_1 \cdots f_\ell$ such that the following properties hold, where we set $f_0 = \varepsilon$:

- $f_i \neq f_j$ for all $i, j, 0 \leq i, j \leq \ell - 1$ with $i \neq j$.
- For all $i, 1 \leq i \leq \ell - 1$ there exist $j, 0 \leq j < i$ and $a \in \Sigma$ such that $f_i = f_j a$.
- $f_\ell = f_i$ for some $0 \leq i \leq \ell - 1$.

Note that the LZ78-factorization is unique for each word w. To compute it, the LZ78 algorithm needs ℓ steps performed by a single left-to-right pass. In the kth step ($1 \leq k \leq \ell - 1$) it chooses the factor f_k as the shortest prefix of the unprocessed suffix $f_k \cdots f_\ell$ such that $f_k \neq f_i$ for all $i < k$. If there is no such prefix, then the end of w is reached and the algorithm sets f_ℓ to the (possibly empty) unprocessed suffix of w.

The factorization $f_1, \ldots, f_\ell$ yields an SLP for w of size at most 3ℓ as described in the following example:

Example 5. The LZ78-factorization of $w = aabaaababababaa$ is a, ab, aa, aba, b, $abab$, aa and leads to an SLP with the following rules:

- $S \to F_1 F_2 F_3 F_4 F_5 F_6 F_3$
- $F_1 \to a$, $F_2 \to F_1 b$, $F_3 \to F_1 a$, $F_4 \to F_2 a$, $F_5 \to b$, $F_6 \to F_4 b$

We have a nonterminal F_i for each factor f_i ($1 \leq i \leq 6$) such that $\text{val}_\mathbb{A}(F_i) = f_i$. The last factor aa is represented in the start rule by the nonterminal F_3.

The LZ78-factorization of a^n ($n > 0$) is $a^1, a^2, \ldots, a^m, a^k$, where $k \in \{0, \ldots, m\}$ such that $n = k + \sum_{i=1}^{m} i$. Note that $m \in \Theta(\sqrt{n})$ and thus $\alpha_{\text{LZ78}}(1, n) \in \Theta(\sqrt{n}/\log n)$. The following bounds for the worst-case approximation ratio of LZ78 were shown in [4, Theorems 3 and 4]:

$$\alpha_{\text{LZ78}}(2, n) \in \Omega(n^{2/3}/\log n) \tag{3}$$

$$\alpha_{\text{LZ78}}(n) \in O((n/\log n)^{2/3}) \tag{4}$$

For ternary (or larger) alphabets, we will improve the lower bound so that it matches the upper bound in (4).

Theorem 6. *For every $k \geq 3$ we have $\alpha_{\text{LZ78}}(k, n) \in \Theta((n/\log n)^{2/3})$.*

Proof. Due to (4) it suffices to show $\alpha_{\text{LZ78}}(3, n) \in \Omega((n/\log n)^{2/3})$. For $k \geq 2, m \geq 1$, let $u_{m,k} = (a^k b^m c)^{k(m+2)-1}$ and $v_{m,k} = (\prod_{i=1}^{m} b^i a^k)^{k^2}$. We now analyse the approximation ratio of LZ78 on the words

$$s_{m,k} = a^{k(k+1)/2} \, b^{m(m+1)/2} \, u_{m,k} \, v_{m,k}.$$

For example we have $u_{2,4} = (a^4 b^2 c)^{15}$, $v_{2,4} = (ba^4 b^2 a^4)^{16}$ and $s_{2,4} = a^{10} b^3 u_{2,4} v_{2,4}$.

Claim 1. The SLP produced by LZ78 on input $s_{m,k}$ has size $\Theta(k^2 m)$.

We consider the LZ78-factorization $f_1, \ldots, f_\ell$ of $s_{m,k}$. The prefix $a^{k(k+1)/2}$ produces the factors $f_i = a^i$ for every $i, 1 \leq i \leq k$ and the substring $b^{m(m+1)/2}$ produces the factors $f_{k+i} = b^i$ for every $i, 1 \leq i \leq m$.

We next show that the substring $u_{m,k}$ then produces (among other factors) all factors $a^i b^j$, where $1 \leq i \leq k$, $1 \leq j \leq m$. All other factors produced by $u_{m,k}$ contain the letter c and therefore do not affect the factorization of the final suffix $v_{m,k} \in \{a, b\}^*$.

The first factors of $u_{m,k}$ in $s_{m,k}$ are $f_{k+m+1} = a^k b$ and $f_{k+m+2} = b^{m-1}c$, which together form the first occurrence of $a^k b^m c$. The next two factors are $a^k b^2$ and $b^{m-2}c$. This pattern continues and the prefix $(a^k b^m c)^m$ of $u_{m,k}$ yields the next $2m$ factors $f_{k+m+2i-1} = a^k b^i$ and $f_{k+m+2i} = b^{m-i}c$ for every $i, 1 \leq i \leq m$. The factorization of $u_{m,k}$ continues with $f_{k+3m+1} = a^k b^m c$ followed by $f_{k+3m+2} = a^k b^m ca$. Next, we have $f_{k+3m+3} = a^{k-1}b$ and $f_{k+3m+4} = b^{m-1}ca$, which is the beginning of a similar pattern as we discovered for $(a^k b^m c)^m$. Therefore, the next $2m$ factors are $f_{k+3m+2i+1} = a^{k-1}b^i$ and $f_{k+3m+2i+2} = b^{m-i}ca$ for every $i, 1 \leq i \leq m$. The next two factors are $f_{k+5m+3} = a^{k-1}b^m c$ followed by $f_{k+5m+4} = a^k b^m ca^2$. The iteration of these arguments yields k (consecutive) blocks of $2m + 2$ factors (resp. $2m + 1$ in the last block) in $u_{m,k}$:

1st	block:	$\prod_{i=1}^{m} ($	$a^k b^i$	$b^{m-i}c$ $)$	$a^k b^m c$	$a^k b^m ca$
2nd	block:	$\prod_{i=1}^{m} ($	$a^{k-1}b^i$	$b^{m-i}ca$ $)$	$a^{k-1}b^m c$	$a^k b^m ca^2$
...						
$(k-1)^{th}$	block:	$\prod_{i=1}^{m} ($	$a^2 b^i$	$b^{m-i}ca^{k-2}$ $)$	$a^2 b^m c$	$a^k b^m ca^{k-1}$
k^{th}	block:	$\prod_{i=1}^{m} ($	ab^i	$b^{m-i}ca^{k-1}$ $)$	$ab^m c$	

We will show that the remaining suffix $v_{m,k}$ of $s_{m,k}$ produces then the set of factors

$$\{a^i b^p a^j \mid 0 \leq i \leq k-1,\ 1 \leq j \leq k,\ 1 \leq p \leq m\}.$$

Let $x = k + m + k(2m + 2) - 1$ and note that this is the number of factors that we have produced so far. The factorization of $v_{m,k}$ in $s_{m,k}$ slightly differs whether m is even or is odd. We now assume that m is even and explain the difference to the other case afterwards. The first factor of $v_{m,k}$ in $s_{m,k}$ is $f_{x+1} = ba$. We already have produced the factors $a^{k-1}b^i$ for every $i, 1 \leq i \leq m$ and hence $f_{x+i} = a^{k-1}b^i a$ for every $i, 2 \leq i \leq m$ and $f_{x+m+1} = a^{k-1}ba$. The next m factors are $f_{x+m+i} = a^{k-1}b^i a^2$ if i is even, $f_{x+m+i} = a^{k-2}b^i a$ if i is odd $(2 \leq i \leq m)$ and $f_{x+2m+1} = a^{k-2}ba$. This pattern continues: The next m factors are $f_{x+2m+i} = a^{k-1}b^i a^3$ if i is even, $f_{x+2m+i} = a^{k-3}b^i a$ if i is odd $(2 \leq i \leq m)$ and $f_{x+3m+1} = a^{k-3}ba$ and so on. Hence, we get the following sets of factors for $(\prod_{i=1}^{m} b^i a^k)^k$:

(i) $\{a^{k-i}b^p a \mid 1 \leq i \leq k,\ 1 \leq p \leq m,\ p \text{ is odd}\}$ for $f_{x+1}, f_{x+3} \cdots, f_{x+km-1}$

(ii) $\{a^{k-1}b^p a^j \mid 1 \leq j \leq k,\ 1 \leq p \leq m,\ p \text{ is even}\}$ for $f_{x+2}, f_{x+4}, \ldots, f_{x+km}$

The remaining word then starts with the factor $f_{y+1} = ba^2$, where $y = x + km$. Now the former pattern can be adapted to the next k repetitions of $\prod_{i=1}^{m} b^i a^k$ which gives us the following factors:

(i) $\{a^{k-i}b^p a^2 \mid 1 \leq i \leq k,\ 1 \leq p \leq m,\ p \text{ is odd}\}$ for $f_{y+1}, f_{y+3} \cdots, f_{y+km-1}$
(ii) $\{a^{k-2}b^p a^j \mid 1 \leq j \leq k,\ 1 \leq p \leq m,\ p \text{ is even}\}$ for $f_{y+2}, f_{y+4}, \ldots, f_{y+km}$

The iteration of this process then reveals the whole pattern and thus yields the claimed factorization of $v_{m,k}$ in $s_{m,k}$ into factors $a^i b^p a^j$ for every $i, 0 \leq i \leq k-1$, $j, 1 \leq j \leq k$ and $p, 1 \leq p \leq m$. If m is odd then the patterns in (1) and (2) switch after each occurrence of $\prod_{i=1}^{m} b^i a^k$, which does not affect the result but makes the pattern slightly more complicated. But the case that m is even suffices in order to derive the lower bound from the theorem.

We conclude that there are exactly $k + m + k(2m + 2) - 1 + k^2 m$ factors (ignoring $f_\ell = \varepsilon$) and hence the SLP produced by LZ78 on input $s_{m,k}$ has size $\Theta(k^2 m)$.

Claim 2. A smallest SLP producing $s_{m,k}$ has size $O(\log k + m)$.

We will combine the points stated in Lemma 1 to prove this claim. Points 2 and 3 yield an SLP of size $O(\log k + \log m)$ for the prefix $a^{k(k+1)/2} b^{m(m+1)/2} u_{m,k}$ of $s_{m,k}$. To bound the size of an SLP for $v_{m,k}$ note at first that there is an SLP of size $O(\log k)$ producing a^k by point 2 of Lemma 1. Applying point 4 and again point 2, it follows that there is an SLP of size $O(\log k) + g(v'_{m,k})$ producing $v_{m,k}$, where $v'_{m,k} = \prod_{i=1}^{m} b^i x$ for some fresh letter x. To get a small SLP for $v'_{m,k}$, we can introduce m nonterminals $B_1, \ldots, B_m$ producing $b^1, \ldots, b^m$ by adding rules $B_1 \to b$ and $B_{i+1} \to B_i b$ $(1 \leq i \leq m - 1)$. This is enough to get an SLP of size $O(m)$ for $v'_{m,k}$ and therefore an SLP of size $O(\log k + m)$ for $v_{m,k}$. Together with our first observation and point 3 of Lemma 1 this yields an SLP of size $O(\log k + m)$ for $s_{m,k}$.

Claim 1 and 2 imply $\alpha_{\mathsf{LZ78}}(s_{m,k}) \in \Omega(k^2 m/(\log k + m))$. Let us now fix $m = \lceil \log k \rceil$. We get $\alpha_{\mathsf{LZ78}}(s_{m,k}) \in \Omega(k^2)$. Moreover, for the length $n = |s_{m,k}|$ of $s_{m,k}$ we have $n \in \Theta(k^3 m + k^2 m^2) = \Theta(k^3 \log k)$. We get $\alpha_{\mathsf{LZ78}}(s_{m,k}) \in \Omega((n/\log k)^{2/3})$ which together with $\log n \in \Theta(\log k)$ finishes the proof. □

It remains open whether also $\alpha_{\mathsf{LZ78}}(2, n) \in \Theta((n/\log n)^{2/3})$ holds. In contrast to BISECTION it is not clear how to encode a ternary alphabet into a binary alphabet while preserving the approximation ratio for LZ78.

4   LZ77 and Composition Systems

The LZ77-*factorization* of a non-empty word $w \in \Sigma^+$ is $w = f_1 f_2 \cdots f_m$, where for every $i, 1 \leq i \leq m$, f_i is (i) the longest non-empty prefix of $f_i f_{i+1} \cdots f_m$ which is a factor of $f_1 f_2 \cdots f_{i-1}$ or (ii) the first symbol of $f_i f_{i+1} \cdots f_m$ if such a prefix does not exist. Let $g_{\mathsf{LZ77}}(w) = m$ be the number of factors in the LZ77-factorization of w.

Example 7. The LZ77-factorization of $w = aabaaababababaa$ is a, a, b, aa, aba, ba, $baba$, a and we have $g_{LZ77}(w) = 8$.

We are interested in the following ratios, where $1 \leq k \leq n$:

$$\beta_{LZ77}(k, n) = \max\{g(w)/g_{LZ77}(w) \mid w \in [1, k]^n\} \text{ and } \beta_{LZ77}(n) = \beta_{LZ77}(n, n).$$

For a word w over a unary alphabet one has $g_{LZ77}(w) \in \Theta(\log |w|)$ and therefore $\beta_{LZ77}(1, n) \in \Theta(1)$. Rytter proved that for every word w, $g(w) \geq g_{LZ77}(w)$ and hence $\beta_{LZ77}(k, n) \geq 1$ for all $k, 1 \leq k \leq n$ [14].[1] Moreover, in the same paper, he constructed for a word w an SLP of size $O(g_{LZ77}(w) \cdot \log |w|)$. This yields $\beta_{LZ77}(n) \in O(\log n)$. Using Kolmogorov complexity we prove the lower bound $\beta_{LZ77}(2, n) \in \Omega(\log n / \log \log n)$.

For a partial recursive function $\phi : \{0, 1\}^* \rightarrow \{0, 1\}^*$ and a word $w \in \{0, 1\}^*$ let $C_\phi(w) = \min\{|p| \mid p \in \{0, 1\}^*, \phi(p) = w\}$ (where we define $\min(\emptyset) = \infty$) be the Kolmogorov complexity of w with respect to ϕ. The invariance theorem of Kolmogorov complexity states that there is a partial recursive surjective function $U : \{0, 1\}^* \rightarrow \{0, 1\}^*$ such that for every partial recursive function $\phi : \{0, 1\}^* \rightarrow \{0, 1\}^*$ there is a constant $c \geq 0$ with $C_U(w) \leq C_\phi(w) + c$ for all w. We fix such a function U (it can be obtained from a universal Turing machine) and define the Kolmogorov complexity of w as $C(w) := C_U(w)$. It is well known that for every $n \geq 0$ there is a word $w \in \{0, 1\}^n$ with $C(w) \geq n$ (such a word is called Kolmogorov random). See [11] for further details.

Theorem 8. $\beta_{LZ77}(2, n) \in \Omega(\log n / \log \log n)$.

Proof. Let $m \in \mathbb{N}$, $w \in \{0, 1\}^*$, $|w| = m^2$ and $C(w) \geq m^2$. We factorize w as $w = w_1 \cdots w_m$ where $|w_i| = m$ for every $i, 1 \leq i \leq m$. We encode every w_i into a binary number of size $\Theta(2^m)$ using the following (ranking) function $p : \{0, 1\}^* \rightarrow \mathbb{N}$: We define $p(u) = i$ if and only if u is the ith word in the length-lexicographic enumeration of all words from $\{0, 1\}^*$ (where $p(\varepsilon) = 0$). This is a computable bijection from $\{0, 1\}^*$ to $\mathbb{N}$ such that $p(x) \in \Theta(2^{|x|})$. Let $N_i = p(w_i)$ for every $i, 1 \leq i \leq m$. Thus, we have $N_i \in \Theta(2^m)$. Let $N = \max\{N_1, \ldots, N_m\} \in \Theta(2^m)$ and define the word $v = a^N \# a^{N_1} \# \ldots \# a^{N_m} \#$ over the alphabet $\{a, \#\}$. Let $\mathbb{A}$ be a smallest SLP for v. Note that v and hence $\mathbb{A}$ uniquely encodes the word w. Since an SLP of size k can be encoded by a bit string of size $O(k \log k)$ [16] and $C(w) \geq m^2$, it follows that $|\mathbb{A}| \cdot \log |\mathbb{A}| \in \Omega(m^2)$. Note that this is the point where the Kolmogorov randomness of w is applied. Moreover, there exists an SLP for v of size $O(m \cdot \log N) = O(m^2)$. Thus, $|\mathbb{A}| \in O(m^2)$, which together with $|\mathbb{A}| \cdot \log(|\mathbb{A}|) \in \Omega(m^2)$ implies $|\mathbb{A}| \in \Omega(m^2 / \log m)$ and hence $g(v) \in \Omega(m^2 / \log m)$. On the other hand, the LZ77-factorization of v has $O(m)$ factors: The prefix $a^N \#$ of v contributes $O(\log N) = O(m)$ factors. Because $N = \max\{N_1, \ldots, N_m\}$, every $a^{N_i} \#$, where $1 \leq i \leq m$, contributes at most one

---

[1] It is shown in [14] that every SLP in Chomsky normal form for w has at least $g_{LZ77}(w)$ many nonterminals. But the number of nonterminals in a smallest Chomsky normal form SLP for w is bounded by $g(w)$.

additional factor. Altogether, we get $g_{\mathsf{LZ77}}(v) \in O(m)$. Let $n = |v| \in \Theta(m \cdot 2^m)$, which implies $\log n \in \Theta(m)$. We get

$$\beta_{\mathsf{LZ77}}(2, n) \geq \frac{g(v)}{g_{\mathsf{LZ77}}(v)} \in \Omega\left(\frac{m^2}{m \log m}\right) = \Omega\left(\frac{m}{\log m}\right) = \Omega\left(\frac{\log n}{\log \log n}\right).$$

This concludes the proof.   □

It remains open, whether the lower bound in Theorem 8 can be raised to $\Omega(\log n)$.

A common generalization of SLPs and LZ77-factorizations are so called *composition systems* [6] or *Cut-SLP* [12] (briefly CSLP), which we define next. For a word $w = a_1 \cdots a_n \in \Sigma^*$ and $1 \leq i \leq j \leq n$ we define $w[i : j] = a_i \cdots a_j$. A CLSP $\mathbb{C} = (N, \Sigma, P, S)$ is defined analogously to an SLP but in addition may contain rules of the form $A \to B[i : j]$ for $A, B \in N$ and $1 \leq i \leq j \leq |\mathrm{val}(B)|$. We then define $\mathrm{val}(A) = \mathrm{val}(B)[i : j]$. The size of a right-hand side $B[i : j]$ is set to $|B[i : j]| = 1$ and the size of a CSLP is $|\mathbb{C}| = \sum_{(A \to w) \in P} |w|$. We denote by $g_{\mathsf{CSLP}}(w)$ the size of a smallest CSLP $\mathbb{C}$ such that $\mathrm{val}(\mathbb{C}) = w$ and define $\beta_{\mathsf{CSLP}}(k, n) = \max\{g(w)/g_{\mathsf{CSLP}}(w) \mid w \in [1, k]^n\}$ and $\beta_{\mathsf{CSLP}}(n) = \beta_{\mathsf{CSLP}}(n, n)$.

Note that if a non-empty word w has an LZ77-factorization $w = f_1 f_2 \cdots f_m$ of length m then $g_{\mathsf{CSLP}}(w) \leq 3m$: We introduce for every $i, 1 \leq i \leq m$ a nonterminal A_i which evaluates to $f_1 \cdots f_i$. For this, we set $A_1 \to f_1$ (f_1 must be a single symbol). For every $i, 2 \leq i \leq m$ we set $A_i \to A_{i-1} f_i$ if f_i is a single symbol and $A_i \to A_{i-1} B_i$, $B_i \to A_{i-1}[j : k]$ if $f_i = (f_1 \cdots f_{i-1})[j : k]$. Together with Theorem 8 this yields the lower bound in the following theorem. The upper bound follows easily using the techniques from [14].

Theorem 9. *We have* $\beta_{\mathsf{CSLP}}(2, n) \in \Omega(\log n / \log \log n)$ *and* $\beta_{\mathsf{CSLP}}(n) \in O(\log n)$.

5 Hardness of Grammar-Based Compression for Binary Alphabets

The goal of this section is to prove the following result:

Theorem 10. *Let* $c \geq 1$ *be a constant. If there exists a polynomial time grammar-based compressor* $\mathcal{C}$ *with* $\alpha_{\mathcal{C}}(2, n) \leq c$ *then there exists a polynomial time grammar-based compressor* $\mathcal{D}$ *with* $\alpha_{\mathcal{D}}(n) \leq 6c$.

For a factor $24 + \varepsilon$ (with $\varepsilon > 0$) instead of 6 this result was shown in [1] using a more complicated block encoding.

We split the proof of Theorem 10 in two lemmas that state translations between SLPs over arbitrary alphabets and SLPs over a binary alphabet. For the rest of this section fix the alphabets $\Sigma = \{a_0, \ldots, a_{k-1}\}$ and $\Sigma_2 = \{a, b\}$. To translate between these two alphabets, we define an injective homomorphism $\varphi : \Sigma^* \to \Sigma_2^*$ by

$$\varphi(a_i) = a^i b \quad (0 \leq i \leq k - 1). \tag{5}$$

Lemma 11. *Let $w \in \Sigma^*$ such that every symbol from Σ occurs in w. From an SLP $\mathbb{A}$ for w one can construct in polynomial time an SLP $\mathbb{B}$ for $\varphi(w)$ of size at most $3 \cdot |\mathbb{A}|$.*

Proof. To translate $\mathbb{A}$ into an SLP $\mathbb{B}$ for $\varphi(w)$, we first add the productions $A_0 \to b$ and $A_i \to aA_{i-1}$ for every $i, 1 \le i \le k-1$.

Finally, we replace in $\mathbb{A}$ every occurrence of $a_i \in \Sigma$ by A_i. This yields an SLP $\mathbb{B}$ for $\varphi(w)$ of size $|\mathbb{A}| + 2k - 1$. Because $k \le |\mathbb{A}|$ (since every symbol from Σ occurs in w), we obtain $|\mathbb{B}| \le 3 \cdot |\mathbb{A}|$. $\square$

Lemma 12. *Let $w \in \Sigma^*$ such that every symbol from Σ occurs in w. From an SLP $\mathbb{B}$ for $\varphi(w)$ one can construct in polynomial time an SLP $\mathbb{A}$ for w of size at most $2 \cdot |\mathbb{B}|$.*

Proof. A factor of a word from $\varphi(\Sigma^*)$ is of the form $s = a^{i_1}b \cdots a^{i_n}ba^{i_{n+1}}$ for some $n \ge 0$, and $0 \le i_1, \ldots, i_{n+1} \le k-1$. Take new symbols $\tilde{a}_i$, $0 \le i \le k-1$. Intuitively, $\tilde{a}_i$ is an abbreviation for a^i (whereas a_i is an abbreviation for $a^i b$). The symbols $\tilde{a}_i$ are only used during the construction for clarification, and disappear at the end. For the word $s = a^{i_1}b \cdots a^{i_n}ba^{i_{n+1}}$ define $\ell(s) \in \Sigma \cup \{\varepsilon\}$, $m(s) \in \Sigma^*$, and $r(s) \in \{\tilde{a}_i \mid 0 \le i \le k-1\}$ as follows:

$$\ell(s) = \begin{cases} a_{i_1} & \text{if } n \ge 1 \\ \varepsilon & \text{if } n = 0 \end{cases} \qquad m(s) = a_{i_2} \cdots a_{i_n} \qquad r(s) = \tilde{a}_{i_{n+1}}$$

Note that $\ell(s) = \varepsilon$ implies that $m(s) = \varepsilon$ as well. Finally, let

$$\psi(s) = \underbrace{a_{i_1}}_{\ell(s)} \underbrace{a_{i_2} \cdots a_{i_n}}_{m(s)} \underbrace{\tilde{a}_{i_{n+1}}}_{r(s)}.$$

Note that for every word $w \in \Sigma^*$ we have $\psi(\varphi(w)) = w\tilde{a}_0$.

Let $w \in \Sigma^*$ and $\mathbb{B} = (N, \Sigma_2, P, S)$ be an SLP for $\varphi(w)$. For a nonterminal $A \in N$ we define $\ell(A), m(A), r(A)$ as $\ell(\text{val}(A)), m(\text{val}(A)), r(\text{val}(A))$. We now define an SLP $\mathbb{A}'$ that contains for every nonterminal $A \in N$ a nonterminal A' such that $\text{val}(A') = m(A)$. Moreover, the algorithm also computes $\ell(A)$ and $r(A)$.

We define the productions of $\mathbb{A}'$ inductively over the structure of $\mathbb{B}$. Consider a production $(A \to \alpha) \in P$, where $\alpha = w_0A_1w_1A_2 \cdots w_{n-1}A_nw_n$ with $n \ge 0$, $A_1, \ldots, A_n \in N$, and $w_0, w_1, \ldots, w_n \in \Sigma_2^*$. Let $\ell_i = \ell(A_i)$ and $r_i = r(A_i)$. The right-hand side for A' is obtained as follows. We start with the word

$$\psi(w_0)\, \ell_1\, A_1'\, r_1\, \psi(w_1)\, \ell_2\, A_2'\, r_2 \cdots \psi(w_{n-1})\, \ell_n\, A_n'\, r_n\, \psi(w_n). \tag{6}$$

Note that each of the factors $\ell_i A_i' r_i$ produces (by induction) $\psi(\text{val}(A_i))$. Next we remove every A_i' that derives the empty word (which is equivalent to $m(A_i) = \varepsilon$). After this step, every occurrence of a symbol $\tilde{a}_i$ is either the last symbol of the word or it is followed by another symbol $\tilde{a}_j$ or a_j. This allows us to eliminate all occurrences of symbols $\tilde{a}_i$ except for the last symbol using the two reduction rules $\tilde{a}_i \tilde{a}_j \to \tilde{a}_{i+j}$ (which corresponds to $a^i a^j = a^{i+j}$) and $\tilde{a}_i a_j \to a_{i+j}$ (which corresponds to $a^i a^j b = a^{i+j} b$). If we perform these rules as long as possible (the order of applications is not relevant since these rules form a confluent and terminating system), only a single occurrence of a symbol $\tilde{a}_i$ at the end of the right-hand side will remain. The resulting word α' produces $\psi(A)$. Hence, we obtain the right-hand side for the nonterminal A' by removing the first symbol of α' if it is of the form a_i (this symbol is then $\ell(A)$) and the last symbol of α', which must be of the form $\tilde{a}_j$ (this symbol is $r(A)$).

Note that for the start variable S of $\mathbb{B}$ we must have $r(S) = \tilde{a}_0$ since val(S) belongs to the image of φ. Let $S' \to \sigma$ be the production for S' in $\mathbb{A}'$. We obtain the SLP $\mathbb{A}$ by replacing this production by $S' \to \ell(S)\sigma$. Since $\text{val}_{\mathbb{A}'}(S') = m(S)$ and $\text{val}_{\mathbb{B}}(S) = \varphi(w)$ we have $\text{val}_{\mathbb{A}}(S') = \ell(S)m(S) = w$.

To bound the size of $\mathbb{A}$ note that the length of the word in (6) is at most $|\alpha| + 2n$. But when forming the right-hand side of A', all symbols $r_1, \ldots, r_n$ are removed from (6). Hence, $|\mathbb{A}'|$ is bounded by the size of $\mathbb{B}$ plus the total number of occurrences of nonterminals in right-hand sides of $\mathbb{B}$, which is at most $2|\mathbb{B}| - 1$ (there is at least one terminal occurrence in a right-hand side). Since $|\mathbb{A}| = |\mathbb{A}'| + 1$ we get $|\mathbb{A}| \leq 2|\mathbb{B}|$.

It is easy to observe that the runtime of the algorithm is linear.   □

Example 13. Consider the production $A \to a^3 b a^5 A_1 a^3 A_2 a^2 b^2 A_3 a^2$ and assume that val$(A_1) = a^2$, val$(A_2) = aba^3ba$ and val$(A_3) = ba^2ba^3$. Hence, when we produce the right-hand side for A' we have: val$(A_1') = \varepsilon$, val$(A_2') = a_3$, val$(A_3') = a_2$, $\ell_1 = \varepsilon$, $r_1 = \tilde{a}_2$, $\ell_2 = a_1$, $r_2 = \tilde{a}_1$, $\ell_3 = a_0$, $r_3 = \tilde{a}_3$. We start with the word

$$a_3 \tilde{a}_5 A_1' \tilde{a}_2 \tilde{a}_3 a_1 A_2' \tilde{a}_1 a_2 a_0 \tilde{a}_0 a_0 A_3' \tilde{a}_3 \tilde{a}_2.$$

Then we replace A_1' by ε and obtain $a_3 \tilde{a}_5 \tilde{a}_2 \tilde{a}_3 a_1 A_2' \tilde{a}_1 a_2 a_0 \tilde{a}_0 a_0 A_3' \tilde{a}_3 \tilde{a}_2$. Applying the reduction rules finally yields $a_3 a_{11} A_2' a_3 a_0 a_0 A_3' \tilde{a}_5$. Hence, we have $\ell(A) = a_3$, $r(A) = \tilde{a}_5$ and the production for A' is $A' \to a_{11} A_2' a_3 a_0 a_0 A_3'$.

Proof of Theorem 10. Let $\mathcal{C}$ be an arbitrary grammar-based compressor working in polynomial time such that $\alpha_\mathcal{C}(2, n) \leq c$. The grammar-based compressor $\mathcal{D}$ works for an input word w over an arbitrary alphabet as follows: Let $\Sigma = \{a_0, \ldots, a_{k-1}\}$ be the set of symbols that occur in w and let φ be defined as in (5). Using $\mathcal{C}$, one first computes an SLP $\mathbb{B}$ for $\varphi(w)$ such that $|\mathbb{B}| \leq c \cdot g(\varphi(w))$. Then, using Lemma 12, one computes from $\mathbb{B}$ an SLP $\mathbb{A}$ for w such that $|\mathbb{A}| \leq 2c \cdot g(\varphi(w))$. Lemma 11 implies $g(\varphi(w)) \leq 3 \cdot g(w)$ and hence $|\mathbb{A}| \leq 6c \cdot g(w)$, which proves the theorem.   □

6   Open Problems

Several open problems arise from this paper. First of all, it would be nice to prove (or disprove) the lower bound $\Omega((n/\log n)^{2/3})$ for the approximation ratio of LZ78 also for a binary alphabet. Our proof needs a ternary alphabet. Another interesting question arises from the gap between the lower bound

$\Omega(\log n / \log \log n)$ and the upper bound $O(\log n)$ for $\beta_{\mathsf{LZ77}}(n)$ (worst case size of a smallest SLP in relation to the number of $\mathsf{LZ77}$-factors). It is open whether the factor $1/\log \log n$ in the lower bound is necessary. Finally, one should try to narrow also the gaps between the lower and upper bounds for the other grammar-based compressors analyzed in [4]. In particular, for the so called global algorithms from [4] these gaps are quite large.

Acknowledgment. The work in this paper was supported by the DFG grant LO 748/10-1.

References

1. Arpe, J., Reischuk, R.: On the complexity of optimal grammar-based compression. In: Proceedings of the DCC 2006, pp. 173–182. IEEE Computer Society (2006)
2. Berstel, J., Brlek, S.: On the length of word chains. Inf. Process. Lett. **26**(1), 23–28 (1987)
3. Casel, K., Fernau, H., Gaspers, S., Gras, B., Schmid, M.L.: On the complexity of grammar-based compression over fixed alphabets. In: Proceeding ICALP 2016, LNCS. Springer, Heidelberg (2016, to appear)
4. Charikar, M., Lehman, E., Lehman, A., Liu, D., Panigrahy, R., Prabhakaran, M., Sahai, A., Shelat, A.: The smallest grammar problem. IEEE Trans. Inf. Theory **51**(7), 2554–2576 (2005)
5. Diwan, A.A.: A New Combinatorial Complexity Measure for Languages. Tata Institute, Bombay (1986)
6. Gasieniec, L., Karpinski, M., Plandowski, W., Rytter, W.: Efficient algorithms for Lempel-Ziv encoding (extended abstract). In: Karlsson, R., Lingas, A. (eds.) SWAT 1996. LNCS, vol. 1097, pp. 392–403. Springer, Heidelberg (1996)
7. Jeż, A.: Approximation of grammar-based compression via recompression. In: Fischer, J., Sanders, P. (eds.) CPM 2013. LNCS, vol. 7922, pp. 165–176. Springer, Heidelberg (2013)
8. Kieffer, J.C., Yang, E.-H.: Grammar-based codes: a new class of universal lossless source codes. IEEE Trans. Inf. Theory **46**(3), 737–754 (2000)
9. Kieffer, J.C., Yang, E.-H., Nelson, G.J., Cosman, P.C.: Universal lossless compression via multilevel pattern matching. IEEE Trans. Inf. Theory **46**(4), 1227–1245 (2000)
10. Larsson, N.J., Moffat, A.: Offline dictionary-based compression. In: Proceedings of the DCC 1999, pp. 296–305. IEEE Computer Society (1999)
11. Li, M., Vitányi, P.: An Introduction to Kolmogorov Complexity and Its Applications, 3rd edn. Springer, Heidelberg (2008)
12. Lohrey, M.: The Compressed Word Problem for Groups. Springer, Heidelberg (2014)
13. Nevill-Manning, C.G., Witten, I.H.: Identifying hierarchical structure in sequences: a linear-time algorithm. J. Artif. Intell. Res. **7**, 67–82 (1997)
14. Rytter, W.: Application of Lempel-Ziv factorization to the approximation of grammar-based compression. Theor. Comput. Sci. **302**(1–3), 211–222 (2003)
15. Storer, J.A., Szymanski, T.G.: Data compression via textual substitution. J. ACM **29**(4), 928–951 (1982)

16. Tabei, Y., Takabatake, Y., Sakamoto, H.: A succinct grammar compression. In: Fischer, J., Sanders, P. (eds.) CPM 2013. LNCS, vol. 7922, pp. 235–246. Springer, Heidelberg (2013)
17. Ziv, J., Lempel, A.: Compression of individual sequences via variable-rate coding. IEEE Trans. Inf. Theory **24**(5), 530–536 (1977)

Efficient and Compact Representations of Some Non-canonical Prefix-Free Codes

Antonio Fariña[1], Travis Gagie[2(✉)], Giovanni Manzini[3,4], Gonzalo Navarro[5], and Alberto Ordóñez[6]

[1] Database Laboratory, University of A Coruña, A Coruña, Spain
[2] Department of Computer Science,
Helsinki Institute for Information Technology (HIIT),
University of Helsinki, Helsinki, Finland
travis.gagie@gmail.com
[3] Institute of Computer Science, University of Eastern Piedmont, Alessandria, Italy
[4] IIT-CNR, Pisa, Italy
[5] Department of Computer Science, University of Chile, Santiago, Chile
[6] Yoop SL, A Coruña, Spain

Abstract. For many kinds of prefix-free codes there are efficient and compact alternatives to the traditional tree-based representation. Since these put the codes into canonical form, however, they can only be used when we can choose the order in which codewords are assigned to characters. In this paper we first show how, given a probability distribution over an alphabet of σ characters, we can store a nearly optimal alphabetic prefix-free code in $o(\sigma)$ bits such that we can encode and decode any character in constant time. We then consider a kind of code introduced recently to reduce the space usage of wavelet matrices (Claude, Navarro, and Ordóñez, *Information Systems*, 2015). They showed how to build an optimal prefix-free code such that the codewords' lengths are non-decreasing when they are arranged such that their reverses are in lexicographic order. We show how to store such a code in $\mathcal{O}\left(\sigma \log L + 2^{\epsilon L}\right)$ bits, where L is the maximum codeword length and ϵ is any positive constant, such that we can encode and decode any character in constant time under reasonable assumptions. Otherwise, we can always encode and decode a codeword of ℓ bits in time $\mathcal{O}(\ell)$ using $\mathcal{O}(\sigma \log L)$ bits of space.

Funded in part by European Union's Horizon 2020 research and innovation programme under the Marie Skłodowska-Curie grant agreement No 690941 (project BIRDS). The first author was supported by: MINECO (PGE and FEDER) grants TIN2013-47090-C3-3-P and TIN2015-69951-R; MINECO and CDTI grant ITC-20151305; ICT COST Action IC1302; and Xunta de Galicia (co-founded with FEDER) grant GRC2013/053. The second author was supported by Academy of Finland grants 268324 and 250345 (CoECGR). The fourth author was supported by Millennium Nucleus Information and Coordination in Networks ICM/FIC P10-024F, Chile.

S. Inenaga et al. (Eds.): SPIRE 2016, LNCS 9954, pp. 50–60, 2016.
DOI: 10.1007/978-3-319-46049-9_5

1 Introduction

Binary prefix-free codes can be represented as binary trees whose leaves are labelled with the characters of the source alphabet, so that the ancestor at depth d of the leaf labelled x is a left child if the dth bit of the codeword for x is a 0, and a right child if it is a 1. To encode a character, we start at the root and descend to the leaf labelled with that character, at each step writing a 0 if we go left and a 1 if we go right. To decode an encoded string, we start at the root and descend according to the bits of the encoding until we reach a leaf, at each step going left if the next bit is a 0 and right if it is a 1. Then we output the character associated with the leaf and return to the root to continue decoding. Therefore, a codeword of length ℓ is encoded/decoded in time $\mathcal{O}(\ell)$. This all generalizes to larger code alphabets, but for simplicity we consider only binary codes in this paper.

There are, however, faster and smaller representations of many kinds of prefix-free codes. If we can choose the order in which codewords are assigned to characters then, by the Kraft Inequality [8], we can put any prefix-free code into canonical form [13]— i.e., such that the codewords' lexicographic order is the same as their order by length, with ties broken by the lexicographic order of their characters—without increasing any codeword's length. If we store the first codeword of each length as a binary number then, given a codeword's length and its rank among the codewords of that length, we can compute the codeword via a simple addition. Given a string prefixed by a codeword, we can compute that codeword's length and its rank among codewords of that length via a predecessor search. If the alphabet consists of σ characters and the maximum codeword length is L, then we can build an $\mathcal{O}(\sigma \log L)$-bit data structure with $\mathcal{O}(\log L)$ query time that, given a character, returns its codeword's length and rank among codewords of that length, or vice versa. If L is at most a constant times the size of a machine word (which it is when we are considering, e.g., Huffman codes for strings in the RAM model) then in theory we can make the predecessor search and the data structure's queries constant-time, meaning we can encode and decode in constant time [5].

There are applications for which there are restrictions on the codewords' order, however. For example, in alphabetic codes the lexicographic order of the codewords must be the same as that of the characters. Such codes are useful when we want to be able to sort encoded strings without decoding them (because the lexicographic order of two encodings is always the same as that of the encoded strings) or when we are using data structures that represent point sets as sequences of coordinates [10], for example. Interestingly, since the mapping between symbols and leaves is fixed, alphabetic codes need only to store the tree topology, which can be represented more succinctly than optimal prefix-free codes, in $2\sigma + o(\sigma)$ bits [9], so that encoding and decoding can still be done in time $\mathcal{O}(\ell)$. There are, however, no equivalents to the faster encoding/decoding methods used on canonical codes [5].

In Sect. 2 we show how, given a probability distribution over the alphabet, we can store a nearly optimal alphabetic prefix-free code in $o(\sigma)$ bits such that we can encode and decode any character in constant time. We note that we can still use our construction even if the codewords must be assigned to the characters according to some non-trivial permutation of the alphabet, but then we must store that permutation such that we can evaluate and invert it quickly.

In Sect. 3 we consider another kind of non-canonical prefix-free code, which Claude, Navarro, and Ordóñez [1] introduced recently to reduce the space usage of their wavelet matrices. (Wavelet matrices are alternatives to wavelet trees [6,10] that are more space efficient when the alphabet is large.) They showed how to build an optimal prefix-free code such that the codewords' lengths are non-decreasing when they are arranged such that their reverses are in lexicographic order. They represent the code in $\mathcal{O}(\sigma L)$ bits, and encode and decode a codeword of length ℓ in time $\mathcal{O}(\ell)$. We show how to store such a code in $\mathcal{O}(\sigma \log L)$ bits, and still encode and decode any character in $\mathcal{O}(\ell)$ time. We also show that, by using $\mathcal{O}(\sigma \log L + 2^{\epsilon L})$ bits, where ϵ is any positive constant, we can encode and decode any character in constant time when L is at most a constant times the size of a machine word. Our first variant is simple enough to be implementable. We show experimentally that it uses 23–30 times less space than a classical implementation, at the price of being 10–21 times slower at encoding and 11–30 at decoding.

2   Alphabetic Codes

Our approach to storing an alphabetic prefix code compactly has two parts: first, we show that we can build such a code such that the expected codeword length is at most a factor of $(1 + \mathcal{O}(1/\sqrt{\log n}))^2 = 1 + \mathcal{O}(1/\sqrt{\log n})$ greater than optimal, the code-tree has height at most $\lg \sigma + \sqrt{\lg \sigma} + 3$, and each subtree rooted at depth $\lceil \lg \sigma - \sqrt{\lg \sigma} \rceil$ is completely balanced; then, we show how to store such a code-tree in $o(\sigma)$ bits such that encoding and decoding take constant time.

Evans and Kirkpatrick [2] showed how, given a binary tree on n leaves, we can build a new binary tree of height at most $\lceil \lg n \rceil + 1$ on the same leaves in the same left-to-right order, such that the depth of each leaf in the new tree is at most 1 greater than its depth in the original tree. We can use their result to restrict the maximum codeword length of an optimal alphabetic prefix code, for an alphabet of σ characters, to be at most $\lg \sigma + \sqrt{\lg \sigma} + 3$, while forcing its expected codeword length to increase by at most a factor of $1 + \mathcal{O}(1/\sqrt{\log \sigma})$. To do so, we build the tree T_{opt} for an optimal alphabetic prefix code and then rebuild, according to Evans and Kirkpatrick's construction, each subtree rooted at depth $\lceil \sqrt{\lg \sigma} \rceil$. The resulting tree, T_{lim}, has height at most $\lceil \sqrt{\lg \sigma} \rceil + \lceil \lg \sigma \rceil + 1$ and any leaf whose depth increases was already at depth at least $\lceil \sqrt{\lg \sigma} \rceil$.

There are better ways to build a tree T_{lim} with such a height limit. Itai [7] and Wessner [14] independently showed how, given a probability distribution over an alphabet of σ characters, we can build an alphabetic prefix code T_{lim} that has maximum codeword length at most $\lg \sigma + \sqrt{\lg \sigma} + 3$ and is optimal among all such codes. Our construction in the previous paragraph, even if not optimal, shows that the expected codeword length of T_{lim} is at most $1 + \mathcal{O}(1/\sqrt{\log \sigma})$ times that of an optimal code with no length restriction.

Further, let us take T_{lim} and completely balance each subtree rooted at depth $\lceil \lg \sigma - \sqrt{\lg \sigma} \rceil$. The height remains at most $\lg \sigma + \sqrt{\lg \sigma} + 3$ and any leaf whose depth increases was already at depth at least $\lceil \lg \sigma - \sqrt{\lg \sigma} \rceil$, so the expected codeword length increases by at most a factor of

$$\frac{\lg \sigma + \sqrt{\lg \sigma} + 3}{\lceil \lg \sigma - \sqrt{\lg \sigma} \rceil} = 1 + \mathcal{O}\left(1/\sqrt{\log \sigma}\right) .$$

Let T_{bal} be the resulting tree. Since the expected codeword length of T_{lim} is in turn at most a factor of $1 + \mathcal{O}(1/\sqrt{\log n})$ larger than that of T_{opt}, the expected codeword length of T_{bal} is also at most a factor of $(1 + \mathcal{O}(1/\sqrt{\log n}))^2 = 1 + \mathcal{O}(1/\sqrt{\log n})$ larger than the optimal. T_{bal} then describes our suboptimal code.

To represent T_{bal}, we store a bitvector $B[1..\sigma]$ in which $B[i] = 1$ if and only if the codeword for the ith character in the alphabet has length at most $\lceil \lg \sigma - \sqrt{\lg \sigma} \rceil$, or if the ith leaf in T is the leftmost leaf in a subtree rooted at depth $\lceil \lg \sigma - \sqrt{\lg \sigma} \rceil$. With Pătraşcu's implementation [12] for B this takes a total of $\mathcal{O}\left(2^{\lg \sigma - \sqrt{\lg \sigma}} \log \sigma + \sigma / \log^c \sigma\right) = \mathcal{O}(\sigma / \log^c \sigma)$ bits for any constant c, and allows us to perform in constant time $\mathcal{O}(c)$ the following operations on B: (1) access, that is, inspecting any $B[i]$; (2) rank, that is, $rank(B, i)$ counts the number of 1s in any prefix $B[1..i]$; and select, that is, $select(B, j)$ is the position of the jth 1 in B, for any j.

Let us for simplicity assume that the alphabet is $[1..\sigma]$. For encoding in constant time we store an array $S[1..2^{\lceil \lg \sigma - \sqrt{\lg \sigma} \rceil}]$, which stores the explicit code assigned to the leaves of T_{bal} where $B[i] = 1$, in the same order of B. That is, if $B[i] = 1$, then the code assigned to the character i is stored at $S[rank(B, i)]$, using $\lg \sigma + \sqrt{\lg \sigma} + 3 = \mathcal{O}(\log \sigma)$ bits. Therefore S requires $\mathcal{O}\left(2^{\lg \sigma - \sqrt{\lg \sigma}} \log \sigma\right) = o(\sigma / \log^c \sigma)$ bits of space, for any constant c. We can also store the length of the code within the same asymptotic space.

To encode the character i, we check whether $B[i] = 1$ and, if so, we simply look up the codeword in S as explained. If $B[i] = 0$, we find the preceding 1 at $i' = select(B, rank(B, i))$, which marks the leftmost leaf in the subtree rooted at depth $\lceil \lg \sigma - \sqrt{\lg \sigma} \rceil$ that contains the ith leaf in T. Since the subtree is completely balanced, we can compute the code for the character i in constant time from that of the character i': The size of the balanced subtree is $r = i'' - i'$, where $i'' = select(B, rank(B, i') + 1)$, and its height is $h = \lceil \lg r \rceil$. Then the first $2r - 2^h$ codewords are of the same length of the codeword for i', and the last $2^h - r$ have one bit less. Thus, if $i - i' < 2r - 2^h$, the codeword for i' is $S[rank(B, i')] + i - i'$, of the same length of that of i; otherwise it is one bit shorter, $(S[rank(B, i')] + 2r - 2^h)/2 + i - i' - (2r - 2^h) = S[rank(B, i')]/2 + i - i' - (r - 2^{h-1})$.

To be able to decode quickly, we store an array $A[1..2^{\lceil \lg \sigma - \sqrt{\lg \sigma} \rceil}]$ such that, for $1 \le j \le 2^{\lceil \lg \sigma - \sqrt{\lg \sigma} \rceil}$, if the $\lceil \lg \sigma - \sqrt{\lg \sigma} \rceil$-bit binary representation of $j - 1$ is prefixed by the ith codeword, then $A[j]$ stores i and the length of that codeword. If, instead, the $\lceil \lg \sigma - \sqrt{\lg \sigma} \rceil$-bit binary representation of j is the path label to the root of a subtree of T_{bal} with size more than 1, then $A[j]$ stores the position

i' in B of the leftmost leaf in that subtree (thus $B[i'] = 1$). Again, A takes $\mathcal{O}\left(2^{\log \sigma - \sqrt{\log \sigma}} \log \sigma\right) = o(\sigma/\log^c \sigma)$ bits, for any constant c.

Given a string prefixed by the ith codeword, we take the prefix of length $\lceil \lg \sigma - \sqrt{\lg \sigma}\rceil$ of that string (padding with 0s on the right if necessary), view it as the binary representation of a number j, and check $A[j]$. This either tells us immediately i and the length of the ith codeword, or tells us the position i' in B of the leftmost leaf in the subtree containing the desired leaf. In the latter case, since the subtree is completely balanced, we can compute i in constant time: We find i'', r, and h as done for encoding. We then take the first h bits of the string (including the prefix we had already read, and padding with a 0 if necessary), and interpret it as the number j'. Then, if $d = j' - S[rank(B, i')] < 2r - 2^h$, it holds $i = i' + d$. Otherwise, the code is of length $h - 1$ and the decoded symbol is $i = i' + 2r - 2^h + \lfloor(d - (2r - 2^h))/2\rfloor = i' + r - 2^{h-1} + \lfloor d/2\rfloor$.

Theorem 1. *Given a probability distribution over an alphabet of σ characters, we can build an alphabetic prefix code whose expected codeword length is at most a factor of $1 + \mathcal{O}\left(1/\sqrt{\log \sigma}\right)$ more than optimal and store it in $\mathcal{O}(\sigma/\log^c \sigma)$ bits, for any constant c, such that we can encode and decode any character in constant time $\mathcal{O}(c)$.*

3   Codes for Wavelet Matrices

As we mentioned in Sect. 1, in order to reduce the space usage of their wavelet matrices, Claude, Navarro, and Ordóñez [1] recently showed how to build an optimal prefix code such that the codewords' lengths are non-decreasing when they are arranged such that their reverses are in lexicographic order. Specifically, they first build a normal Huffman code and then use the Kraft Inequality to build another code with the same codeword lengths with the desired property. They store an $\mathcal{O}(\sigma L)$-bit mapping between characters and their codewords, where again σ is the alphabet size and L is the maximum length of any codeword, which allows them to encode and decode codewords of length ℓ in time $\mathcal{O}(\ell)$. (In the wavelet matrices, they already spend $\mathcal{O}(\ell)$ time in the operations associated with encoding and decoding.)

Assume we are given a code produced by Claude et al.'s construction. We reassign the codewords of the same length such that the lexicographic order of the reversed codewords of that length is the same as that of their characters. This preserves the property that codeword lengths are non-decreasing with their reverse lexicographic order. The positive aspect of this reassignment is that all the information on the code can be represented in $\sigma \lg L$ bits as a sequence $D = d_1, \ldots, d_\sigma$, where d_i is the depth of the leaf encoding character i in the code-tree T. We can then represent D using a wavelet tree [6], which uses $\mathcal{O}(\sigma \log L)$ bits and supports the following operations on D in time $\mathcal{O}(\log L)$: (1) access any $D[i]$, which gives the length ℓ of the codeword of character i; (2) compute $r = rank_\ell(D, i)$, which gives the number of occurrences of ℓ in $D[1..i]$, which if $D[i] = \ell$ gives the position (in reverse lexicographic order) of the leaf representing character i among those of codeword length ℓ; and (3) compute $i = select_\ell(D, r)$, which gives the position in D of the rth occurrence of ℓ, or which is the same, the character i corresponding to the rth codeword of length ℓ (in reverse lexicographic order).

If, instead of $\mathcal{O}(\log L)$ time, we wish to perform the operations in time $\mathcal{O}(\ell)$, where ℓ is the length of the codeword involved in the operation, we can simply give the wavelet tree of D the same shape of the tree T. We can even perform the operations in time $\mathcal{O}(\log \ell)$ by using a wavelet tree shaped like the trie for the first σ codewords represented with Elias γ- or δ-codes [4, Observation 1]. The size stays $\mathcal{O}(\sigma \log L)$ if we use compressed bitmaps at the nodes [6,10].

We are left with two subproblems. For decoding the first character encoded in a binary string, we need to find the length ℓ of the first codeword and the lexicographic rank r of its reverse among the reversed codewords of that length, since then we can decode $i = select_\ell(D, r)$. For encoding a character i, we find its length $\ell = D[i]$ and the lexicographic rank $r = rank_\ell(D, i)$ of its reverse among the reversed codewords of length ℓ, and then we must find the codeword given ℓ and r. We first present a solution that takes $\mathcal{O}(L \log \sigma) = \mathcal{O}(\sigma \log L)$ further bits[1] and works in $\mathcal{O}(\ell)$ time. We then present a solution that takes $\mathcal{O}(2^{\epsilon L})$ further bits and works in constant time.

Let T be the code-tree and, for each depth d between 0 and L, let $\mathsf{nodes}(d)$ be the total number of nodes at depth d in T and let $\mathsf{leaves}(d)$ be the number of leaves at depth d. Let v be a node other than the root, let u be v's parent, let r_v be the lexicographic rank (counting from 1) of v's reversed path label among all the reversed path labels of nodes at v's depth, and let r_u be defined analogously for u. Notice that since T is optimal it is strictly binary, so half the nodes at each positive depth are left children and half are right children. Moreover, the reversed path labels of all the left children at any depth are lexicographically less than the reversed path labels of all the right children at the same depth (or, indeed, at any depth). Finally, the reversed path labels of all the leaves at any depth are lexicographically less than the reversed path labels of all the internal nodes at that depth. It follows that

- v is u's left child if and only if $r_v \leq \mathsf{nodes}(\mathsf{depth}(v))/2$,
- if v is u's left child then $r_v = r_u - \mathsf{leaves}(\mathsf{depth}(u))$,
- if v is u's right child then $r_v = r_u - \mathsf{leaves}(\mathsf{depth}(u)) + \mathsf{nodes}(\mathsf{depth}(v))/2$.

Of course, by rearranging terms we can also compute r_u in terms of r_v.

Suppose we store $\mathsf{nodes}(d)$ and $\mathsf{leaves}(d)$ for d between 0 and L. With the three observations above, given a codeword of length ℓ, we can start at the root and in $\mathcal{O}(\ell)$ time descend in T until we reach the leaf v whose path label is that codeword, then return its depth ℓ and the lexicographic rank $r = r_v$ of its reverse path label among all reversed path labels of nodes at that depth.[2] Then we compute i from ℓ and r as described, in further $\mathcal{O}(\log \ell)$ time. For encoding i,

---

[1] Since the code tree has height L and σ leaves, it follows that $L < \sigma$.

[2] This descent is conceptual; we do not have a concrete node v at each level, but we do know r_v.

we obtain as explained its length ℓ and the rank $r = r_v$ of its reversed codeword among the reversed codewords of that length. Then we use the formulas to walk up towards the root, finding in each step the rank r_u of the parent u of v, and determining if v is a left or right child of u. This yields the ℓ bits of the codeword of i in reverse order (0 when v is a left child of u and 1 otherwise), in overall time $\mathcal{O}(\ell)$. This completes our first solution, which we evaluate experimentally in Sect. 4.

Theorem 2. *Suppose we are given an optimal prefix code in which the code-words' lengths are non-decreasing when they are arranged such that their reverses are in lexicographic order. We can store such a code in $\mathcal{O}(\sigma \log L)$ bits — possibly after swapping characters' codewords of the same length — where σ is the alphabet size and L is the maximum codeword length, such that we can encode and decode any character in $\mathcal{O}(\ell)$ time, where ℓ is the corresponding codeword length.*

If we want to speed up descents, we can build a table that takes as arguments a depth and several bits, and returns the difference between r_u and r_v for any node u at that depth and its descendant v reached by following edges corresponding to those bits. Notice that this difference depends only on the bits and the numbers of nodes and leaves at the intervening levels. If the table accepts t bits as arguments at once, then it takes $L2^t \log \sigma$ bits and we can descend in $\mathcal{O}(L/t)$ time. Setting $t = \epsilon L/2$, and since $L \geq \lg \sigma$, we use $\mathcal{O}(2^{\epsilon L})$ space and descend from the root to any leaf in constant time.

Speeding up ascents is slightly more challenging. Consider all the path labels of a particular length that end with a particular suffix of length t: the lexico-graphic ranks of their reverses form a consecutive interval. Therefore, we can partition the nodes at any level by their r values, such that knowing which part a node's r value falls into tells us the last t bits of that node's path label, and the difference between that node's r value and the r value of its ancestor at depth t less. For each depth, we store the first r value in each interval in a predecessor data structure, implemented as a trie with degree $\sigma^{\epsilon/3}$; since there are at most 2^t intervals in the partition for each depth and $L \geq \lg \sigma$, setting $t = \epsilon L/2$ again we use a total of $\mathcal{O}(L2^{\epsilon L/2}\sigma^{\epsilon/3} \log \sigma) \subset \mathcal{O}(2^{\epsilon L})$ bits and ascend from any leaf to the root in constant time.

Finally, the operations on the wavelet tree can be made constant-time by using a balanced multiary variant [3].

Theorem 3. *Suppose we are given an optimal prefix code in which the code-words' lengths are non-decreasing when they are arranged such that their reverses are in lexicographic order. Let L be the maximum codeword length, so that it is at most a constant times the size of the machine word. Then we can store such a code in $\mathcal{O}(\sigma \log L + 2^{\epsilon L})$ bits — possibly after swapping characters' codewords of the same length — where ϵ is any positive constant, such that we can encode and decode any character in constant time.*

4   Experiments

We have run experiments to compare the solution of Theorem 2 (referred to as WMM in the sequel, for Wavelet Matrix Model) with the only previous encoding, that is, the one used by Claude et al. [1] (denoted by TABLE). Note that our codes are not canonical, so other solutions [5] do not apply.

Claude et al. [1] use for encoding a single table of σL bits storing the code of each symbol, and thus they easily encode in constant time. For decoding, they have tables separated by codeword length ℓ. In each such table, they store the codewords of that length and the associated character, sorted by codeword. This requires $\sigma(L + \lg \sigma)$ further bits, and permits decoding binary searching the codeword found in the wavelet matrix. Since there are at most 2^ℓ codewords of length ℓ, the binary search takes time $\mathcal{O}(\ell)$.

For the sequence D used in our WMM, we use binary Huffman-shaped wavelet trees with plain bitmaps. The structures for supporting rank/select efficiently require 37.5 % space overhead, so the total space is $1.37\,\sigma\mathcal{H}_0(D)$, where $\mathcal{H}_0(D) \leq \lg L$ is the per-symbol zero-order entropy of the sequence D. We also add a small index to speed up select queries [11] (that is, decoding), which can be parameterized with a sampling value that we set to $\{16, 32, 64, 128\}$. Finally, we store the values leaves and nodes, which add an insignificant L^2 bits in total.

We used a prefix of three datasets in http://lbd.udc.es/research/ECRPC. The first one, EsWiki, contains a sequence of word identifiers generated by using the Snowball algorithm to apply stemming to the Spanish Wikipedia. The second one, EsInv, contains a concatenation of differentially encoded inverted lists extracted from a random sample of the Spanish Wikipedia. The third dataset, Indo was created with the concatenation of the adjacency lists of Web graph Indochina-2004 available at http://law.di.unimi.it/datasets.php. In Table 1 we provide some statistics about the datasets. We include the number of symbols in the dataset (n) and the alphabet size (σ). Assuming P is the relative frequency of the alphabet symbols, $\mathcal{H}(P)$ indicates (in bits per symbol) the empirical entropy of the sequence. This approximates the average ℓ value of queries. Finally we show L, the maximum code length, and the zero-order entropy of the sequence D, $\mathcal{H}_0(D)$, in bits per symbol. The last column is then a good approximation of the size of our Huffman-shaped wavelet tree for D.

Table 1. Main statistics of the texts used.

Collection	Length (n)	Alphabet size (σ)	Entropy ($\mathcal{H}(P)$)	Max code length(L)	Entropy of level entries ($\mathcal{H}_0(D)$)
EsWiki	200,000,000	1,634,145	11.12	28	2.24
EsInv	300,000,000	1,005,702	5.88	28	2.60
Indo	120,000,000	3,715,187	16.29	27	2.51

Our test machine has a Intel(R) Core(tm) i7-3820@3.60 GHz CPU (4 cores/8 siblings) and 64 GB of DDR3 RAM. It runs Ubuntu Linux 12.04 (Kernel 3.2.0-99-generic). The compiler used was g++ version 4.6.4 and we set compiler optimization flags to −O9. All our experiments run in a single core and time measures refer to CPU *user-time*.

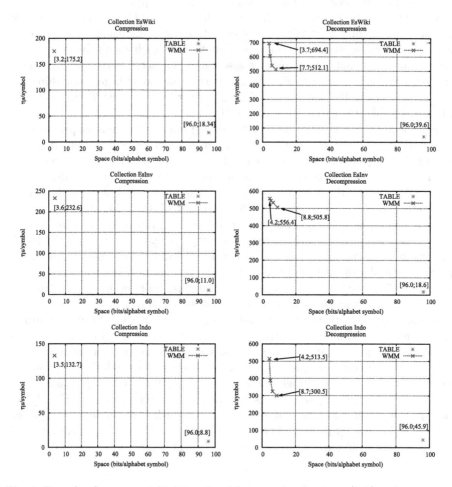

Fig. 1. Size of code representations versus either compression time (left) or decompression time (right). Time is measured in nanoseconds per symbol.

Figure 1 compares the space required by both code representations and their compression and decompression times. As expected, the space per character of our new code representation, WMM, is close to $1.37\,\mathcal{H}_0(D)$, whereas that of TABLE is close to $2L + \lg \sigma$. This explains the large difference in space between both representations, a factor of 23–30 times. For decoding we show the mild effect of adding the structure that speeds up select queries.

The price of our representation is the encoding and decoding time. While the TABLE approach encodes using a single table access, in 8–18 ns, our representation needs 130–230, which is 10 to 21 times slower. For decoding, the binary search performed by TABLE takes 20–50 ns, whereas our WMM representation requires 510–700 in the slowest and smallest variant (i.e., 11–30 times slower). Our faster variants require 300–510 ns, which is still several times slower.

5 Conclusions

A classical prefix code representation uses $\mathcal{O}(\sigma L)$ bits, where σ is the alphabet size and L the maximum codeword length, and encodes in constant time and decodes a codeword of length ℓ in time $\mathcal{O}(\ell)$. Canonical prefix codes can be represented in $\mathcal{O}(\sigma \log L)$ bits, so that one can encode and decode in constant time under reasonable assumptions. In this paper we have considered two families of codes that cannot be put in canonical form. Alphabetic codes can be represented in $\mathcal{O}(\sigma)$ bits, but encoding and decoding take time $\mathcal{O}(\ell)$. We gave an approximation that worsens the average code length by a factor of $1 + \mathcal{O}(1/\sqrt{\log \sigma})$, but in exchange requires $o(\sigma)$ bits and encodes and decodes in constant time. We then consider a family of codes that are canonical when read right to left. For those we obtain a representation using $\mathcal{O}(\sigma \log L)$ bits and encoding and decoding in time $\mathcal{O}(\ell)$, or even in $\mathcal{O}(1)$ time under reasonable assumptions if we use $\mathcal{O}(2^{\epsilon L})$ further bits, for any constant $\epsilon > 0$.

We have implemented the simple version of these right-to-left codes, which are used for compressing wavelet matrices, and shown that our encodings are significantly smaller than classical ones in practice (up to 30 times), albeit also slower (up to 30 times). For the journal version of the paper, we plan to implement the wavelet tree of D with a shape that lets it operate in time $\mathcal{O}(\ell)$ or $\mathcal{O}(\log \ell)$, as used to prove Theorem 2; currently we gave it Huffman shape in order to minimize space. Since there are generally more longer than shorter codewords, the Huffman shape puts them higher in the wavelet tree of D, so the longer codewords perform faster and the shorter codewords perform slower. This is the opposite effect as the one sought in Theorem 2. Therefore, a faithful implementation may lead to a slightly larger but also faster representation.

An interesting challenge is to find optimal alphabetic encodings that can encode and decode faster than in time $\mathcal{O}(\ell)$, even if they use more than $\mathcal{O}(\sigma)$ bits of space. Extending our results to other non-canonical prefix codes is also an interesting line of future work.

Acknowledgements. This research was carried out in part at University of A Coruña, Spain, while the second author was visiting and the fifth author was a PhD student there. It started at a StringMasters workshop at the Research Center on Information and Communication Technologies (CITIC) of the university. The workshop was partly funded by EU RISE project BIRDS (Bioinformatics and Information Retrieval Data Structures). The authors thank Nieves Brisaboa and Susana Ladra.

References

1. Claude, F., Navarro, G., Ordóñez, A.: The wavelet matrix: an efficient wavelet tree for large alphabets. Inf. Syst. **47**, 15–32 (2015)
2. Evans, W., Kirkpatrick, D.G.: Restructuring ordered binary trees. J. Algorithms **50**, 168–193 (2004)
3. Ferragina, P., Manzini, G., Mäkinen, V., Navarro, G.: Compressed representations of sequences, full-text indexes. ACM Trans. Algorithm **3**(2), 20 (2007)
4. Gagie, T., He, M., Munro, J.I., Nicholson, P.K.: Finding frequent elements in compressed 2D arrays and strings. In: Grossi, R., Sebastiani, F., Silvestri, F. (eds.) SPIRE 2011. LNCS, vol. 7024, pp. 295–300. Springer, Heidelberg (2011)
5. Gagie, T., Navarro, G., Nekrich, Y., Ordóñez, A.: Efficient and compact representations of prefix codes. IEEE Trans. Inf. Theory **61**(9), 4999–5011 (2015)
6. Grossi, R., Gupta, A., and Vitter, J.S.: High-order entropy-compressed text indexes. In: Proceedings SODA, pp. 841–850 (2003)
7. Itai, A.: Optimal alphabetic trees. SIAM J. Comp. **5**, 9–18 (1976)
8. Kraft, L.G.: A device for quantizing, grouping, and coding amplitude modulated pulses. M.Sc. thesis, EE Dept., MIT (1949)
9. Munro, J.I., Raman, V.: Succinct representation of balanced parentheses and static trees. SIAM J. Comp. **31**(3), 762–776 (2001)
10. Navarro, G.: Wavelet trees for all. J. Discr. Algorithm **25**, 2–20 (2014)
11. Navarro, G., Providel, E.: Fast, small, simple rank/select on bitmaps. In: Klasing, R. (ed.) SEA 2012. LNCS, vol. 7276, pp. 295–306. Springer, Heidelberg (2012)
12. Pătraşcu, M.: Succincter. In: Proceedings FOCS, pp. 305–313 (2008)
13. Schwartz, E.S., Kallick, B.: Generating a canonical prefix encoding. Commun. ACM **7**, 166–169 (1964)
14. Wessner, R.L.: Optimal alphabetic search trees with restricted maximal height. Inf. Proc. Lett. **4**, 90–94 (1976)

Parallel Lookups in String Indexes

Anders Roy Christiansen[1(✉)] and Martín Farach-Colton[2]

[1] The Technical University of Denmark, Kongens Lyngby, Denmark
aroy@dtu.dk
[2] Rutgers University, New Brunswick, USA
farach@cs.rutgers.edu

Abstract. Recently, the first PRAM algorithms were presented for looking up a pattern in a suffix tree. We improve the bounds, achieving optimal results.

Keywords: Parallel · Pattern matching · Suffix trees · PRAM

1 Introduction

Looking up a pattern string in an index is one of the most basic primitives in stringology, and the suffix tree (and its suffix array representation) is among the most basic indexes. It is therefore surprising that, until recently, there were no known PRAM algorithms for looking up an m-character pattern P in a suffix tree of an n-character text T. This contrasts sharply with the rich PRAM literature for the problem of finding all occurrences of P in T in the case where P can be preprocessed, optimal solutions of which are known for the full range of PRAM models [5,8,15].

Recently Jekovec and Brodnik [10] considered the problem of parallel lookups in an index, specifically suffix trees and quadratic-space suffix tries. They achieved work-time optimal $O(m)$ work and $O(\log m)$ time for suffix trie lookups in the CREW PRAM, although the preprocessing involves quadratic work and space. For suffix tree lookups, they achieve $O(m \log m)$ work and $O(\log m)$ time by augmenting the $O(n)$-size suffix tree with further data structures[1] that increase the size to $O(n \log n)$. These bounds are time-optimal due to the $\Omega(\log n)$ time lower bound for computing the OR of n-bits [4] in the CREW PRAM.

Fischer et al. [7] gave an CREW PRAM algorithm using the suffix array and some additional compact data structures requiring a total of $n \log n + O(n)$ bits (i.e. $n + o(n)$ words), thus improving the space. Their algorithm uses $O(\log \log m \log \log n + \log m)$ time and $O(m + \min(m, \log n)(\log m + \log \log m \log \log n))$ work. Additionally they considered the approximate pattern lookup problem and lookups in compressed suffix arrays.

In this paper, we improve the bounds for looking up a pattern in an index in several ways. First, we provide an algorithm that matches the time-work

---

[1] Suffix trees of subsets of characters, hash tables, etc.

© Springer International Publishing AG 2016
S. Inenaga et al. (Eds.): SPIRE 2016, LNCS 9954, pp. 61–67, 2016.
DOI: 10.1007/978-3-319-46049-9_6

optimal bounds of $O(\log m)$ time, $O(m)$ work while achieving $O(n)$ space. Also, our algorithm runs on the EREW PRAM, thus improving on the earlier CREW PRAM algorithms. As in the previous algorithms, we use randomization, but only in the preprocessing, whereas the pattern matching phase is deterministic[2].

We consider two variants of the pattern lookup problem: exact matching and prefix matching. In exact matching, we find the place in the suffix tree where the complete pattern matches. In prefix matching, we find the location in the suffix tree which matches the longest possible prefix of the pattern.

Our main result is:

Theorem 1. *Given a suffix tree of a string T of length n and a pattern P of length m, then parallel prefix pattern lookup in the suffix tree takes worst-case $O(\log m)$ time and $O(m)$ work, after $O(\log n \log^* n)$ time and $O(n)$ work preprocessing w.h.p. requiring $O(n)$ additional space. All bounds are on the EREW PRAM model.*

In order to present this result, we first present a simpler but similar method that does more work during preprocessing and does not support prefix pattern lookups. Both results augment suffix trees with Karp-Rabin fingerprints [12] and perfect hashing [2]. The final result is obtained by reducing the number of strings that must be guaranteed to have collision-free KR fingerprints by discarding possible false-positives during a query. We note that the techinque of combining indexes with Karp-Rabin fingerprints for efficient pattern lookups was introduced in [1], but in that case it was to improve sequential dictionary pattern matching.

Furthermore we include a simple algorithm for parallel prefix pattern lookup in a suffix array because it is deterministic in both the query and preprocessing phases and works on general alphabets whereas the first works on integer alphabets, at the cost of some running time. The result is summarized below:

Theorem 2. *Prefix pattern matching in a suffix array with LCP-values can be performed in $O(\log n)$ time and $O(m + \log n)$ work on the CRCW PRAM model with no other preprocessing than computing the suffix array and the LCP array.*

2   Preliminaries

Denote by T a text of length n of letters from an alphabet Σ. Call the corresponding suffix tree S. For an edge $e \in S$ denote by $T(e)$ the string of letters on the path from the root to e and including the letters on e. Similarly let pre(e) denote the string of letters on the path from the root to e including only the first letter on e. Let parent(e) be the edge that shares a node with e and is on the e-root path. Let $T[i]$ be the ith character of T and $T[i,j]$ be the substring of T from the ith character to the jth character, both inclusive.

---

[2] Both earlier results involve hashing, as does ours. We give our bounds using fast, randomized perfect hashing, rather than slow, deterministic perfect hashing.

In this paper we will be working in the PRAM model see [9] for details. We present all results based on the work-time presentation framework described in [9] (i.e. without having the number of processors as a parameter).

We will be using the following lemmas throughout our solutions:

Lemma 1 (Follows from list rank in [3]). *Given a set of linked lists represented by a table of length n of next-pointers and the index of a head element one can compute which elements are in the linked list that contains the head. This can be done in $O(\log n)$ time and $O(n)$ work in the EREW PRAM model.*

Lemma 2 ([6]). *Given a table B of n bits, one can find the leftmost 1-bit in B in $O(1)$ time and $O(n)$ work in the CRCW PRAM model.*

Lemma 3 (Follows from Prefix Sum [13]). *Given a string T of length n, all prefix Karp-Rabin fingerprints [12] $\phi(T[1,1]), \phi(T[1,2]), \ldots$ can be computed in $O(\log n)$ time and $O(n)$ work in the EREW PRAM model.*

Lemma 4 (From [2], adapted to EREW). *Given a (multi-)set of n integers a perfect hash table of size n can be computed in time $O(\log n \log^* n)$ using $O(n)$ work and space w.h.p. in the EREW PRAM model.*

3   Simple Fingerprint-Based Pattern Lookup

The main idea in this solution is to use a combination of Karp-Rabin fingerprints and perfect hash tables to avoid doing an actual traversal of the suffix tree from the root. We first show a simplified version of this solution, and then extend it to reduce preprocessing time and to support prefix lookups.

Data Structure. Let ϕ be a Karp-Rabin based fingerprint function that is collision free for all substrings in T. We store the string T, the suffix tree S for T, and a perfect hash table H_d for each $d = 1 \ldots n$ mapping $H_{|\mathrm{pre}(e)|}[\phi(\mathrm{pre}(e))] \to e$ for each edge e in S. These structures use $O(n)$ space in total.

Query. Given a pattern P, first compute the prefix fingerprints of P using Lemma 3. In parallel, look up a fingerprint $\phi(P[1,d])$ in hash table H_d, for all $d = 1 \ldots m$. If there is a match, let $M[d] = H_d[\phi(P[1,d])]$, and otherwise let $M[d] = \bot$. Since all lookups are in different hash tables there are no read conflicts. Find the rightmost non-$\bot$ value in M and call it e_c. If P occurs in T then this match must be on e_c in the suffix tree. Match P character-by-character to $T(e_c)[1, m]$. If there are no differences, report that P exists on e_c in T; otherwise, report that P does not occur.

Since all characters of P are compared to a substring of T before reporting an occurrence, no false positives are reported. We need this verification part because our fingerprints are only guaranteed to be collision-free on T, not on P.

If P does occur in T, then the fingerprint function is guaranteed to be collision-free in both P and T, and so we will find a single maximal e_c so that a prefix of P matches with $\mathrm{pre}(e_c)$. The brute-force matching phase then extends the match length down the edge e_c.

The bottleneck of the query is the $O(\log m)$-time, $O(m)$-work of computing the fingerprints (Lemma 3), and the same time and work to verify a match. We conclude that these are overall work-time bounds.

Preprocessing. We assume the suffix tree is given[3]. In $O(\log n)$ time and $O(n)$ work we can compute all prefix fingerprints of T using Lemma 3. From these prefix fingerprints the fingerprint of an arbitrary substring of T can be computed in constant time and work.

Validate that ϕ is collision-free for the substrings of T by computing all possible fingerprints. Since there are $\Theta(n^2)$ different substrings this takes $O(n^2)$ work. They can all be calculated independently, but $O(n)$ fingerprints might depend on the same fingerprint prefix which means the algorithm might need to read the same memory cell at the same time. Since a CREW algorithm can be simulated as an EREW algorithm with $O(\log n)$ time overhead per step [9], this takes $O(\log n)$ time. Construct a hash table over all the fingerprints using Lemma 4 to check for duplicates - if there are any duplicates, start over with a new random Karp-Rabin fingerprint function. In total this takes $O(\log n \log^* n)$ time and $O(n^2)$ work w.h.p.

Finally constructing the n different hash tables with a total of $O(n)$ elements can be done in $O(\log n \log^* n)$ time and $O(n)$ work w.h.p. using Lemma 4.

Overall preprocessing takes $O(\log n \log^* n)$ time and $O(n^2)$ work, both w.h.p.

4   Better Fingerprint-Based Pattern Lookup

We now show how to improve the above solution such that the preprocessing work will be $O(n)$ w.h.p. instead of $O(n^2)$. Furthermore, this method will support general prefix pattern lookups. These improvements are achieved by reducing the number of substrings of T that must be guaranteed to have collision-free fingerprints from $O(n^2)$ to $O(n)$, and instead taking care of possible false positives during the query.

Data Structure. The data structure used is the same as above with the difference that the fingerprint function ϕ is only guaranteed to be collision free for the substrings pre(e) for all $e \in S$, of which there are $O(n)$.

Query. Given a pattern P, first compute its prefix fingerprints. In parallel, look up the fingerprint $\phi(P[1, d])$ in the respective hash table H_d for all $d = 1 \ldots m$. If there is a match set $M[d] = H_d[\phi(P[1, d])]$ otherwise $M[d] = \bot$. If $M[1] = \bot$ then P does not occur in T, so in this case stop and report no match. The edges contained in M form a set of disjoint paths in S (see proof below). Consider each of these paths to be a linked list of edges. Let $N[i]$ be a table describing the next-pointers i.e. which edge $M[N[i]]$ follows $M[i]$. Define $e = M[d]$ and $e' = parent(e)$, and set $N[|pre(e')|] = d$ if $e \neq \bot$ and $M[|pre(e')|] = e'$. Let $N[d] = \bot$ denote unset entries. Use Lemma 1 to compute which edges are in

---

[3] Though, in fact, the suffix tree of T can be constructed in $O(\log^2 n)$ time, $O(n \log n)$ work and $O(n)$ space [11] for general alphabets.

the same linked list as $M[1]$, let d be the index of the right-most of these. Now $e_c = M[d]$ is our candidate edge. In parallel find the longest prefix of the strings P and $T(e_c)[1, m]$ that matches. Report the result.

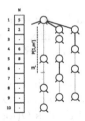

Fig. 1. An illustration of a small part of the suffix tree S. Green edges represent the edges in M. As illustrated they all form disjoint monotone paths. A prefix of the pattern P of length m' occurs on the left-most path in the illustration. An example of the N-array is included. The y-position of a node represents the string depth. (Color figure online)

Before reporting any results we verify by comparing P to a substring of T, so that no false-positives are reported.

We focus on proving that we always find a (prefix) match of P in T if it exists. So assume a non-empty prefix of P exists somewhere in T. In this case there is a path $\hat{P}$ from the root spelling out this prefix of P. We now need to show $\hat{P}$ is a prefix of the path $\mathcal{P}$ from the root to the edge e_c our algorithm picks as the candidate edge for verification.

Consider the set of edges the algorithm finds in M. All edges $e \in \hat{P}$ are in this set as $P[1, |\text{pre}(e)|] = \text{pre}(e) \Rightarrow \phi(P[1, |\text{pre}(e)|]) = \phi(\text{pre}(e))$. If the fingerprint function were collision-free (even with P), then this set of edges would be exactly the edges on $\mathcal{P}$. Unfortunately, this is not the case for the restricted-collision-free fingerprint function we are using. In our case the set of edges form a disjoint set of monotone paths in S as illustrated in Fig. 1. To prove this, we show that at most one outgoing edge of a node can be in M. Assume to the contrary that e_1 and e_2 are both outgoing edges of a node with string depth d and they are both in M. Then $\phi(P[1, d + 1]) = \phi(\text{pre}(e_1))$ and $\phi(P[1, d + 1]) = \phi(\text{pre}(e_2))$, which implies that $\phi(\text{pre}(e_1)) = \phi(\text{pre}(e_2))$. This contradicts that the fingerprint function is collision free for strings $\text{pre}(e)$ where $e \in S$.

Since all edges in $\mathcal{P}$ are in M and any node can have at most one outgoing edge, the path we are interested in is the one containing the root of S. All other paths can safely be discarded. Therefore we use Lemma 1 to remove all edges of M not connected to the root. Since all the edges on $\hat{P}$ are on this path and we pick the deepest, $\hat{P}$ is a prefix of $\mathcal{P}$. This completes the proof.

Preprocessing. All steps of the preprocessing are similar to the steps of the preprocessing before with the only exception we only need to verify our Karp-Rabin fingerprint function is collision free on a set of $O(n)$ strings. As this was the bottleneck on the work before, the work is now reduced to $O(n)$ w.h.p.

5   Parallel Suffix Array Pattern Lookup

Here we describe a parallelization of [14], which has the advantage of working for any alphabet and of being deterministic in both query and preprocessing. The query run time is slower.

Manber's algorithm performs a binary search over the suffix array. It maintains an interval $[L, R] \subseteq [1, n]$ of the suffix array wherein potential matches lie. In each round the middle element M in $[L, R]$ is found, and it is determined if the search should continue in the interval $[L, M]$ or $[M, R]$. This is accomplished by matching P to $T[\mathrm{SA}[M], \mathrm{SA}[M] + m]$. Finding the leftmost mismatch between the two strings in parallel takes $O(1)$ time and $O(m)$ work using Lemma 2. There are $O(\log n)$ rounds, so the overall time is $O(\log n)$ and the work is $O(m \log n)$.

This method can be generalized to the algorithm that uses the LCP-array as well. If we just keep comparing the current suffix with the entire part of P that has not yet been matched we will obtain the same time and work bounds as above. By a small modification, the work can be reduced to $O(m)$ as follows. Instead of comparing all of the pattern to the current suffix the algorithm should perform the comparison in chunks of size $\frac{m}{\log n}$.

In rounds where no more than $\frac{m}{\log n}$ characters match, the total work is $O(m + \log n)$. In the remaining rounds, the total work is $O(m)$. Thus the overall time is still $O(\log n)$ but the work is reduced to $O(m + \log n)$.

References

1. Amir, A., Farach, M., Matias, Y.: Efficient randomized dictionary matching algorithms. In: Proceedings of the 3rd CPM, pp. 262–275 (1992)
2. Bast, H., Hagerup, T.: Fast and reliable parallel hashing. In: Proceedings of the Third Annual ACM Symposium on Parallel Algorithms and Architectures, pp. 50–61. ACM (1991)
3. Cole, R., Vishkin, U.: Approximate and exact parallel scheduling with applications to list, tree and graph problems. In: Proceedings of the IEEE 27th Annual Symposium on Foundations of Computer Science, pp. 478–491 (1986)
4. Cook, S., Dwork, C., Reischuk, R.: Upper and lower time bounds for parallel random access machines without simultaneous writes. SIAM J. Comput. **15**(1), 87–97 (1986)
5. Czumaj, A., Galil, Z., Gasieniec, L., Park, K., Plandowski, W.: Work-time-optimal parallel algorithms for string problems. In: Proceedings of the Twenty-Seventh Annual ACM Symposium on Theory of Computing, pp. 713–722. ACM (1995)
6. Fich, F.E., Ragde, P., Wigderson, A.: Relations between concurrent-write models of parallel computation. SIAM J. Comput. **17**(3), 606–627 (1988)
7. Fischer, J., Köppl, D., Kurpicz, F.: On the benefit of merging suffix array intervals for parallel pattern matching. In: Proceedings of the 27th CPM (2016)
8. Galil, Z.: A constant-time optimal parallel string-matching algorithm. J. ACM (JACM) **42**(4), 908–918 (1995)
9. JáJá, J.: An Introduction to Parallel Algorithms. Addison Wesley, Redwood City (1992)

10. Jekovec, M., Brodnik, A.: Parallel query in the suffix tree. arXiv preprint arXiv:1509.06167 (2015)
11. Kärkkäinen, J., Sanders, P., Burkhardt, S.: Linear work suffix array construction. J. ACM **53**(6), 918–936 (2006)
12. Karp, R.M., Rabin, M.O.: Efficient randomized pattern-matching algorithms. IBM J. Res. Dev. **31**(2), 249–260 (1987)
13. Ladner, R.E., Fischer, M.J.: Parallel prefix computation. J. ACM (JACM) **27**(4), 831–838 (1980)
14. Manber, U., Myers, G.: Suffix arrays: a new method for on-line string searches. SIAM J. Comput. **22**(5), 935–948 (1993)
15. Vishkin, U.: Optimal parallel pattern matching in strings. Inf. Control **67**(1–3), 91–113 (1985)

Fast Classification of Protein Structures by an Alignment-Free Kernel

Taku Onodera[1,2(✉)] and Tetsuo Shibuya[1,2]

[1] Human Genome Center, Institute of Medical Science,
The University of Tokyo, Shirokanedai, Minato-ku, Tokyo, Japan
{tk-ono,tshibuya}@hgc.jp
[2] CREST, JST, Tokyo, Japan

Abstract. Alignment is the most fundamental algorithm that has been widely used in numerous research in bioinformatics, but its computation cost becomes too expensive in various modern problems because of the recent explosive data growth. Hence the development of alignment-free algorithms, i.e., alternative algorithms that avoid the computationally expensive alignment, has become one of the recent hot topics in algorithmic bioinformatics.

Analysis of protein structures is a very important problem in bioinformatics. We focus on the problem of predicting functions of proteins from their structures, as the functions of proteins are the keys of everything in the understandings of any organisms and moreover these functions are said to be determined by their structures. But the previous best-known (i.e., the most accurate) method for this problem utilizes alignment-based kernel method, which suffers from the high computation cost of alignments.

For the problem, we propose a new kernel method that does not employ alignments. Instead of alignments, we apply the two-dimensional suffix tree and the contact map graph to reduce kernel-related computation cost dramatically. Experiments show that, compared to the previous best algorithm, our new method runs about 16 times faster in training and about 37 times faster in prediction while preserving comparatively high accuracy.

1 Introduction

Proteins are fundamental biomolecules that work as functional units of biological systems. A protein consists of amino acids connected like a chain. In natural environments, this chain is folded into a three-dimensional structure by physical or chemical forces. Roughly speaking, each sequence is folded into a specific structure, which gives rise to specific functions. Thus, researchers have devoted much energies to identify protein structures and maintain them with annotations.

Since protein structures are notoriously difficult to determine by wet lab. techniques such as X-ray crystallography, once identified, they are often become subjects of "heavy" analyses. However, there are several situations where computationally inexpensive analysis methods are desirable. In particular,

© Springer International Publishing AG 2016
S. Inenaga et al. (Eds.): SPIRE 2016, LNCS 9954, pp. 68–79, 2016.
DOI: 10.1007/978-3-319-46049-9_7

computer simulations, nowadays, are beginning to produce large amounts of protein structure data. This technology is promising because it enables us to study proteins that are not amenable to conventional methods. Also, by simulation, one can observe the dynamic behaviors of proteins in environments such as water. To further this technology, more efficient algorithms for protein structure analysis are needed.

In this article, we propose a method to classify protein structures. For the classification of non-vectorial data such as strings or graphs, kernel methods are known to be effective [21]. We design a kernel function, a kind of similarity measure, for protein structures and plug it in to support vector machines (SVMs) [24]. Previously, researchers have developed *structural alignment*, a group of problem formulations/methods that capture relevant protein structure similarities very well [11]. But structural alignment is often computationally too expensive to apply to classification for two reasons. First, all structural alignment formulations require finding "good" correspondences between one protein's amino acids and those of another. This gives rise to hard combinatorial optimization problems. Second, most formulations involve complex algebraic scores such as the root mean square deviation. This makes it difficult to classify data by the structural similarity even if amino acids correspondence is known.

To overcome these difficulties, we take the approach of *alignment-free* method and apply it to compare protein structures. Alignment-free methods are recently gaining popularity among the community of sequence data analysis as a computationally cheaper and (sometimes) more relevant alternative of sequence alignment [1,5,6,12,23]. The crux of these methods is to avoid the quadratic cost of pairwise sequence alignment or NP-hardness of multiple alignment. We take the same approach for protein structures, for which even pairwise alignment is NP-hard (or it has a high degree polynomial time complexity depending on the exact formulation).

To compare protein structures without alignment, we define a kernel function through *protein contact maps*. The contact map of a protein is a graph with totally ordered vertices representing the amino acids and edges representing the proximity of amino acids in the folded state. It was introduced in the context of structural alignment and researchers have used it mainly to formulate contact map overlap problem, where optimal order preserving correspondence between vertices from different contact maps are sought [10]. We, instead, respect the sequential aspect of proteins more and characterize each protein by the histogram of square submatrices of the adjacency matrix of the contact map. See Fig. 1 for an illustration. The leftmost image is a protein comprising of 3 amino acids. Every pair of amino acids except the 1st and the 3rd is spatially close enough to each other. The middle image is the adjacency matrix of the corresponding contact map. The right image is the resulting feature vector (only some of the coordinates are shown.) Then we define the kernel function of two proteins as the inner product of their feature vectors. This function is positive semidefinite and thus, we can feed it into SVMs.

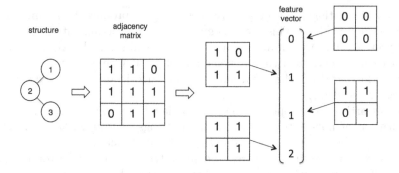

Fig. 1. Example feature vector of the proposed kernel with parameter $k = 2$. Every pair of amino acids except the 1st and the 3rd is spacially close to each other. In the feature vector on the right, only some of the coordinates are shown

Physically, the characterization of proteins described above corresponds to treating contiguous amino acids on the chain as groups and counting the patterns of interactions between such groups. This is natural for two reasons. First, interactions between spatially close amino acids have much more significant impact on the global structure than those between spatially remote amino acids do. Therefore, by representing proteins by contact map instead of, e.g., distance matrix, we lose some information but the loss should be minor. On the other hand, the combinatorial nature of the contact map opens the possibility of efficient solution. Second, amino acids close on the chain are close in space. Thus, we can treat a set of contiguous amino acids as a unit of interaction. In existing work, researchers chose the set of amino acids that are actually spatially close as units of interaction [4,25]. While this is also a reasonable choice, by using the proximity on the chain, and the contact map representation, we can apply the very efficient techniques of *combinatorial pattern matching* for the computation of the kernel function.

The results of this paper are summarized as follows:

- We propose a novel alignment-free kernel function for protein structures that is based on protein contact maps;
- We propose an efficient algorithm to calculate the proposed kernel function. The algorithm is based on the two-dimensional suffix tree [9,15] and runs in $\Theta(n^2)$-time where n is the size of input proteins. This bound matches the time complexity of the fastest existing method [25];
- We also propose a prediction algorithm for the test phase of SVMs based on the kernel we introduce. The time complexity of the algorithm does not depend on the size of the support vectors;
- We experimentally show that the combination of the proposed kernel function and the SVMs achieves classification performance comparative to the most accurate existing method [4] while it runs about 16 times faster in training and 37 times faster in prediction.

Related work. Wang et al. [25] and Bhattacharya et al. [4] worked on the same problem addressed in the current paper. For performance comparison, see Tables 1, 2 and 3. Wang et al. proposed a kernel function for protein structures that incorporates both sequential information, i.e., amino acid types, and structural information. They used the set of amino acids within some distance from an amino acid as a unit of interaction. Their method was the fastest but the least accurate among the methods we tested. Bhattacharya et al., taking the opposite approach to ours, proposed kernel functions based on structural alignment. They used the set of a fixed number of amino acids closest to an amino acid as a unit of interaction. In our experiment, their method gave accurate results but it took very long computation time. Qiu et al. [19] also proposed a kernel function based on structural alignment but in a different context of function annotation. This method may be applicable to the problem considered in the current paper but we did not test it because it is similar to Bhattacharya et al.'s method.

Alignment-free analysis is a well-studied topic among sequence data analysis community [1,5,6,12,23]. In particular, our method can be seen as a two-dimensional analogue of the spectrum kernel [16].

2 Preliminaries

Proteins and contact maps. In this paper, we model a protein as a sequence of three-dimensional coordinates. These coordinates represent the positions of carbon atoms called C_α. Due to the nature of peptide bond, the distance between a pair of neighboring C_α atoms is always about 3.8 Å.

Let $P = (p_1, p_2, \ldots, p_n)$ be a protein. The contact map of P is a graph consisting of n totally ordered vertices $v_1, v_2, \ldots, v_n$. There exists an edge between v_i and v_j iff the distance between p_i and p_j is less than a parameter $t > 0$.

The Isuffix tree. For a string T, we denote the substring from the i-th to the j-th character inclusive by $T[i : j]$. We denote the length of T by $|T|$. We denote the submatrix from the i_1-th row to the i_2-th row and the j_1-th column to the j_2-th column of a matrix M as $M[i_1 : i_2, j_1 : j_2]$. We also write $M[i, j_1 : j_2]$ to mean $M[i : i, j_1 : j_2]$ and write $M[i_1 : i_2, j]$ to mean $M[i_1 : i_2, j : j]$.

The two-dimensional suffix tree for an $m \times n$ matrix M should support that, given an $\ell \times \ell$ square matrix Q as a query, finding all (i, j) s.t. $M[i : i+\ell-1, j : j+\ell-1] = Q$. The Isuffix tree [15] is an instance of the two-dimensional suffix tree. To explain the Isuffix tree, we first need to define Istrings. For an $n \times n$ square matrix M, the Istring of M, denoted by I_M, is a length $2n - 1$ string of strings. The characters of I_M are defined as

$$I_M[i] = \begin{cases} M[1 : (i+1)/2, (i+1)/2] & \text{if } i \text{ is odd} \\ M[i/2 + 1, 1 : i/2] & \text{if } i \text{ is even.} \end{cases}$$

For each i, the string $I_M[i]$ is called an Icharacter. Strings of strings are ordered by naturally generalizing the order on strings. The suffix of an $m \times n$ matrix M at position (i, j), denoted by $M_{i,j}$, is $M[i : i+k, j : j+k]$ where $k = \min\{m-i, n-j\}$.

In other words, $M_{i,j}$ is the largest square submatrix of M whose upper left corner is (i, j).

The Isuffix tree of an $m \times n$ matrix M is a compressed trie storing Istrings of all suffixes of M. See Fig. 2 for an example. One can search a query square matrix Q in M by searching I_Q in the Isuffix tree. Because each edge label is equal to $I_{M_{i',j'}}[i:j]$ for some suffix $M_{i',j'}$ and i, j, it can be represented by the tuple (i', j', i, j). Thus, the Isuffix tree takes $\Theta(mn)$-space. The path label of a node is the concatenation of the edge labels of edges in the path from the root to the node. The depth of a node is the number of Icharacters in the path label of the node. One can guarantee that the leaves and suffixes correspond to each other one-to-one by appending a row at the bottom of M and a column on the right of M consisting of unique elements.

Kim et al. [15] gave an algorithm to construct the Isuffix tree of an $m \times n$ matrix consisting of entries from universe $\{1, 2, \ldots, (mn)^{O(1)}\}$ in $\Theta(mn)$-time.

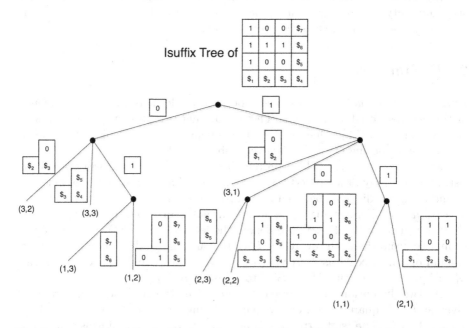

Fig. 2. An example of the Isuffix tree. Length 1 suffixes (those submatrices consisting only of 1 entry $\$_i$ for $1 \le i \le 7$) are not shown

Support vector machines. Support vector machines (SVMs) are supervised learning models for classification and regression [24]. Here, we explain how kernel-based SVMs work.[1] See [21] for the details.

A supervised learning problem consists of training phase and test phase. In the training phase, training data $(x_1, y_1) \ldots, (x_\ell, y_\ell) \in U \times \{-1, 1\}$ are given

---

[1] The SVM described here is called 1-norm soft margin SVM.

where U is the universe of data. The value y_i indicates if x_i belongs to the group of interest or not. In the test phase, test data (x, y) is given but the algorithm cannot access y. The algorithm should predict y from the training data and x.

Kernel-based SVMs do this task as follows. First one specifies a map ϕ from the universe of data U to $\mathbf{R}^d$. The map is called feature map and the resulting vectors are called feature vectors. Then, in the training phase, one solves the following optimization problem:

$$\text{maximize} \quad - \sum_{i,j=1}^{\ell} y_i y_j \alpha_i \alpha_j \phi(x_i)^\top \phi(x_j),$$

$$\text{subject to} \quad \sum_{i=1}^{\ell} y_i \alpha_i = 0, \quad \sum_{i=1}^{\ell} \alpha_i = 1$$

$$\text{and } 0 \le \alpha_i \le C \text{ for } i = 1, 2, \ldots, \ell.$$

The value C is a positive parameter. Let $(\alpha_1^*, \alpha_2^*, \ldots, \alpha_\ell^*)$ be the solution of the problem and let $S := \{1 \le i \le \ell : \alpha_i^* > 0\}$. Training data x_i is said to be a support vector if $i \in S$. In the test phase, one judges $y = 1$ iff

$$\sum_{i \in S} y_i \alpha_i^* \phi(x_i)^\top \phi(x) > b^* \tag{1}$$

where b^* is a constant that is determined in the training phase.

Sometimes, it is possible to compute kernel function $K : U \times U \ni (x_1, x_2) \mapsto \phi(x_1)^\top \phi(x_2) \in R$ without constructing feature vectors $\phi(x_1)$ and $\phi(x_2)$. Such algorithms are usually much faster than the naïve algorithm involving feature vector construction when the dimension of feature vectors is high.

3   Method

3.1   Definition of the Kernel Function

Let P be a protein and A_P be the adjacency matrix of the contact map of P. Let $k > 0$ and $\mathcal{M}_k$ be the space of all $k \times k$ binary matrices. The feature vector $\Phi_k(P)$ is a vector defined as follows:

$$\Phi_k(P) := (\#\{(i, j) : A_P[i : i + k - 1, j : j + k - 1] = H\})_{H \in \mathcal{M}_k}.$$

The kernel function K_k is defined to be the function that takes two proteins P_1 and P_2 and outputs the inner product of $\Phi_k(P_1)$ and $\Phi_k(P_2)$.

3.2   Algorithm to Compute the Kernel Function

Let P_1 and P_2 be two proteins. For brevity, we assume both of them are of length n. We compute $K_k(P_1, P_2)$ as follows. First, we construct the contact maps of P_1 and P_2. From contact maps, we compute adjacency matrices A_{P_1} and A_{P_2}.

We append a row and a column consisting of unique numbers to each of A_{P_1} and A_{P_2} making them $(n+1) \times (n+1)$ matrices. Then we construct the Isuffix tree of the $(n+1) \times (2n+2)$ matrix derived by concatenating A_{P_2} to the right of A_{P_1}. Next, we traverse the Isuffix tree in depth first order following Algorithm 1. During the traversal, the depth of the current node goes up and down. While we are at a node of depth greater than or equal to $2k-1$, we count the number of leaves from A_{P_1} that we have encountered since the last time we were at a node of depth less than $2k-1$. We also count the same number for A_{P_2}. When we climb up from a node of depth greater or equal to $2k-1$ to a node of depth less than $2k-1$ we compute the product of these counts and reset counters to 0. The output is the sum of these products.

The sizes of the adjacency matrices and the Isuffix tree are all $\Theta(n^2)$. The computation of adjacency matrices and the Isuffix tree and the traversal of the Isuffix tree take $\Theta(n^2)$-time. The contact maps can be constructed in $\Theta(n)$-time probabilistically by hashing[2] but to derive $\Theta(n^2)$-time bound on the kernel computation, the trivial deterministic algorithm is enough.

Algorithm 1. Traversal of the Isuffix tree

$a \leftarrow 0$
while traversal of the Isuffix tree in depth first order **do**
  if depth of the current node $< 2k-1$ **then**
    $a \leftarrow a + c_1 c_2$
    $c_1 \leftarrow 0$
    $c_2 \leftarrow 0$
  else
    if the current node is a leaf from A_{P_1} **then**
      $c_1 \leftarrow c_1 + 1$
    if the current node is a leaf from A_{P_2} **then**
      $c_2 \leftarrow c_2 + 1$
  return a

The second column of Table 1 shows the comparison of the time complexities of existing methods and our method. The expression $\mathsf{align}(n)$ is the time needed to align protein structures of length n.[3] The smallest bound on $\mathsf{align}(n)$ we could find in literature was $O(n^4)$ by Poleksic's formulation and algorithm [18].

Therefore, in terms of the dependence on the size of input, the time complexity of our algorithm matches the fastest existing method. Also, it does not depend on the parameters.

---

[2] This bound uses the sparsity of protein contact maps. See the discussion in Sect. 5.
[3] It is impossible to give a single explicit form for $\mathsf{align}(n)$ because unlike sequence alignment, there is no *de facto* standard formulation of structural alignment. Also, practical structural alignment tools do not even formally state the problem it solves.

3.3 Training Algorithm

In training, we compute kernel functions for all pairs of the training data and solve SVM optimization problem. The third column of Table 1 shows time complexities. They involve only kernel computation time because it is not clear how optimization time, which depends on various parameters and heuristic techniques, scales. In our experiment, optimization took much less time than the kernel computation. See also the discussion in Sect. 5.

3.4 Prediction Algorithm

In the test phase of SVM, one needs to evaluate the left hand side of inequality (1). In the case of our kernel, when query protein P is given, one needs to evaluate $\sum_{i \in S} y_i \alpha_i^* \phi(P_i)^\top \phi(P) = \sum_{i \in S} y_i \alpha_i^* K_k(P_i, P)$. A trivial method is to compute each $K_k(P_i, P)$ separately and take the weighted sum. However, it takes time linear to the number of support vectors ℓ', which, in the worst case, can be as large as the size of the training set. Here, we explain another algorithm whose computational cost does not depend on ℓ'.

First observe that

$$\sum_{i \in S} y_i \alpha_i^* K_k(P_i, P) = \sum_{i \in S} y_i \alpha_i^* \sum_{H \in \mathcal{M}_k} occ(H, A_{P_i}) occ(H, A_P)$$

$$= \sum_{H \in \mathcal{M}_k} \left(\sum_{i \in S} y_i \alpha_i^* occ(H, A_{P_i}) \right) occ(H, A_P)$$

$$= \sum_{1 \le s,t \le |P|-k+1} \left(\sum_{i \in S} y_i \alpha_i^* occ(A_P[s : s + k - 1, t : t + k - 1], A_{P_i}) \right).$$

For each $i \in S$, we prepare $n+1$-dimensional square matrix derived by appending a row and a column consisting of unique numbers to the bottom and right of A_{P_i}. Then, we construct the Isuffix tree of the $(n + 1) \times \ell'(n + 1)$ matrix derived by concatenating these ℓ' square matrices. Each leaf of the Isuffix tree corresponds to some Isuffix of A_{P_i} for some $i \in S$. For each $i \in S$, we label each leaf from A_{P_i} by $y_i \alpha_i^*$. We label each internal node by the sum of the labels of its children. Given query P, for each (i, j) s.t. $1 \le s, t \le |P| - k + 1$, we search $P[s : s+k-1, t : t+k-1]$ in the Isuffix tree for all (i, j) s.t. $1 \le s, t \le |P|-k+1$ and take the sum of the labels of the nodes the searches end up at. When a search ends up at somewhere in the middle of an edge, take the label of the upper end point node of the edge.

It takes $O(k^2|P|^2)$-time because each search takes $O(k^2)$-time and there are $O(|P|^2)$ search instances invoked.

4 Experiments

To assess the effectiveness of our algorithm, we tested if it can recover the classification of existing classified databases correctly. We used SCOP database [8] as

Table 1. Comparison of time complexities. Each protein in the training set and the test set is assumed to be of length n. ℓ is the number of proteins for training. K is a parameter of our kernel. ℓ' is the number of the support vectors. $\mathsf{align}(n)$ is the time needed to align protein structures of length n

Method	Pairwise	Training	Prediction
Ours	$O(n^2)$	$O(\ell^2 n^2)$	$O(k^2 n^2)$
[25]	$O(n^2)$	$O(\ell^2 n^2)$	$O(\ell' n^2)$
[4]	$O(\mathsf{align}(n))$	$O(\ell^2 \mathsf{align}(n))$	$O(\ell' \mathsf{align}(n))$

the data source.[4] In SCOP, data are hierarchically organized into a tree structure. The deepest level of the tree based on structural information is called superfamily and the level just above it is called fold. In order to test the ability of structure classifiers, it is reasonable to check if they can recognize the boundaries between superfamilies that are within the same fold. Thus, we made a dataset by extracting all entries in a superfamily that is under the fold c.1 and contains at least 16 entries. We selected c.1 because it is the fold that contains the largest number of superfamilies satisfying the condition. The resulting dataset consists of 9 classes (superfamilies) and contains 383 entries. We used half of the entries as the training set and the other half as the test set. For each class, we classified the test data as either in that class or not. To take both precision and recall into account, we used F-score to measure classification performance.

Our kernel has two parameters, namely, the threshold t for contact maps and the size of submatrix k. We optimized k by performing 2-fold cross validation within the training set. More precisely, we chose k between 3 and 7 that gave the best cross validation result. On the other hand, we treated t as a fixed parameter. We just report different results for different values of t. This is because optimization of k, which can be done by reusing the same two-dimensional suffix tree, is computationally inexpensive while optimization of t is expensive. We optimized the soft margin hyperparameter C of SVM by grid search. We used LIBSVM [7] for SVM optimization.

In terms of implementation details, we used two-dimensional analogue of the suffix array instead of the suffix tree. Since the $\Theta(n^2)$-time algorithm of Kim et al. [15] (and its adoption to two-dimensional suffix array) is very complex, we used ternary sort [3] for Isuffix sorting. For prediction, we used standard binary search instead of the two-dimensional analogue of the search algorithm of Manber and Myers [17]. We run the codes on Xeon E5–2670 v3 processors.

We compared our method with the method of Wang et al. [25] and Bhattacharya et al. [4]. The latter authors proposed many kernels but we only report the results of the kernel called K_1^{Al} in the original paper because we found the

---

[4] Each SCOP entry corresponds to a subregion of a protein called domain. In most applications, domains are not known *a priori*. Nevertheless, SCOP entries are widely used for benchmarking in protein classification studies including the work related to the current one. Thus, we followed the convention.

other kernels were much less accurate than K_1^{Al}. The kernel K_1^{Al} requires structural alignment as input and its performance may depend on the quality of the alignment. In the original paper, the authors used their own structural alignment algorithm but they did not provide the detail. Thus, we instead used a famous structural alignment algorithm called combinatorial extension [22] (CE). For the eigendecomposition needed in K_1^{Al}, we used LAPACK [2].

Table 2 shows the comparison of classification performances. The proposed kernel scored better than Wang et al.'s kernel [25] for most cases. It is difficult to judge which one of the proposed kernel and Bhattacharya et al.'s kernel [4] is better. However, it should be noted that for some classes such as c.1.4 and c.1.12, the dominance of the proposed kernel is clear for all values of t we tested.

Table 3 shows the comparison of runtimes. The numbers of our method were taken from the case when $t = 12$. Runtimes for other values of t were similar. The training time includes SVM optimization and cross validations. The method of Wang et al. [25] runs very fast because it is also alignment-free. However, this method had low classification scores. Compared to the alignment based method [4], our method runs about 16 times faster in training and about 37 times faster in prediction.

Table 2. Classification performance comparison. The performance measure is F-score. The unit of parameter t is Å. The cases when the proposed method had the highest score among all tested methods are indicated by boldface numbers

Method		c.1.2	c.1.4	c.1.7	c.1.8	c.1.9	c.1.10	c.1.11	c.1.12	c.1.18
[25]		0.31	0.14	0.33	0.84	0.40	0.39	0.79	0	0
[4]		0.37	0.27	0.86	0.81	0.84	0.62	0.93	0.13	0.84
Ours	$t = 8$	**0.41**	**0.59**	0.36	**0.86**	0.76	0.58	0.75	**0.40**	0.83
	$t = 9$	**0.48**	**0.40**	0.62	0.79	**0.89**	**0.65**	**0.93**	**0.53**	0.67
	$t = 10$	0.35	**0.48**	0.19	0.80	0.82	0.60	0.84	**0.29**	**0.89**
	$t = 11$	**0.73**	**0.59**	0.50	0.79	0.72	0.50	**0.95**	**0.38**	0.55
	$t = 12$	**0.39**	**0.38**	**0.88**	0.82	**0.92**	**0.64**	0.92	**0.17**	0.59
	$t = 13$	0.26	**0.56**	0.80	**0.84**	0.72	**0.62**	0.86	**0.50**	0.80
	$t = 14$	**0.73**	**0.56**	0.57	**0.84**	**0.86**	0.57	**0.93**	**0.27**	0.77

Table 3. Runtime comparison

Method	Training time	Average prediction time
[25]	4.4 s	0.0168 s
[4]	1182 min	72.4 s
Ours	40 min 14 s	0.619 s

5   Conclusion

We proposed a kernel function for protein structures that is based on a novel use of the protein contact map and an efficient algorithm to compute the kernel function applying the two-dimensional suffix tree. We also experimentally showed that, by using the proposed kernel, one can classify protein structures much faster than the most accurate existing method while achieving a comparable classification performance.

We conclude this paper with some discussion. For large datasets, SVM optimization may become the computational bottleneck of the training phase. There are several researches addressing this problem [14,20]. These results are orthogonal to ours.

We did not consider tuning of parameter t in this paper. If we consider distance between objects into account, there is no need to introduce such parameters as t. Since there are structural alignment problems/algorithms based on distance matrices [13], it may be possible to design a relevant alignment-free similarity measure for protein structures from distance matrices.

When t is independent of n, protein contact maps are sparse because there is a limit on the number of amino acids packed in a certain volume of space. Our algorithm needs quadratic time because it does not take into account this sparsity. Ideally, it is desirable to have relevant protein similarity measures that are computable in linear time.

Acknowledgement. This work was supported by JSPS KAKENHI Grant Numbers 25280002 and 24106007. The super-computing resource was provided by Human Genome Center (the Univ. of Tokyo).

References

1. Aluru, S., Apostolico, A., Thankachan, S.V.: Efficient alignment free sequence comparison with bounded mismatches. In: Przytycka, T.M. (ed.) RECOMB 2015. LNCS, vol. 9029, pp. 1–12. Springer, Heidelberg (2015)
2. Anderson, E., Bai, Z., Bischof, C., Blackford, S., Demmel, J., Dongarra, J., Du Croz, J., Greenbaum, A., Hammarling, S., McKenney, A., Sorensen, D.: LAPACK Users' Guide, 3rd edn. SIAM, Philadelphia (1999)
3. Bentley, J.L., Sedgewick, R.: Fast algorithms for sorting and searching strings. In: Proceedings of the 8th Annual ACM-SIAM Symposium on Discrete Algorithms, pp. 360–369 (1997)
4. Bhattacharya, S., Bhattacharyya, C., Chandra, N.: Structural alignment based kernels for protein structure classification. In: Proceedings of the 24th International Conference on Machine Learning, pp. 73–80 (2007)
5. Bonham-Carter, O., Steele, J., Bastola, D.: Alignment-free genetic sequence comparisons: a review of recent approaches by word analysis. Briefings Bioinform. **15**(6), 890–905 (2014)
6. Břinda, K., Sykulski, M., Kucherov, G.: Spaced seeds improve k-mer-based metagenomic classification. Bioinformatics **31**(22), 3584–3592 (2015)

7. Chang, C.C., Lin, C.J.: LIBSVM: a library for support vector machines. ACM Trans. Intell. Syst. Technol. **2**(3), 27:1–27:27 (2011)

8. Fox, N.K., Brenner, S.E., Chandonia, J.M.: SCOPe: structural classification of proteins-extended, integrating scop and astral data and classification of new structures. Nucleic Acids Res. **42**(D1), D304–D309 (2014)

9. Giancarlo, R.: A generalization of the suffix tree to square matrices, with applications. SIAM J. Comput. **24**(3), 520–562 (1995)

10. Goldman, D., Istrail, S., Papadimitriou, C.H.: Algorithmic aspects of protein structure similarity. In: Proceedings of the 40th Symposium on Foundations of Computer Science, pp. 512–521 (1999)

11. Hasegawa, H., Holm, L.: Advances and pitfalls of protein structural alignment. Curr. Opin. Struct. Biol. **19**(3), 341–348 (2009)

12. Haubold, B.: Alignment-free phylogenetics and population genetics. Briefings Bioinf. **15**(3), 407–418 (2014)

13. Holm, L., Sander, C.: Protein structure comparison by alignment of distance matrices. J. Mol. Biol. **233**(1), 123–138 (1993)

14. Joachims, T.: Training linear SVMs in linear time. In: Proceedings of the 12th ACM SIGKDD International Conference on Knowledge Discovery and Data Mining, pp. 217–226 (2006)

15. Kim, D.K., Na, J.C., Sim, J.S., Park, K.: Linear-time construction of two-dimensional suffix trees. Algorithmica **59**(2), 269–297 (2011)

16. Leslie, C.S., Eskin, E., Noble, W.S.: The spectrum kernel: a string kernel for SVM protein classification. In: Pacific Symposium on Biocomputing, pp. 566–575 (2002)

17. Manber, U., Myers, G.: Suffix arrays: a new method for on-line string searches. In: Proceedings of the 1st Annual ACM-SIAM Symposium on Discrete Algorithms, pp. 319–327 (1990)

18. Poleksic, A.: Algorithms for optimal protein structure alignment. Bioinformatics **25**(21), 2751–2756 (2009)

19. Qiu, J., Hue, M., Ben-Hur, A., Vert, J.P., Noble, W.S.: A structural alignment kernel for protein structures. Bioinformatics **23**(9), 1090–1098 (2007)

20. Severyn, A., Moschitti, A.: Large-scale support vector learning with structural kernels. In: Balcázar, J.L., Bonchi, F., Gionis, A., Sebag, M. (eds.) ECML PKDD 2010, Part III. LNCS, vol. 6323, pp. 229–244. Springer, Heidelberg (2010)

21. Shawe-Taylor, J., Cristianini, N.: Kernel Methods for Pattern Analysis. Cambridge University Press, New York (2004)

22. Shindyalov, I.N., Bourne, P.E.: Protein structure alignment by incremental combinatorial extension (CE) of the optimal path. Protein Eng. **11**(9), 739–747 (1998)

23. Song, K., Ren, J., Reinert, G., Deng, M., Waterman, M.S., Sun, F.: New developments of alignment-free sequence comparison: measures, statistics and next-generation sequencing. Briefings Bioinf. **15**(3), 343–353 (2014)

24. Vapnik, V.N.: Statistical Learning Theory, vol. 1. Wiley, New York (1998)

25. Wang, C., Scott, S.D.: New kernels for protein structural motif discovery and function classification. In: Proceedings of the 22nd International Conference on Machine Learning, pp. 940–947 (2005)

XBWT Tricks

Giovanni Manzini[1,2]([⊠])

[1] Computer Science Institute, University of Eastern Piedmont, Alessandria, Italy
giovanni.manzini@uniupo.it
[2] Institute of Informatics and Telematics, CNR, Pisa, Italy

Abstract. The eXtended Burrows-Wheeler Transform (XBWT) is a data transformation introduced in [Ferragina et al., FOCS 2005] to compactly represent a labeled tree and simultaneously support navigation and path-search operations over its label structure.

A natural application of the XBWT is to store a dictionary of strings. A recent extensive experimental study [Martínez-Prieto et al., Information Systems, 2016] shows that, among the available string dictionary implementations, the XBWT is attractive because of its good tradeoff between small space usage, speed, and support for substring searches. In this paper we further investigate the use of the XBWT for storing a string dictionary. Our first contribution is to show how to add suffix links (aka failure links) to a XBWT string dictionary. For a XBWT dictionary with n internal nodes our suffix links can be traversed in constant time and only take $2n + o(n)$ bits of space.

Our second contribution are practical construction algorithms for the XBWT, including the additional data structure supporting the traversal of suffix links. Our algorithms build on the many well engineered algorithms for Suffix Array and BWT construction and offer different tradeoffs between running time and working space.

1 Introduction

A trie [15] is a fundamental data structure to represent a set of strings. A trie with n nodes takes $\mathcal{O}(n \log n)$ bits of space and supports extremely simple and efficient algorithms to determine whether a string belongs to the set. In this paper we are interested in the "compressed" version of a trie obtained applying to it the eXtended Burrows Wheeler Transform (XBWT): a generalization of the BWT introduced in [6–8] to compactly represent an arbitrary labeled tree. The XBWT represents an n-node trie in $\mathcal{O}(n)$ bits of space still supporting constant time upward and downward navigation.

In a recent comprehensive study of string dictionaries [18], the authors show that in many applications we need to handle dictionaries whose size is larger than the available RAM. In this setting, compression is mandatory to avoid incurring the penalties of external memory access. In the same paper the authors show that, among the available string dictionary implementations, the XBWT-trie is particularly attractive because of its good tradeoff between small space usage, speed, and support for substring searches.

S. Inenaga et al. (Eds.): SPIRE 2016, LNCS 9954, pp. 80–92, 2016.
DOI: 10.1007/978-3-319-46049-9_8

In this paper we present two contributions related to the XBWT-trie. Our first contribution is the observation that we can enrich the XBWT with $2n + o(n)$ additional bits in order to support suffix links. Suffix links, also known as *failure links*, are useful to speedup some search operations as in the classical Aho-Corasick algorithm [9, Sect. 3.4].

Our second contribution is related to the problem of computing the XBWT. For a set of strings $x_1, \ldots, x_k$ of total length m we can compute the XBWT-trie by first building the n-nodes (uncompressed) trie and then applying the XBWT construction algorithm from [8]. This approach takes optimal $\mathcal{O}(m + n)$ time but it may not work well in practice because trie construction may constitute a memory bottleneck. Indeed, as shown by the Suffix-Tree vs Suffix Array debate, pointer based tree structures often have very large multiplicative constants hidden in the $\mathcal{O}$ notation that in practice prevent their use for large datasets. An indirect confirmation of this state of affairs is that in [18] the authors report that they were unable to build the trie for the largest dataset due to excessive memory usage.

In this paper we take advantage of the similarities between XBWT and BWT to derive alternative algorithms for the construction of the XBWT starting from the Suffix Array or the BWT. Our motivation is that the algorithms for constructing these data structures have been widely studied and engineered so there are practical algorithms using very little working space or even designed for external memory, see [3–5,10,12,13] and references therein. Our contribution is to show that given the Suffix Array or BWT we can compute the XBWT, including the data structure supporting suffix links, in $\mathcal{O}(m)$ time. Combining our algorithms with the available (and future!) Suffix Array and BWT construction algorithms we obtain a wide range of tradeoffs between running time and working space for XBWT construction.

2 XBWT Trie Representation

Given a string $x[1, n]$ over a finite ordered alphabet Σ we write $x[i]$ to denote its i-th symbol and $x[i, j]$ to denote the substring $x[i]x[i+1] \cdots x[j]$. We write x^R to denote the string x reversed $x[n] \cdots x[1]$. We write $x \preceq y$ ($x \prec y$) to denote that x is lexicographically (strictly) smaller than y. As usual we assume that if x is a prefix of y then $x \prec y$. Throughout the paper we use the notation $\mathsf{rank}_c(x, i)$ to denote the number of occurrences of c in $x[1, i]$, and $\mathsf{select}_c(x, j)$ to denote the position of the j-th c in x.

Tries [15] are a fundamental data structure for representing a set of k distinct strings $x_1, x_2, \ldots, x_k$. A trie efficiently supports the two basic dictionary operations: $\mathsf{locate}(s)$ returning i if $s = x_i$ for some $i \in [1, k]$ or 0 otherwise, and $\mathsf{extract}(i)$ returning the string x_i given an index $i \in [1, k]$. In addition, it supports the operation $\mathsf{locatePrefix}(s)$ returning the strings which are prefixed by s [18]. To simplify the algorithms, and ensure that no string is the prefix of another one, it is customary to add a special symbol $\$ \notin \Sigma$ at the end of each string x_i. A trie for the set of strings $\{\mathsf{aa}, \mathsf{acaa}, \mathsf{ba}, \mathsf{aba}, \mathsf{aac}, \mathsf{bc}\}$ is shown in Fig. 1.

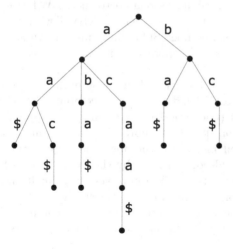

Last	L	Π
0	a	ε
1	b	
0	a	a
0	b	
1	c	
0	$	aa
1	c	
1	$	aaca
1	$	ab
1	$	aba
1	a	aca
0	a	b
1	c	
1	a	ba
1	a	ca
1	$	caa
1	a	cb

Fig. 1. A trie representing the strings and its XBWT representation (the arrays Last and L). The array Π is not stored in the XBWT even if navigation algorithms use it to identify internal nodes.

The eXtended Burrows-Wheeler Transform is a generalization of the BWT designed to compactly represent a labeled tree. We now show how to compute the XBWT of a trie T and obtain two arrays L and Last that compactly represent T. Our description of the XBWT is slightly different (simpler) from the one in [6,8] that takes as input an arbitrary labeled tree.

To each *internal* trie node w we associate the string λ_w obtained by concatenating the symbols in the arcs in the upward path from w to the root of T. Hence, if node w has depth d its associated string has length d. If T has n internal nodes we have n strings overall. Let $\Pi[1, n]$ denote the array containing the above set of n strings sorted lexicographically. Note that $\Pi[1]$ is always the empty string corresponding to the root of T.

For $i = 1, \ldots, n$ let L_i denote the set of symbols in the arcs exiting from the trie node corresponding to $\Pi[i]$. We do not require that the symbols in L_i are in any particular order, but since T is a trie they are distinct. We define the array L as the concatenation of the arrays $L_1, \ldots, L_n$. Clearly if T has n' nodes, then L has $n' - 1$ elements: one for each trie edge. By construction L contain $n - 1$ symbols from Σ and $n' - n$ occurrences of \$. To keep an explicit representation of the intervals $L_1, \ldots, L_n$ we define a binary array Last$[1, n']$ such that Last$[i] = 1$ iff $L[i]$ is the last symbol of some interval L_j. Hence Last contains exactly n 1's. See Fig. 1 for a complete example.

If $L[i] \neq \$$ belongs to the interval L_j then $L[i]$ naturally corresponds to the internal trie node reachable from the node corresponding to $\Pi[j]$ following the arc labeled $L[i]$. Such a node corresponds to the entry $\Pi[i']$ such that $\Pi[i'] =$

$L[i]\Pi[j]$. In other words, there is a bijection between the symbols in L different from \$ and the entries in Π different from the empty string. For historical reasons this bijection is called the LF-map, and we call $LF(i)$ the index in Π of the entry corresponding to $L[i]$. Hence, LF is defined by the relation

$$\Pi[LF(i)] = L[i]\Pi[j]$$

for every i, j with $L[i] \in L_j$ and $L[i] \neq \$$. The following results are a simple restatement of Properties 1–3 in [8] using the notation of this paper.

Lemma 1 (Order preserving property). *For every pair of indices i, k such that $L[i] \neq \$$, $L[k] \neq \$$, it is*

$$L[i] < L[k] \implies LF(i) < LF(k),$$
$$L[i] = L[k] \implies LF(i) < LF(k) \Leftrightarrow i < k.$$

$\square$

For any symbol $c \in \Sigma$ let $C(c)$ denote the index of the first position in Π containing a path starting with symbol c. Lemma 1 makes it possible to compute LF and its inverse LF^{-1} using rank and select operations. In turn, the LF map makes it possible to navigate the XBWT-trie, that is to move from the entry in Π representing a trie node to the entries representing its children and parent.

Lemma 2 (Downward navigation). *Let $c = L[i] \neq \$$. Then*

$$LF(i) = C[c] + \mathsf{rank}_c(L, i - 1).$$

As a consequence, if node w corresponds to $\Pi[j]$ and has a child with label c, then such child corresponds to entry $\Pi[j']$ with

$$j' = C[c] + \mathsf{rank}_c(L, \mathsf{select}_1(\mathsf{Last}, j)).$$

$\square$

Lemma 3 (Upward navigation). *For $i > 1$ let c denote the first symbol of path $\Pi[i]$. Then*

$$LF^{-1}(i) = \mathsf{select}_c(L, 1 + i - C[c]).$$

As a consequence, if node w corresponds to the non empty path $\Pi[j]$ whose first character is c, the parent of w' corresponds to the entry $\Pi[j']$ with

$$j' = 1 + \mathsf{rank}_1(\mathsf{Last}, LF^{-1}(j) - 1) = 1 + \mathsf{rank}_1(\mathsf{Last}, \mathsf{select}_c(L, 1 + j - C[c])).$$

$\square$

Using downward (resp. upward) navigation we can implement the locate (resp. extract) trie operation. As observed in [18] it is convenient to take as the ID of x_i the rank in L of the \$ occurrence that we reach starting from the root and following x_i's symbols. If we reorder the strings in reverse lexicographic order (i.e. so that $x_1^R \prec x_2^R \prec \cdots \prec x_k^R$) then $\mathsf{ID}(x_i) = i$.

The most common representation of the array L is a (possibly compressed) Wavelet tree. We also need a bitarray representation of Last supporting constant time rank_1, select_1 operations, and a suitable representation of the array C (possibly another bitarray). Using a balanced uncompressed Wavelet trees for L the space usage is $\mathcal{O}(n' \log(|\Sigma|))$ bits and each upward or downward step takes $\mathcal{O}(\log |\Sigma|)$ time.

3   Adding Suffix Links

In addition to pointers to their children and parent, trie nodes may store an additional pointer called a *suffix link*. The node corresponding to path α has a suffix link pointing to the node corresponding to the longest proper suffix of α that is also in T. Hence, if we have reached the node corresponding to the path $c_0 c_1 \cdots c_i$ the suffix link makes it possible to reach in constant time the node corresponding to path $c_j \cdots c_i$ where $j > 0$ is the smallest positive integer for which such node exists. Since the root corresponds to the empty string, a suffix link exists for all internal nodes except for the root itself.

In a XBWT-trie internal nodes are identified with their position in Π. Because of the ordering of the paths in Π, the target of the suffix link of node $\Pi[i]$ is the node $\ell < i$ such that $\Pi[\ell]$ is the longest proper *prefix* of $\Pi[i]$ which is in Π.

To emulate suffix links we build a string P of balanced parentheses of length $2n$. We write a pair of parentheses for each internal node so that the parentheses for node j enclose those for i iff $\Pi[j]$ is a prefix of $\Pi[i]$. To build P we start with an empty string and consider $\Pi[i]$ for $i = 1, \ldots, n$. When we reach $\Pi[i]$ first we write a) for every $\ell < i$ such that the closed parenthesis for $\Pi[\ell]$ has not been written and $\Pi[\ell]$ *is not* a prefix of $\Pi[i]$; then we write the (corresponding to $\Pi[i]$. After we have reached $\Pi[n]$ we write a closing parenthesis for all indices ℓ such that the closed parenthesis for $\Pi[\ell]$ has not yet been written. For example, for the XBWT of Fig. 1 it is $P = (((()) (()) ())(()) (()) ())$.

The following lemma shows that to find the suffix link for node $\Pi[i]$ it suffices to find the closest set of parentheses enclosing the (associated to $\Pi[i]$.

Lemma 4. *Let $1 < i \leq n$ and $\alpha = \Pi[i]$. Define $k = \mathsf{select}_{(}(P, i)$ and $j = \mathsf{enclose}(k)$. Then, the longest proper prefix of α in Π is $\alpha' = \Pi[\ell]$ with $\ell = \mathsf{rank}_{(}(P, j)$.*

Proof. First note that enclose is always defined since the pair $P[1] = ($, $P[2n] =)$ corresponding to $\Pi[1]$ encloses every other pair of parentheses.

We need to prove that $\alpha' = \Pi[\ell]$ is the longest proper prefix of α which is in Π. Since the) for $\Pi[\ell]$ is not written when we reach $\Pi[i]$, by construction

$\Pi[\ell]$ is a prefix of $\Pi[i]$. To prove it is the longest prefix assume by contradiction that $\Pi[\ell']$ is also a prefix of $\Pi[i]$ and $|\Pi[\ell']| > |\Pi[\ell]|$. Because of the ordering in Π we would have $\ell < \ell' < i$. Also because of the ordering, for $i' = \ell' \ldots i$ $\Pi[i']$ would be a prefix of $\Pi[i]$. But then the parentheses for ℓ' would enclose those for i, which is a contradiction since by construction ℓ corresponds to the closest enclosing pair. $\square$

Using the range min-max tree from [19] we can represent the balanced parenthesis sequence P in $2n + o(n)$ bits of space and support rank, select, and enclose in $\mathcal{O}(1)$ time. We have therefore established the following result.

Theorem 1. *We can add to the XBWT-trie suffix links traversable in constant time using additional $2n + o(n)$ bits.* $\square$

Since Π only contains internal nodes, the approach described above only provides suffix links for the trie internal nodes. However, it can be extended to the trie leaves if necessary. Since the symbol $ appears only at the end of a string, the suffix link of a leaf can only point to another leaf. Thus, we can build a subsequence Π' of Π containing only the internal nodes which have $ among their children. It is easy to see that the parenthesis array P' build on Π' provides suffix links for the leaves.

4   Alternative Construction Algorithms

In this section we propose new algorithms for computing the XBWT of the trie containing the set of distinct strings $x_1, x_2, \ldots, x_k$. Our algorithms derive the XBWT from the Suffix Array or BWT of the concatenation $t = y_1 \$ y_2 \$ \cdots y_k \$$, where $y_i = x_i^R$ reversed and $ is assumed to be lexicographically smaller than any symbol in Σ. We denote by SA, LCP and BWT respectively the Suffix Array, LCP array, and Burrows Wheeler Transform of the string t (See Fig. 2 for an example). Throughout this section let m denote the length of t, i.e. $m = \sum_i (|x_i| + 1)$.

Let z be a string not containing the symbol $ and such that $z\$$ is a substring of t. We denote by $[b_z, e_z]$ the maximal range of suffix array rows prefixed by $z\$$. For example, in Fig. 2 for $z = \epsilon$ the maximal range is $[1, 6]$, for $z = $ aa the maximal range is $[11, 12]$, and for $z = $ ca the maximal range is $[20, 20]$.

Lemma 5. *Let $[b_z, e_z]$ denote the maximal range for the string z. Then $e_z - b_z + 1$ is equal to the number of strings in $x_1, \ldots, x_k$ which have z as a prefix. In addition it is $\mathsf{LCP}[b_z] < |z|$ and*

$$\mathsf{LCP}[i] \geq |z| + 1 \qquad for\ i = b_z + 1, \ldots, e_z.$$

Proof. By construction the rows prefixed by $z\$$ are in a bijection with the strings y_i's which have z as a suffix. Since $y_i = x_i^R$ the first part of the lemma follows. Since b_z is the first row prefixed by $z\$$ row $b_z - 1$ must be prefixed by a string lexicographically strictly smaller than $z\$$. Since $ is the smallest symbol, row $b_z - 1$ cannot be prefixed by z. $\square$

#	SA	LCP	BWT	RCP	MR	Sorted suffixes
1	22	–	b	–	1	$
2	3	1	a	0	0	$aaca$ababacaacb
3	8	2	a	0	0	ababacaacb$
4	11	3	b	0	0	abacaacb
5	15	1	a	0	0	caacb$
6	19	2	a	0	0	cb
7	2	0	a	0	1	a$aaca$ababacaacb
8	7	3	c	1	0	aababacaacb$
9	14	2	b	1	0	acaacb$
10	18	3	a	1	0	acb
11	1	1	$	1	1	aa$aaca$ababacaacb
12	17	3	c	2	0	aacb
13	4	2	$	2	1	aacaababacaacb$
14	9	1	$	1	1	ababacaacb
15	12	2	$	2	1	abacaacb$
16	5	1	a	1	1	acaababacaacb$
17	21	0	c	0	1	b$
18	10	1	a	1	0	babacaacb
19	13	1	a	1	1	bacaacb$
20	6	0	a	0	1	caababacaacb$
21	16	2	$	2	1	caacb
22	20	1	$	1	1	cb$

Fig. 2. Suffix array, LCP array, MR array, and BWT for the concatenation $t =$ aa$aaca$ababacaacb obtained from the set of strings aac, aa, aba, acaa, ba, bc The arrays MR and RCP will be introduced later.

Lemma 6. *Let T denote the trie representing the strings $x_1, \ldots, x_k$. There is a one-to-one (bijective) correspondence between internal nodes of T and maximal row ranges of SA. Each node w corresponds to a maximal range containing a number of rows equals to the number of leaves in the subtree rooted at w. The correspondence is order preserving in the sense that row range $[b_y, e_y]$ precedes $[b_z, e_z]$ iff the node corresponding to the former interval precedes the node corresponding to the latter in the array Π used to define the XBWT.*

Proof. For each internal node w let λ_w denote the string obtained concatenating the symbols in the upward path from w to the root. The image of node w is the maximal row range associated to λ_w, that is, the set of SA rows prefixed by λ_w\$. As we have already observed, the number of rows in this interval is equal to the number of strings $x_1, \ldots, x_k$ which have λ_w as prefix which coincides with the number of leaves in the subtree rooted at w. The correspondence is order preserving since both in Π and in the suffix array the order is determined by the lexicographic order of λ_w. □

Lemma 7. *Let $[b, e]$ denote the maximal row range associated to the internal node w. Then, the labels on the arcs exiting from w coincide with the set of symbols in the substring BWT$[b, e]$.*

Proof. Let λ_w denote the string containing the symbols in the upward path from w to the root. There is an arc with label $c \in \Sigma$ leaving w iff there is at least a string x_i prefixed by $\lambda_w^R c$. This implies $c\lambda_w$ is a prefix of y_i. If $j \in [b, e]$ is the row prefixed by $\lambda_w \$ y_{i+1} \$ \cdots y_k \$$ it is $\mathsf{BWT}[j] = c$. Viceversa, if $\mathsf{BWT}[h] = c$ for $h \in [b, e]$ then at least one y_i is prefixed by $c\lambda_w$, hence $\lambda_w^R c$ is a prefix of x_i and there must be an arc with label c exiting from node w.

Finally, there is an arc with label \$ leaving w iff $\lambda_w^R = x_h$ for some $h \in [1, k]$. But then there will be one SA row prefixed by $y_h \$ y_{h+1} \$ \cdots y_k \$$ and the corresponding BWT position will contain the symbol \$. □

From Lemma 7 we can derive a simple strategy to compute the XBWT, that is, the arrays L and Last defined in Sect. 2. Assume we are given a binary array MR such that $\mathsf{MR}[i] = 1$ iff row i is the starting position of a maximal row range (see example in Fig. 2). MR encodes the maximal row ranges and by Lemma 6 each maximal row range corresponds to an element in the array Π. In Sect. 2 we have logically partitioned the array L into $L_1, \ldots, L_n$ where L_i contains the labels in the arcs leaving the internal node associated to $\Pi[i]$. We compute the subarrays $L_1, \ldots, L_n$ in that order. We scan the array MR starting from its first position until we find an index j_1 such that $\mathsf{MR}[j_1 + 1] = 1$. We know that $[1, j_1]$ is the maximal row range corresponding to $\Pi[1]$. In $\mathcal{O}(j_1)$ time we compute the set of distinct symbols in $\mathsf{BWT}[1, j_1]$ and we write them to L. By Lemma 7 we have just computed L_1 and we complete this phase by writing $\mathbf{0}^{|L_1|-1}\mathbf{1}$ to Last. Next we restart the scanning of MR until we find an index j_2 such that $\mathsf{MR}[j_2 + 1] = 1$. By construction $[j_1 + 1, j_2]$ is the maximal range corresponding to $\Pi[2]$ so from $\mathsf{BWT}[j_1 + 1, j_2]$ we can derive L_2 and so on. The above algorithm takes $\mathcal{O}(m)$ time and only requires the arrays BWT and MR.

The bit array MR can be derived from the SA and LCP arrays. However a faster alternative is to modify one of the algorithms computing the LCP from the SA so that, instead of the LCP, it computes the RCP (Reduced Common Prefix) array storing the lengths of the common prefix among lexicographically consecutive suffixes assuming that all instances of the \$ symbol are different. See again Fig. 2 for an example.[1] The linear time LCP construction algorithms in [11,14,17] can all be easily modified to compute the RCP values instead of LCP values. The MR array can be computed along with the RCP array observing that $\mathsf{MR}[i] = \mathbf{0}$ iff $\mathsf{LCP}[i] > \mathsf{RCP}[i]$. The latter condition can be verified even without knowing the LCP values by testing whether $t[\mathsf{SA}[i] + \mathsf{RCP}[i]] = t[\mathsf{SA}[i - 1] + \mathsf{RCP}[i]] = \$$. Indeed, the RCP array satisfies the following lemma which is an immediate consequence of Lemma 5.

Lemma 8. *Let $[b_z, e_z]$ denote the maximal range for the string $z \in \Pi$. It is*

$$\mathsf{RCP}[b_z + 1] = \mathsf{RCP}[b_z + 2] = \cdots = \mathsf{RCP}[e_z] = |z|$$

and $\mathsf{RCP}[b_z] = \mathsf{lcp}(z, z')$ where z' is the string immediately preceding z in the Π array. □

---

[1] The RCP array coincides with the LCP array if we build the concatenation t inserting a different symbol $\$_i$ at the end of each string x_i. However, this approach is not practical since would increase significantly the size of the alphabet.

Compute P
1:   $S \leftarrow$ empty stack; $P \leftarrow$ empty string
2:   **for** $i = 1, \ldots m$ **do**
3:       **if** MR$[i]$ == 1 **do**   // *beginning of maximal row range*
4:           $\ell \leftarrow$ RCP$[i]$
5:           **while** $(\ell_{top} \geq \ell)$ **do**
6:               $S.pop()$ // *pop if not prefix of the new string*
7:           **if** $t[\text{SA}[i_{top}] + \ell + 1] \neq \$$ **do**
8:               $\ell \leftarrow \ell_{top}$
9:               $S.pop()$
10:          $S.push(i, \ell)$
11: **while** $(S$ not empty$)$ **do**
12:      $S.pop()$

Fig. 3. Algorithm for computing the parenthesis array P given t, SA, RCP and MR. An open parenthesis is written to P at each *push* operation, and a closed parenthesis at each *pop* operation. (i_{top}, ℓ_{top}) represents the pair currently at the top of the stack.

Note that computing the RCP array is faster than computing the LCP array (the common prefixes are shorter) and its storage takes less space since each entry takes at most $\lceil \log(\max_i |x_i|) \rceil$ bits.

We have established that with a single scan of the BWT and MR array we can compute the arrays L and Last. We now show that using the RCP array we can also compute the parenthesis string P that supports suffix links emulation as described in Sect. 3. The algorithm for computing P is described in Fig. 3. To prove its correctness we first establish the following Lemma.

Lemma 9. *In the algorithm of Fig. 3 let $(i_1, \ell_1), (i_2, \ell_2), \ldots, (i_h, \ell_h)$ denote the pairs stored in the stack at any given moment, and let $z_1, z_2, \ldots, z_h$ denote the corresponding strings, i.e. z_j corresponds to the maximal row range $[i_j, e_j]$. Then, for $i = 2, \ldots, h$ we have that z_{i-1} is a proper prefix of z_i and $|z_{i-1}| = \ell_i$.*

Proof. Initially the stack is empty so the hypothesis is true. Assume now the stack $(i_1, \ell_1), (i_2, \ell_2), \ldots, (i_h, \ell_h)$ satisfies the hypothesis and we have reached position i which is the beginning of the next maximal row range which corresponds to the string z. Note that i_h is the starting point of the immediately preceding row range. Hence, setting $\ell = \text{RCP}[i]$ we have $\ell = \text{lcp}(z_h, z)$. In addition, for $j < h$ since z_j is a prefix of z_h it is $\text{lcp}(z_j, z) = \min(\ell, |z_j|)$. Clearly if $\ell_j \geq \ell$, z_j cannot be a prefix of z since

$$|z_j| > |z_{j-1}| = \ell_j \geq \ell \geq \text{lcp}(z_j, z)$$

so it is correct to remove (i_j, ℓ_j) from the stack at Line 6. If $\ell_j < \ell$ then z_j is a prefix of z iff $\ell = |z_j|$ which is the condition tested at Line 7. If this is the case we push (i, ℓ) to the stack and the invariant is maintained. If z_j is not a prefix of z then z_{j-1} certainly is, since it is a proper prefix of $z_j[1, \ell] = z[1, \ell]$, and we add to the stack (i, ℓ_j) after having removed (i_j, ℓ_j) thus maintaining the invariant.   □

Theorem 2. *The algorithm of Fig. 3 correctly computes the array P in $\mathcal{O}(m)$ time.*

Proof. Because of the order preserving correspondence between maximal row ranges and paths in Π, scanning the array MR is equivalent to scanning the array Π. Lemma 9 ensures that we write an open parenthesis for each path $\Pi[i]$ and that the corresponding closed parenthesis is written immediately before the opening parenthesis of the first path $\Pi[h]$ with $h > i$ such that $\Pi[i]$ is not a prefix of $\Pi[h]$. This is exactly how P is defined in Sect. 3 and the correctness follows.

To see that the running time is $\mathcal{O}(m)$ observe that in addition to the outer loop we only have push and pop operations on the stack. Since we push one pair (i, ℓ) for each 1 in MR, and once popped from the stack pairs are discarded, the overall time is $\mathcal{O}(m)$.   □

For the construction of P, in addition to the input arrays, the algorithm needs extra storage only for the stack. Since the values in the stack are strictly increasing, it uses at most $\mathcal{O}(\ell \log \ell)$ bits where $\ell = \max_i \mathsf{RCP}[i]$. Summing up, we are able to compute the XBWT with simple sequential scans using the SA and LCP (actually RCP) arrays. Since there are many well engineered algorithms for computing the SA and LCP array, we believe our solution is the most practical choice when the working space is not an issue. Indeed, its working space is dominated by the space required for the storage and computation of the SA which is still $\mathcal{O}(m \log m)$ bits but in practice it could be much less than the space required for storing a pointer based representation of the trie T.

We now describe an alternative XBWT construction algorithm that only uses the BWT of the string $t = y_1\$y_2\$ \cdots y_k\$$. Since the BWT takes $m \log |\Sigma|$ bits and can be computed using $o(m \log m)$ bits of working space, our algorithm provides new time/space trade-offs for XBWT construction. In addition, our algorithm works without modification if the BWT of t is replaced by the Multi String BWT [10] of $\{x_1, \ldots, x_k\}$. Although BWT algorithms have been studied for a longer time, Multi String BWT algorithms are potentially faster and have recently received much attention, see [1, 10, 16] and references therein.

The idea of our algorithm is to compute the MR array emulating a depth first visit of the trie T using the BWT. Since each internal trie node corresponds to a maximal row range, the visit will give us all maximal row ranges, i.e., the bit array MR. Our solution is inspired by the algorithm in [2] that computes the LCP array emulating a breadth first visit on the suffix trie using the BWT.

Assuming the BWT is stored in a balanced Wavelet Tree we can use the algorithm getInterval from [2] to compute, given the maximal row range corresponding to an internal node w, the maximal row ranges corresponding to w's children. This computation takes $\mathcal{O}(d \log |\Sigma|)$ time, where d is the number of w's children. Using getInterval, the computation of the MR array can be done by the algorithm in Fig. 4 whose running time is $\mathcal{O}(m + n \log |\Sigma|)$ where n is the

---

Compute MR (lightweight)
1:    **for** $i = 1, \ldots m$
2:        $MR[i] \leftarrow 0$ // *Clear the MR array*
3:    **df_visit**$(1, k, \epsilon)$

df_visit(b_z, e_z, z)
1:    $MR[e_z] \leftarrow 1$ // *mark the endpoint of the maximal row range*
2:    **foreach** $c \neq \$ $ in $BWT[b_z, e_z]$ **do**
3:        $[b_{cz}, e_{cz}] \leftarrow$ maximal row range for cz
4:        **df_visit**(b_{cz}, e_{cz}, cz)

Fig. 4. Algorithm for computing the MR array given the BWT.

Compute P (lightweight)
1:    $S \leftarrow$ empty stack; $P \leftarrow$ empty string
2:    **for** $i = 1, \ldots n$ **do**
3:        $\ell \leftarrow RCP'[i]$
4:        **while** $(\ell_{top} \geq \ell)$ **do**
5:            $S.pop()$ // *pop if not prefix of the new one*
6:        **if** $LEN'[i_{top}] \neq \ell$ **do**
7:            $\ell \leftarrow \ell_{top}$
8:            $S.pop()$
9:        $S.push(i, \ell)$
10:    **while** $(S$ not empty$)$ **do**
11:        $S.pop()$

Fig. 5. Algorithm for computing the parenthesis array P given RCP' and LEN'.

number of internal trie nodes. The working space of the algorithm, in addition to the BWT and MR arrays, is dominated by the stack for the depth first visit which takes $(\max_i |x_i|)|\Sigma|$ words. After the computation of MR, the arrays L and Last can be obtained in $\mathcal{O}(m + |\Sigma|)$ time as described above.

To compute also the parenthesis array P we use the following approach. Our starting point is the observation that the algorithm in Fig. 3 only uses the RCP values for the entries i such that $MR[i] = 1$. In addition, the SA is only used at Line 7 to check if the string that prefixes row i_{top} is a prefix of the string that prefixes row i. This property can be tested also by checking if the length of the string at i_{top} is equal to $RCP[i]$.

This observation suggests that after the computation of MR we count the number of **1**'s in it: this gives us the number n of internal trie nodes. Then, we allocate two length-n arrays RCP' and LEN' where we store the RCP and the length of the entries in Π with $MR[i] = 1$. These arrays take $\mathcal{O}(n \log(\max_i |x_i|))$ bits and can be computed in $\mathcal{O}(n \log |\Sigma|)$ time using a straightforward modification of the LCP construction algorithm from [2]. Using RCP' and LCP' we can compute the parenthesis array P using the algorithm of Fig. 5 which is derived from the one in Fig. 3 but has a simpler structure since, instead of scanning MR skipping the **0** entries, it scans directly RCP' and LEN'.

5   Concluding Remarks

With the advent of applications that use very large string dictionaries the XBWT-trie becomes a valid alternative for their storage. In this paper we have presented two contributions that can increase the practical appeal of this data structure. We believe there are other improvements to the original XBWT-trie design that can make this data structure even more appealing to practitioners. For example, it is relatively simple to support the contraction of unary paths. The computation of the XBWT also deserves further investigations: we have shown how to compute it from the SA or the BWT but we are currently working on the design of efficient and lightweight direct construction algorithms.

References

1. Bauer, M.J., Cox, A.J., Rosone, G.: Lightweight algorithms for constructing and inverting the BWT of string collections. Theor. Comput. Sci. **483**, 134–148 (2013)
2. Beller, T., Gog, S., Ohlebusch, E., Schnattinger, T.: Computing the longest common prefix array based on the Burrows-Wheeler transform. J. Discrete Algorithms **18**, 22–31 (2013)
3. Beller, T., Zwerger, M., Gog, S., Ohlebusch, E.: Space-efficient construction of the Burrows-Wheeler transform. In: Kurland, O., Lewenstein, M., Porat, E. (eds.) SPIRE 2013. LNCS, vol. 8214, pp. 5–16. Springer, Heidelberg (2013)
4. Crochemore, M., Grossi, R., Kärkkäinen, J., Landau, G.M.: Computing the Burrows-Wheeler transform in place and in small space. J. Discrete Algorithms **32**, 44–52 (2015)
5. Ferragina, P., Gagie, T., Manzini, G.: Lightweight data indexing and compression in external memory. Algorithmica **63**, 707–730 (2012)
6. Ferragina, P., Luccio, F., Manzini, G., Muthukrishnan, S.: Structuring labeled trees for optimal succinctness, and beyond. In: Proceedings of the 46th IEEE Symposium on Foundations of Computer Science (FOCS), pp. 184–193 (2005)
7. Ferragina, P., Luccio, F., Manzini, G., Muthukrishnan, S.: Compressing and searching XML data via two zips. In: Proceedings of the 15th International World Wide Web Conference (WWW), pp. 751–760 (2006)
8. Ferragina, P., Luccio, F., Manzini, G., Muthukrishnan, S.: Compressing and indexing labeled trees, with applications. J. ACM, **57** (2009)
9. Gusfield, D.: Algorithms on Strings, Trees, and Sequences: Computer Science and Computational Biology. Cambridge University Press, Cambridge (1997)
10. Holt, J., McMillan, L.: Constructing Burrows-Wheeler transforms of large string collections via merging. In: BCB, pp. 464–471. ACM (2014)
11. Kärkkäinen, J., Manzini, G., Puglisi, S.J.: Permuted longest-common-prefix array. In: Kucherov, G., Ukkonen, E. (eds.) CPM 2009 Lille. LNCS, vol. 5577, pp. 181–192. springer, Heidelberg (2009)
12. Kärkkäinen, J., Kempa, D.: Engineering a lightweight external memory suffix array construction algorithm. In: Proceedings of CEUR Workshop, ICABD, vol. 1146, pp. 53–60 (2014). http://CEUR-WS.org

13. Kärkkäinen, J., Kempa, D., Puglisi, S.J.: Parallel external memory suffix sorting. In: Cicalese, F., Porat, E., Vaccaro, U. (eds.) CPM 2015. LNCS, vol. 9133, pp. 329–342. Springer, Heidelberg (2015)

14. Kasai, T., Lee, G.H., Arimura, H., Arikawa, S., Park, K.: Linear-time longest-common-prefix computation in suffix arrays and its applications. In: Amir, A., Landau, G.M. (eds.) CPM 2001. LNCS, vol. 2089, pp. 181–192. Springer, Heidelberg (2001)

15. Knuth, D.E.: Sorting and Searching. The Art of Computer Programming, 2nd edn. Addison-Wesley, Reading (1998)

16. Li, H.: Fast construction of FM-index for long sequence reads. Bioinformatics **30**, 3274–3275 (2014)

17. Manzini, G.: Two space saving tricks for linear time LCP array computation. In: Hagerup, T., Katajainen, J. (eds.) SWAT 2004. LNCS, vol. 3111, pp. 372–383. Springer, Heidelberg (2004)

18. Martínez-Prieto, M.A., Brisaboa, N.R., Cánovas, R., Claude, F., Navarro, G.: Practical compressed string dictionaries. Inf. Syst. **56**, 73–108 (2016)

19. Navarro, G., Sadakane, K.: Fully-functional static and dynamic succinct trees. ACM Trans. Algorithms **10** (2014). Article 16

Maximal Unbordered Factors of Random Strings

Patrick Hagge Cording[1]($\boxtimes$) and Mathias Bæk Tejs Knudsen[2]

[1] DTU Compute, Technical University of Denmark, Kongens Lyngby, Denmark
phaco@dtu.dk
[2] Department of Computer Science,
University of Copenhagen, Copenhagen, Denmark

Abstract. A border of a string is a non-empty prefix of the string that is also a suffix of the string, and a string is unbordered if it has no border. Loptev, Kucherov, and Starikovskaya [CPM 2015] conjectured the following: If we pick a string of length n from a fixed alphabet uniformly at random, then the expected length of the maximal unbordered factor is $n - O(1)$. We prove that this conjecture is true by proving that the expected value is in fact $n - \Theta(\sigma^{-1})$, where σ is the size of the alphabet. We discuss some of the consequences of this theorem.

1 Introduction

A string S is a finite sequence of n characters from an alphabet Σ of size σ. $S[i, j]$, $1 \leq i \leq j \leq n$, is the sequence of characters of S starting in i and j, both indices included. We denote $S[i, j]$ a *factor* of S. The factor $S[1, j]$ is a prefix of S and $S[i, n]$ is a suffix. A *border* of a string is a non-empty prefix of the string that is also a suffix of the string. If $S = \alpha\beta = \lambda\alpha$, for non-empty strings β and λ, then α is a border of S with length $|\alpha|$. The maximal border of S is the longest border among all borders of S. S is unbordered if it does not have a border. The maximal unbordered factor is the longest factor that does not have a border. A string is periodic if it can be written as $S = \alpha^k \alpha'$, where α^k is the string α repeated $k > 0$ times and α' is a prefix of α.

Borders were first studied by Ehrenfeucht and Silberger [2] with emphasis on the relationship between the maximal unbordered factor of a string and its minimal period. This relationship has since received more attention in the litterature [1, 4, 5].

Loptev, Kucherov, and Starikovskaya [10] prove that for $\sigma \geq 2$ the expected length of the maximal unbordered factor is at least $n(1 - \xi(\sigma) \cdot \sigma^{-4}) + O(1)$, where $\xi(\sigma)$ converges to 2 as σ grows. When $\sigma \geq 5$ and n is sufficiently large this implies that the expected length of the maximal unbordered factor is at least

P.H. Cording—Supported by the Danish Research Council under the Sapere Aude Program (DFF 4005-00267).

M.B.T. Knudsen—Research partly supported by Mikkel Thorup's Advanced Grant from the Danish Council for Independent Research under the Sapere Aude research career programme and the FNU project AlgoDisc - Discrete Mathematics, Algorithms, and Data Structures.

© Springer International Publishing AG 2016
S. Inenaga et al. (Eds.): SPIRE 2016, LNCS 9954, pp. 93–96, 2016.
DOI: 10.1007/978-3-319-46049-9_9

$0.99n$. Supported by experimental results, the authors of [10] conjectured that the expected length of maximal unbordered factor is $n - O(1)$. We prove that this conjecture is true and obtain the following theorem.

Theorem 1. *Let S be a string of length n, where each character is chosen i.i.d. uniformly from an alphabet A of size $\sigma \geq 2$. The expected length of the maximal unbordered factor is $n - O(\sigma^{-1})$.*

The problem of computing the maximal unbordered factor of a string has been studied by Loptev et al. [10] and Gawrychowski et al. [3], who give algorithms with average-case running times $O(\frac{n^2}{\sigma^4} + n)$ and $O(n \log n)$, respectively. It can be decided in $O(n)$ time if a string of length n has a border by computing the *border array* (also known as the failure function, made famous by the KMP pattern matching algorithm [7,11]). Entry i of the border array B of a string S contains the length of the maximal border of the prefix $S[1, i]$. If $B[n] = 0$ then S is unbordered. Let B_j be the border array for the suffix $S[j, n]$. If $B_j[i] = 0$ it means that the factor $S[j, i]$ is unbordered. Computing B_j for $j = 1 \ldots n$ and scanning these to find the maximal unborderd factor of S takes $O(n^2)$ time. As mentioned in [10], we can compute the B_j's in decreasing order of the suffix length and stop the algorithm once $n - j$ is smaller than the currently longest unbordered factor and obtain an algorithm with average-case running time $O((n - \mu + 1)n)$ where μ is the expected length of the maximal unbordered factor. With our new bound on the expected length of the maximal unbordered factor, we therefore get the following corollary.

Corollary 1. *There is an algorithm with average-case running time $O(n)$ that finds the maximal unbordered factor.*

This improves the previously best known average-case bounds for finding the maximal unbordered factor of a string.

Related Work. The worst-case running time of the above mentioned algorithm is still $O(n^2)$. Gawrychowski et al. [3] give an algorithm with worst-case running time $O(n^{1.5})$.

Holub and Shallit [6] investigated the expected length of the maximal border of a random word.

Data structures for answering a border query have also been developed. A border query takes two indices i and j and the answer is the maximal border of the factor $S[i, j]$. Kociumaka et al. [8] show several time-space trade-offs for this problem. For one of these, their data structure can answer border queries in $O(\log^{1+\epsilon} n)$ time and uses $O(n)$ space. Kociumaka et al. [9] improved this to $O(1)$ time for answering border queries while using $O(n)$ space.

2 The Proof of Theorem 1

Fix A and $\sigma \geq 2$. Let X_n be the expected length of the maximal unbordered factor of a random string of length n. We define $X_0 = 0$, and we let $Y_n = n - X_n$.

We prove in the following that $Y_n \leq c$, where c is given by:

$$c = \frac{2\sigma}{(\sigma - 1)^2(1 - \sigma^{-1} - \sigma^{-2})}$$

Since $c \leq \frac{32}{\sigma}$ this will prove the theorem. This follows from $\sigma \geq 2$ and the following calculation:

$$c = \frac{2}{\sigma\left(1 - \sigma^{-1}\right)^2\left(1 - \sigma^{-1} - \sigma^{-2}\right)} \leq \frac{2}{\sigma\left(1 - 2^{-1}\right)^2\left(1 - 2^{-1} - 2^{-2}\right)} = \frac{32}{\sigma}$$

We will prove the claim by induction on n. By definition this is true whenever $n \leq 1$. So fix some n and assume that $Y_m \leq c$ for all $m < n$.

Let S be a random string of length n. Let $f = f(S)$ be the smallest positive integer $< n$ such that $S[1, f] = S[n - f + 1, n]$. If no such integer exists we let $f = 0$. We note that if $f > 0$ then $f \leq \frac{n}{2}$, since if $f > \frac{n}{2}$ then $f' = 2f - n$ satisfies $S[1, f'] = S[n - f' + 1, n]$ as well and $f' < f$ which is impossible. Let $L = L(S)$ be the length of the maximal unbordered factor of S. Then:

$$Y_n = n - X_n = n - E(L) = \sum_{\ell=0}^{\lfloor n/2 \rfloor} P(f = \ell)(n - E(L \mid f = \ell)) \qquad (1)$$

If $1 \leq \ell < \frac{n}{2}$ then $S[\ell + 1, n - \ell]$ is independent of the event $f = \ell$, since $f = \ell$ is determined by $S[1, \ell]$ and $S[n - \ell + 1, n]$. The longest unbordered factor in $S[\ell + 1, n - \ell]$ is also an unbordered factor in S and hence for $\ell < \frac{n}{2}$:

$$E(L \mid f = \ell) \geq X_{n-2\ell} \qquad (2)$$

If n is odd (2) holds for all integers $\ell \in \left\{1, 2, \ldots, \lfloor \frac{n}{2} \rfloor\right\}$. If n is even we see that if $\ell = \lfloor \frac{n}{2} \rfloor$ the right hand side of (2) is 0 and hence it also holds for all integers $\left\{1, 2, \ldots, \lfloor \frac{n}{2} \rfloor\right\}$. If $f = 0$, then S is an unbordered factor, and therefore $E(L \mid f = 0) = n$. So we can use this observation together with the inequality (2) to upper bound Y_n in (1) by:

$$Y_n \leq \sum_{\ell=1}^{\lfloor n/2 \rfloor} P(f = \ell)(n - X_{n-2\ell}) = \sum_{\ell=1}^{\lfloor n/2 \rfloor} 2\ell P(f = \ell) + \sum_{\ell=1}^{\lfloor n/2 \rfloor} P(f = \ell)Y_{n-2\ell} \qquad (3)$$

Nielsen [12] proved the following lower bound on the probability that S is unbordered. Since S is unbordered iff $f = 0$ we get:

Theorem 2 (Nielsen[12]).

$$P(f = 0) \geq 1 - \sigma^{-1} - \sigma^{-2}$$

Using Theorem 2 together with the fact that $Y_{n-2\ell} \leq c$ we get:

$$\sum_{\ell=1}^{\lfloor n/2 \rfloor} P(f = \ell)Y_{n-2\ell} \leq c \sum_{\ell=1}^{\lfloor n/2 \rfloor} P(f = \ell) = c(1 - P(f = 0)) \leq c(\sigma^{-1} + \sigma^{-2}) \qquad (4)$$

If $f = \ell$ then $S[1, \ell] = S[n - \ell + 1, n]$. After fixing $S[1, \ell]$ there are σ^ℓ ways to choose $S[n - \ell + 1, n]$ and hence $P(f = \ell) \leq \sigma^{-\ell}$. Using this we get:

$$\sum_{\ell=1}^{\lfloor n/2 \rfloor} 2\ell P(f = \ell) \leq \sum_{\ell=1}^{\infty} 2\ell \sigma^{-\ell} = \frac{2\sigma}{(\sigma - 1)^2} \tag{5}$$

Inserting (4) and (5) into (3) gives:

$$Y_n \leq \frac{2\sigma}{(\sigma - 1)^2} + c(\sigma^{-1} + \sigma^{-2}) = c$$

which finishes the induction and the proof.   □

References

1. Duval, J.-P.: Relationship between the period of a finite word and the length of its unbordered segments. Discrete Math. **40**(1), 31–44 (1982)
2. Ehrenfeucht, A., Silberger, D.: Periodicity and unbordered segments of words. Discrete Math. **26**(2), 101–109 (1979)
3. Gawrychowski, P., Kucherov, G., Sach, B., Starikovskaya, T.: Computing the longest unbordered substring. In: Iliopoulos, C., Puglisi, S., Yilmaz, E. (eds.) SPIRE 2015. LNCS, vol. 9309, pp. 246–257. Springer, Heidelberg (2015). doi:10. 1007/978-3-319-23826-5_24
4. Harju, T., Nowotka, D.: Periodicity and unbordered words: a proof of the extended duval conjecture. J. ACM (JACM) **54**(4), 20 (2007)
5. Holub, Š., Nowotka, D.: The ehrenfeucht-silberger problem. J. Comb. Theor. Ser. A **119**(3), 668–682 (2012)
6. Holub, Š., Shallit, J.: Periods and borders of random words. In: 33rd Symposium on Theoretical Aspects of Computer Science (2016)
7. Knuth, D.E., Morris Jr., J.H., Pratt, V.R.: Fast pattern matching in strings. SIAM J. Comput. **6**(2), 323–350 (1977)
8. Kociumaka, T., Radoszewski, J., Rytter, W., Waleń, T.: Efficient data structures for the factor periodicity problem. In: Calderón-Benavides, L., González-Caro, C., Chávez, E., Ziviani, N. (eds.) SPIRE 2012. LNCS, vol. 7608, pp. 284–294. Springer, Heidelberg (2012). doi:10.1007/978-3-642-34109-0_30
9. Kociumaka, T., Radoszewski, J., Rytter, W., Waleń, T.: Internal pattern matching queries in a text and applications. In: Proceedings of the Twenty-Sixth Annual ACM-SIAM Symposium on Discrete Algorithms, pp. 532–551. SIAM (2015)
10. Loptev, A., Kucherov, G., Starikovskaya, T.: On maximal unbordered factors. In: Cicalese, F., Porat, E., Vaccaro, U. (eds.) CPM 2015. LNCS, vol. 9133, pp. 343–354. Springer, Heidelberg (2015). doi:10.1007/978-3-319-19929-0_29
11. Morris Jr., J.H., Pratt, V.R.: A linear pattern-matching algorithm (1970)
12. Nielsen, P.T.: A note on bifix-free sequences (corresp.). IEEE Trans. Inf. Theor. **19**(5), 704–706 (1973)

Fragmented BWT: An Extended BWT for Full-Text Indexing

Masaru Ito[1(✉)], Hiroshi Inoue[2], and Kenjiro Taura[1]

[1] Department of Information and Communication Engineering,
Graduate School of Information Science and Technology,
The University of Tokyo, Tokyo, Japan
{mito,tau}@eidos.ic.i.u-tokyo.ac.jp
[2] IBM Research, Tokyo, Japan
inouehrs@jp.ibm.com

Abstract. This paper proposes Fragmented Burrows Wheeler Transform (FBWT), an extension to the well-known BWT structure for full-text indexing and searching. A FBWT consists of a number of BWT fragments each covering only a subset of all the suffixes of the original string. As constructing FBWT does not entail building the BWT of the whole string, it is faster than constructing BWT. On the other hand, searching with FBWT can be more costly than that with BWT, since searching the former requires searching all fragments; its amount of work is $O(dp + occ \log^{1+\epsilon} n)$ as opposed to $O(p + occ \log^{1+\epsilon} n)$ of regular BWT, where p is the length of the query string, n the length of the original text, occ the occurrences of the query string, and d the number of fragments. To compensate the search cost, searching with FBWT can be accelerated with SIMD instructions by searching multiple fragments in parallel. Experiments show that building FBWT is about twice as fast as building BWT via a state of the art algorithm (SA-IS); and that FBWT's search performance compared to BWT's depends on the number of occurrences, ranging from four times slower than BWT (when there are few occurrences), to twice as fast as BWT (when there are many).

Keywords: Suffix array · Burrows- Wheeler transform · Full-text indexing

1 Introduction

Full-text searching is a fundamental topic for text processing applications. It is used in many applications such as genome analysis, data mining, search engine, and so on. As the amount of text data has been rapidly increasing in recent years, many efforts have been made to index and search texts faster. Suffix array [12] and Burrows Wheeler Transform (BWT) [1] are popular data structures to accomplish the task and are widely used by genome analysis applications such as BWA [10], SOAPv2 [11], bowtie [9]. For constructing suffix arrays and BWTs, many algorithms have been developed such as SA-IS [13], DC3 [7] and

© Springer International Publishing AG 2016
S. Inenaga et al. (Eds.): SPIRE 2016, LNCS 9954, pp. 97–109, 2016.
DOI: 10.1007/978-3-319-46049-9_10

parallelized [6] to accelerate constructing them. For full-text searching, many data structures have been proposed such as Ferragina and Manzini [3,4], wavelet tree [5], and wavelet matrix [2].

We propose an extension of BWT, named *Fragmented BWT (FBWT)*. This data structure consists of a number of BWT fragments. Each fragment is essentially a BWT covering only a subset of the original string. An FBWT can be constructed more cheaply than BWT because it needs to build only a part of the suffix array. On the other hand, searching with FBWT can be more costly than that with BWT, as it needs to search all fragments individually. Fortunately, the cost can be reduced by searching them in parallel with SIMD instructions. Experiments show that we can build FBWT twice as fast as building the regular BWT by a state of the art algorithm (SA-IS). The search performance of FBWT depends on the number of occurrences (*occ*) and it is from about four times slower to two times faster than searching with BWT, after vectorization applied.

2   Related Work

Suffix array [12] is a fundamental data structure whose construction and searching have been extensively studied. For the construction, Manber and Myers proposed the first $O(n \log n)$-time algorithm [12]. $O(n)$-time algorithms have been studied [7,13] and the state of the art algorithm is SA-IS introduced by Nong et al. [13]. As suffix array needs $O(n \log n)$ bits for n characters, researchers have proposed more compact data structures that still allow full text searching. A representative method is due to Sadakane et al. [14] and Ferragina and Manzini [3]. Sadakane, et al. uses Ψ array which stores the position of the next suffix on the original string in the suffix array. This method has $O(p \log n + occ \log^\epsilon n)$ time complexity, where n is length of the original text. Ferragina and Manzini introduced a new data structure using Burrows Wheeler Transform [1]. This method consists of two steps. The first step finds *ranks* of matching suffixes in time complexity $O(p)$, where p is the length of the query string, and the second step locates them in the original reference string, in $O(occ \log^{1+\epsilon} n)$-time, where *occ* is the number of occurrences. The present work is largely based on Ferragina and Manzini's, with the main difference being that we build a suffix array of a string d times smaller than the entire string. Our method also makes the second step of the searching algorithm more efficient, although it makes the first step more expensive.

3   Background

3.1   Notational Preliminaries

Throughout this paper, T is an original text we build the index of. As is customary we assume T ends with a special character, written as '$\$$', which is the lexicographically smallest character. n is the number of characters in T including

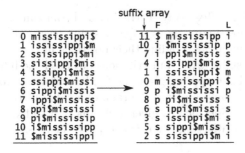

suffix array

```
      0 mississippi$        11 $ mississipp i
      1 ississippi$m        10 i $mississip p
      2 ssissippi$mi         7 i ppi$missis s
      3 sissippi$mis         4 i ssippi$mis s
      4 issippi$miss         1 i ssissippi$ m
      5 ssippi$missi         0 m ississippi $
      6 sippi$missis   →     9 p i$mississi p
      7 ippi$mississ         8 p pi$mississ i
      8 ppi$mississi         6 s ippi$missi s
      9 pi$mississip         3 s issippi$mi s
     10 i$mississipp         5 s sippi$miss i
     11 $mississippi         2 s sissippi$m i
```

Fig. 1. Burrows Wheeler transform for mississippi$

this end character. $T[i]$ is the character at index i of T (i starting from zero). $T[a:b]$ is the substring of T starting from its ath character to bth character. $T[a:]$ is an abbreviation of $T[a:n]$, the suffix starting from position a. We write $s \prec t$ to mean string s is lexicographically smaller than t.

3.2 Suffix Array and Burrows Wheeler Transform

Given a reference string T, its suffix array [12] is a sorted array of its all suffixes. Each suffix is actually represented by its starting index in T, so the suffix array of T, SA, is a permutation of $\{0, \ldots, n-1\}$, such that for all i and j, $0 \leq i < j < n \Rightarrow T[SA[i]:] \prec T[SA[j]:]$.

Burrows Wheeler transform of T, BWT, is an array of n characters defined as follows.

$$BWT[i] = \begin{cases} T[SA[i] - 1] & (SA[i] \neq 0) \\ T[n-1] & (SA[i] = 0) \end{cases} \tag{1}$$

That is, BWT's ith element is the character ahead of ith smallest suffix (for convenience, we consider the character ahead of $T[0]$ is $T[n-1]$). In other words, BWT is a permutation of the reference string, sorted by lexicographic order of the suffix following each character.

This is illustrated in Fig. 1, which shows the BWT of a reference string "mississippi$". On the left is all the suffixes of the reference string. Each suffix is represented as a cyclic shift of the original string. Note that the order between cyclic shifts is equivalent to that between suffixes, as '$' is smaller than any other character. On the right is its lexicographical sort, signifying the first and the last characters of each cyclic shift in the F and L columns, respectively. BWT is the sequence of characters in the L column.

3.3 LF-Mapping and Backward Search

Searching for a string using a suffix array amounts to finding suffixes that have the query string as a prefix. As suffixes are lexicographically sorted, such suffixes form a contiguous interval in the suffix array. If the suffix array of the entire string were available, searching for a string could be done by a straightforward binary search on the suffix array. Ferragina and Manzini [3] have shown that BWT can answer how many times a query string occurs in the original string without using a suffix array. Finding their positions in the original string needs a sparsely sampled suffix array, but still does not need the full suffix array. The following function, LF-mapping, gives the fundamental tool to accomplish this.

LF-mapping is a mapping from the rank of a suffix to the rank of the suffix one character longer. Specifically, given the rank of a suffix, i, its LF-mapping, $LF(i)$ is defined as j satisfying $SA[j] = SA[i] - 1$. That is,

$$LF(i) \overset{\text{def}}{=} j \text{ such that } SA[j] = SA[i] - 1. \tag{2}$$

Remarkably, this can be computed without materializing SA, as follows.

$$LF(i) = C[c] + occ(c, i - 1), \tag{3}$$

where c is the character ahead of the given suffix (that is, $c = BWT[i]$), $C[x]$ the number of characters smaller than x in the original string (or equivalently in BWT), and $occ(x, p)$ the number of character x's in $BWT[0:p]$.

We sketch its correctness for the sake of extending it for our method. Recall that we are trying to find the rank of suffix "c" $\parallel T[SA[i]:]$ (the symbol $\parallel$ means a string concatenation). As it starts with c, all suffixes starting from characters smaller than c have ranks lower than this suffix, which are counted by the term $C[c]$. Now we count suffixes starting from c and lexicographically smaller than "c" $\parallel T[SA[i]:]$. As they all share the same first character (c), this amounts to counting the number of suffixes lexicographically smaller than $T[SA[i]:]$ *and* immediately following c, which is the number of suffixes among $T[SA[0]:], T[SA[1]:], \cdots, T[SA[i-1]:]$ that immediately follow c. This is exactly the number of times c appears in $BWT[0:i-1]$, which is $occ(c, i-1)$.

The above argument can be extended to searching for an *arbitrary* query string, q. In order to search for a string q, we start from an empty query string and extend it ahead one character at a time, keeping track of the minimum/maximum ranks of suffixes that have the query string in their prefixes. The procedure is illustrated as follows.

```
1   search_bwt(q) {
2       s = 0; e = n - 1;
3       for (i = |q| - 1; i ≥ 0; i--) {
4           c = q[i];
5           s = C[c] + occ(c, s - 1);
6           e = C[c] + occ(c, e) - 1;
7           if (s > e) break;
8       }
9       return (s, e);
10  }
```

3.4 Locating Suffixes in the Reference String

This backward search procedure immediately gives the number of occurrences of q, which is $(e - s + 1)$. To locate them in the original string, however, we need an extra step for each suffix we have found. This would be trivial, again if the entire suffix array were available; we merely return $SA[j]$ for each $j \in [s, e]$. A cleverer technique was invented to accomplish this without holding the entire suffix array in memory. The key idea is to sparsely sample the suffix array, to build a (partial) mapping from ranks to locations in the original string. Samples are taken so that they are regularly interspaced in the original string. By taking a sample from every $\log^{(1+\epsilon)} n$ consecutive elements, the space usage becomes $n \log n / \log^{(1+\epsilon)} n \in o(n)$. Once such sampled suffix array is available, finding the position of the suffix can be done by extending the suffix backward in the manner of LF-mapping, until we reach a suffix that is mapped in the sampled suffix array. The time complexity thus becomes $O(\log^{1+\epsilon} n)$ for each occurrence A sampled suffix array is actually implemented as a hash table mapping from integers (rank) to integers (positions).

```
1   /* SSA : a sampled suffix array (hash table)
2        j : rank of the suffix we want to locate in the reference string */
3   get_position (j, SSA) {
4       backward_count = 0;
5       while (j not found in SSA) {
6           c = L[i];
7           j = C[c] + occ(c, j) − 1;
8           backward_count++;
9       }
10      return SSA[j] + backward_count;
11  }
```

4 Fragmented BWT

In this section, we describe our index data structure and searching method. We assume the special character '$' appears at the end of T at least once and so many times that the length of T becomes a multiple of the number of fragments in the fragmented BWT we are to build.

Definition 1. *For a string T, (r/d)-fragment of T's suffix array, $SA_{r/d}$, is a sorted array of T's suffixes that start from positions that are r modulo d. (r/d)-fragment of T's BWT, $BWT_{r/d}$, is defined similarly to Eq. (1):*

$$BWT_{r/d}[i] = \begin{cases} T[SA_{r/d}[i] - 1] & (SA_{r/d}[i] \neq 0) \\ T[n - 1] & (SA_{r/d}[i] = 0) \end{cases} \tag{4}$$

The fragmented suffix array (or BWT) of T is the collection of all (r/d)-fragments ($r = 0, \ldots, d - 1$). We show how to construct a fragmented BWT of a string in Sect. 4.1 and how to search using a fragmented BWT in Sects. 4.2 and 4.3.

4.1 Constructing FBWT

A straightforward approach to constructing a fragmented BWT of a string T is to build the full suffix array of T, scan it to split it into fragments (put elements that are r modulo d into the (r/d)-fragment), and then derive BWT fragments. This method, obviously, does not bring any benefit in construction time. Alternatively, we could build each suffix array fragment individually, by sorting only suffixes that belong to the fragment we would like to build, and derive the corresponding BWT fragment. In essence, instead of sorting an array of n elements, we sort d arrays each having n/d elements. The latter may be faster than the former but the benefit will be marginal, especially when we use fast $O(n)$ suffix array construction algorithms, as we are merely replacing sorting n elements with sorting n/d elements d times.

Our method constructs only the $(0/d)$-fragment from the original string. Other fragments are obtained from fragments already built, as shown below. We call d consecutive characters *a wide character*. The original string T of n characters can be naturally viewed as an array of (n/d) wide characters, each starting from a position that is a multiple of d. Define the lexicographical order between wide characters in the obvious manner. The procedure for obtaining all BWT fragments, $L_0, L_1, \cdots, L_{d-1}$, is the following. Figure 2 shows the example of FBWT construction for "cock-a-doodle-doo".

```
 1   build_FBWT(T, d) {
 2      /* T : array of n characters including  trailing $'s
 3          d : the number of fragments (divides n) */
 4      W = array of (n/d) wide characters in T;
 5      /* i.e. W[i] = T[id, (i + 1)d − 1] for i ∈ [0, n/d) */
 6      /* get the first fragment L₀ */
 7      L = BWT of W;
 8      L₀ = array of (n/d) characters, obtained by
 9              collecting (d − 1)th character of each wide character in L;
10      /* i.e., L₀[i] = (d − 1)th character of L[i]  (i ∈ [0, n/d)) */
11      /* derive other fragments */
12      for (r = d − 1; r ≥ 1; r−−) {
13          stably sort L by rth character;
14          Lᵣ = array of (n/d) characters, obtained by
15                  collecting (r − 1)th character of each wide character in L;
16          /* i.e., Lᵣ[i] = (r − 1)th character of L[i]  (i ∈ [0, n/d)) */
17      }
18      return {L₀, L₁, ..., L_{d−1}} ;
19   }
```

The $(0/d)$-fragment of BWT, L_0, can be obtained by making the BWT of the wide character array W and then collecting the last character from each wide character in W's BWT. From this fragment, we work backward, first obtaining L_{d-1}, then L_{d-2}, ..., until finally obtaining L_1. When obtaining L_{d-1} from L_0, we *stably* sort the wide characters in the just obtained BWT of W, L, by its

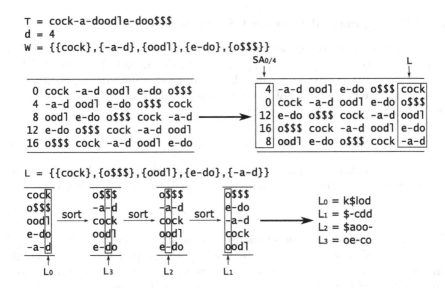

Fig. 2. FBWT for "cock-a-doodle-doo$$$"

$(d-1)$th character and then collect their $(d-2)$th characters. We repeat this process $(d-1)$ times; each iteration obtains L_r from L_{r+1} ($L_d = L_0$ for notational convenience) by stably sorting wide characters in L by their rth characters and then collecting their $(r-1)$th characters in the sorted L.

We will show the correctness of the above procedure. Consider an (imaginary) array L' whose ith element $L'[i]$ is a pair $\langle L[i]$, the suffix that follows $L[i] \rangle$ and it is sorted in the same way as L is sorted along the way. Below, we denote the second element of the ith pair as $S[i]$.

Lemma 1. *At the beginning of iteration r ($d-1, d-2, \ldots, 0$), the imaginary array L' is sorted by the lexicographic order of $(L[i][r+1:] \parallel S[i])$.*

$L[i]$ is a wide character (d characters) and $L[i][r+1:]$ is thus a suffix of it starting from its $(r+1)$th character.

Proof. By induction. Recall that initially ($r = d-1$), L is the BWT of W. By the definition of BWT, L' is sorted by the lexicographical order of the $S[i]$'s. As $r = d-1$, each $L[i][r+1:]$ is an empty string, so the order between $S[i]$'s is equivalent to the lexicographical order of $(L[i][r+1:] \parallel S[i])$. This shows that the claim initially holds for $r = d-1$. Assume the claim holds at the beginning of iteration r. In this iteration, it stably sorts L (and the imaginary L') with rth characters. As L' is already sorted by $(L[i][r+1:] \parallel S[i])$, stably sorting them with rth character is equivalent to sorting them by $(L[i][r:] \parallel S[i])$, establishing the claim in the next iteration $(r-1)$.

Given the lemma, it is easy to observe that at the end of iteration r, taking their $(r-1)$th character will give the desired fragment L_r.

4.2   LF-Mapping and Backward Search with FBWT

Let us extend LF-mapping and backward search for fragmented BWT. Let the fragments be $\{L_0, L_1, \cdots, L_{d-1}\}$ (L_r is (r/d)-fragment). The extended LF-mapping maps the rank (i) of a suffix, known to be starting from a position that is r modulo d, to the rank (j) of the suffix one character longer. A rank is a local rank computed within each fragment. That is,

$$LF(r, i) \stackrel{\text{def}}{=} j \text{ such that } SA_{r-1/d}[j] = SA_{r/d}[i] - 1. \tag{5}$$

(for $r = 0$, we define $SA_{-1/d} = SA_{d-1/d}$ for convenience). Analogously to the regular LF-mapping, this can be computed with BWT, which is now fragmented.

$$LF(r, i) = C_r[c] + occ_r(c, i - 1) \tag{6}$$

where c is the character ahead of the suffix ranked ith in (r/d)-fragment; $C_r[x]$ is the number of characters smaller than x in positions that are $(r - 1)$ modulo d in the reference string, which is equivalent to the number of characters smaller than x in L_r ((r/d)-fragment of BWT); $occ_r(x, p)$ is the number of x's $L_r[0 : p]$. Due to space limitation, we omit the correctness proof of the above procedure, which is analogous to that of the original LF-mapping.

The extended LF-mapping can be similarly extended to searching for arbitrary strings. The procedure below finds occurrences of q starting from positions that are r modulo d in the reference string. It returns an interval in (r/d)-fragment of the suffix array. That is, (s, e) such that

$$s \leq j \leq e \iff q \text{ is a prefix of } T[SA_{r/d}[j]:].$$

```
1   backward_search_fragment(q, r) {
2       s = 0; e = n − 1;
3       for (i = |q| − 1; i ≥ 0; i−−) {
4           /* extend search one character ahead */
5           x = (r + i + 1) mod d;
6           s = Cₓ[q[i]] + occₓ(q[i], s − 1);
7           e = Cₓ[q[i]] + occₓ(q[i], e) − 1;
8           if (s > e) break;
9       }
10      return (s, e); }
```

To find all occurrences in the original string, we must search all fragments; this is roughly d times more costly than backward-searching a single BWT covering the entire string. Fortunately, however, we can search multiple fragments in parallel using SIMD instructions.

4.3   Locating Suffixes with FBWT

The second step is similar to get_position of Ferragina and Manzini. Starting from the rank of a suffix whose position in the reference string is unknown, we repeat applying LF-mapping, until we find a suffix that is sampled and whose position in the original string is thus known. In this process, we enjoy a benefit of splitting the BWT into fragments, as explained below. As we go backward with LF-mapping, we need to check if the current suffix is sampled. We take samples only from a single fragment (specifically, $(0/d)$-fragment). In this setting, the check is redundant if we know a suffix does not start at a position not a multiple of d. The get-position procedure by Ferragina and Manzini cannot take advantage of this fact, as such information is not readily available. In our algorithm, in contrast, *suffixes are already classified into fragments by their starting positions modulo d*. Thus we can easily know that the current suffix ever has a chance to have been sampled. This can reduce the number of costly checks to see if a suffix is in the sampled array roughly by a factor of d (see line 6 below). As a result, this step is faster than that in the ordinary BWT and the entire search faster when it is dominated by this step (i.e. there are many occurrences).

```
1  /* return the position in the reference string of suffix whose rank is
2       j in (r/d)−fragment of suffix array. in other words, it returns
3       SA_{r/d}[j], without fully materializing SA_{r/d}. */
4  get_position_fragment(r, j, SSA) {
5      backward_count = 0;
6      while (r ≠ 0 or j not found in SSA) {
7          c = L_r[j];
8          j = C_r[c] + occ_r(c, j) − 1;
9          backward_count++;
10     }
11     return SSA[j] + backward_count;
12 }
```

Finally, the toplevel procedure to search for q in the entire string is below.

```
1  /* search the entire reference string to find
2       occurrences of q, no matter where they start. */
3  search(q) {
4      ranks = {};  /* empty set */
5      /* get interval in suffix array of all fragments having prefix q */
6      for (r = 0; r < d; r++) {
7          s, e = backward_search_fragment(q, r);
8          ranks = ranks ∪ {(r, s, e)};
9      }
10     /* get positions in the reference string */
11     I = {};  /* empty set */
12     for each ((r, s, e) ∈ ranks) {
13         for (j = s; j ≤ e; j++) {
14             I = I ∪ { get_position_fragment(r, j, SSA) };
15         }
16     }
17     return I;
18 }
```

5 Implementation

We describe implementation details not described so far. For the number of fragments, we chose $d = 8$, a natural choice because the size of a wide character then fits a single 64 bits register. In the backward search procedure, we used 256 bits AVX2 SIMD instructions to search all fragments in parallel. We assumed an array index fits 32 bits, so the choice of $d = 8$ allows us to process all d fragments in parallel with the 256 bits instruction set. If we desire to handle strings larger than 2^{32} characters, the natural choice would be $d = 4$.

To initially construct the suffix array of the wide characters, we combined DC3 [7] and SA-IS [13]; we implemented DC3 that can handle eight bytes wide characters by ourselves. For SA-IS, we used an implementation by Mori (https://sites.google.com/site/yuta256/sais/). The toplevel procedure is DC3 which, upon the recursive call, switches to SA-IS if the character size becomes smaller than 2^{32}, the size the SA-IS implementation we used can handle. In practice, DC3 is used only in the first level, where we have eight bytes wide characters. We build a sampled suffix array of $SA_{0/d}$, taking every other element from it (equivalent to taking a sample from every 16 elements of the entire suffix array). We used unordered_map of C++ Standard Template Library to implement a sampled suffix array. occ_r function is implemented by building a wavelet matrix of $BWT_{r/d}$. In this implementation, our index is almost the same size as BWT method.

6 Evaluation

Our evaluation platform has Xeon E5-2699 v3 processor running Linux 3.1.6 kernel. In creating suffix array and fragmented BWT, we compare our method and SA-IS, the state of the art algorithm. Figure 3 shows index construction time for genome sequence (http://ftp.ensembl.org/pub/release-75/fasta/homo_sapiens/dna/) over various string lengths. This shows that our method is consistently two times faster against a range of data sizes.

To evaluate search performance, we took substrings of various lengths from the reference string and use them as queries. We varied query lengths from 10 to 100 and for each length, we extracted 1000 strings from the head of the string, skipping the very first region that entirely consists of many 'N's.

Figure 4 shows query execution time including both steps (backward search and locating suffixes in the reference string), against a range of the number of occurrences. As seen in Fig. 4, our method is four times slower when there are few occurrences; our method starts outperforming regular BWT backward search when the number of occurrences becomes as many as 16 and becomes twice as fast when there are many occurrences.

To better understand this result, Fig. 5 shows execution time of the backward search step against query string lengths. As shown, despite the use of SIMD instructions to search many fragments in parallel, our method is about three times slower than the regular backward search. Ideally, we like to expect searching all the eight fragments in parallel with eight-way SIMD instructions would

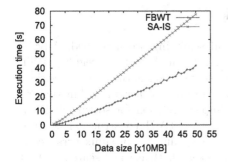

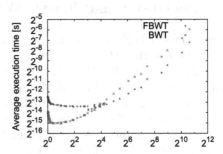

Fig. 3. Construction time of FBWT and SA-IS

Fig. 4. Search time

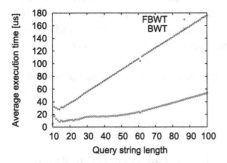

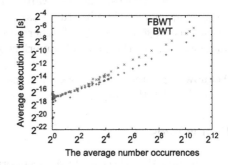

Fig. 5. Backward search time

Fig. 6. Time to locate suffixes found

result in roughly the same execution time as the ordinary backward search. This did not happen because of the following reasons.

- Fundamentally, the backward search can quit before reaching the head of the query string, as soon as one of its suffixes turns out not to occur at all, but by searching eight fragments in parallel, the search continues until *all* the eight fragments quit.
- Even if eight fragments quit with roughly the same number of iterations, searching eight fragments involve some overheads associated with the use of SIMD instructions. In particular, we use gather instructions to access multiple locations in parallel, but this does not have the same latency/throughput as the scalar load instruction.

Figure 6 shows execution time for locating suffixes (get_position_fragment) in the reference string, which seems consistently roughly twice faster than the ordinary procedure (get_position). This is because the former does not need to check if the current query string is found in the sampled suffix array in every iteration; it knows which fragments the current suffix belongs to and checks its presence only when the suffix is known to be from $(0/d)$-fragment.

7 Conclusion and Future Work

Fragmented BWT for full-text indexing and searching are described. Advantages include that (1) its construction is faster than that of the regular BWT and (2) searching with FBWT is faster in locating suffixes in the reference string. To accelerate to construct FBWT more, we can apply other algorithm such as parallel construction [8] or parallelize our implementation for sorting in DC3. We can expect FBWT construction be faster than BWT construction because the space consumption is smaller thanks to fewer number of suffix array construction. It is slower in the first step of the search (backward search to find ranks of suffixes), but this can be mitigated by SIMD instructions. Experiments show that the construction becomes roughly twice as fast as the state of the art method (SA-IS), locating suffixes roughly twice as fast as the same procedure for monolithic BWTs. The backward search becomes third to four times slower (despite applying SIMD instructions). Our future work includes applying SIMD to the second step as well to further accelerate the search procedure and better understanding of the cause of imperfect speedup with SIMD instructions.

Acknowledgement. This work was in part supported by Grant-in-Aid for Scientific Research (A) 16H01715.

References

1. Burrows, M., Wheeler, D.: A block-sorting lossless data compression algorithm. Algorithm Data Compression (124), p. 18 (1994)
2. Claude, F., Navarro, G.: The wavelet matrix. In: SPIRE, pp. 167–179 (2012)
3. Ferragina, P., Manzini, G.: Indexing compressed text. J. ACM **52**(4), 552–581 (2000)
4. Ferragina, P., Manzini, G., Mäkinen, V., Navarro, G.: Compressed representations of sequences and full-text indexes. ACM Trans. Algorithms **3**(2), 20 (2007)
5. Grossi, R., Gupta, A., Vitter, S.: High-order entropy-compressed text indexes. In: Proceedings of the Fourteenth Annual ACM-SIAM Symposium on Discrete Algorithms, pp. 841–850 (2003)
6. Hayashi, S., Taura, K.: Parallel and memory-efficient Burrows-Wheeler transform. In: Proceedings - 2013 IEEE International Conference on Big Data, pp. 43–50 (2013)
7. Kärkkäinen, J., Sanders, P.: Simple linear work suffix array construction. In: Colloquium on Automata, Languages and Programming, pp. 943–955 (2003)
8. Kärkkäinen, J., K.D., S., P.: Parallel external memory suffix sorting. In: CPM 2015, pp. 329–342 (2015)
9. Langmead, B., Trapnell, C., Pop, M., Salzberg, S.: Ultrafast and memory-efficient alignment of short DNA sequences to the human genome. Genome Biol. **10**(3), 1 (2009)
10. Li, H., Durbin, R.: Fast and accurate short read alignment with Burrows-Wheeler transform. Bioinformatics **25**, 1754–1760 (2009)
11. Li, R., Yu, C., Li, Y., Lam, W., Yiu, M., Kristiansen, K., Wang, J.: SOAP2: an improved ultrafast tool for short read alignment. Bioinformatics **25**, 1966–1967 (2009)

12. Manber, U., Myers, G.: Suffix string arrays: a new searches method for on-line. In: Proceedings of the First Annual ACM-SIAM Symposium on Discrete Algorithms, pp. 319–327 (1990)
13. Nong, G., Zhang, S., Chan, H.: Linear suffix array construction by almost pure induced-sorting. In: 2009 Data Compression Conference, pp. 193–202 (2009)
14. Sadakane, K.: New text indexing functionalities of the compressed suffix arrays. J. Algorithms **48**(2), 294–313 (2003)

AC-Automaton Update Algorithm
for Semi-dynamic Dictionary Matching

Diptarama$^{(\boxtimes)}$, Ryo Yoshinaka, and Ayumi Shinohara

Graduate School of Information Sciences, Tohoku University,
6-6-05 Aramaki Aza Aoba, Aoba-ku, Sendai, Japan
diptarama@shino.ecei.tohoku.ac.jp, {ry,ayumi}@ecei.tohoku.ac.jp

Abstract. Given a set of pattern strings called a dictionary and a text
string, dictionary matching is the problem to find the occurrences of the
patterns on the text. Dynamic dictionary matching is dictionary matching where patterns may dynamically be inserted into and deleted from
the dictionary. The problem is called semi-dynamic dictionary matching when we only consider the insertion. An AC-automaton is a data
structure which enables us to solve dictionary matching in $O(d \log \sigma)$
preprocessing time and $O(n \log \sigma)$ matching time, where d denotes the
total length of the patterns in the dictionary, n denotes the length of the
text, and σ denotes the alphabet size. In this paper we propose an efficient algorithm that dynamically updates an AC automaton for insertion
of a new pattern by using a directed acyclic word graph.

Keywords: Semi-dynamic dictionary matching · AC-automaton ·
DAWG

1 Introduction

The pattern matching problem is one of the fundamental problems in string
processing. Given a pattern string and a text string, output all occurrence positions of the pattern in the text. Pattern matching algorithms can be applied in
data mining, search engines, text editors, etc.

An extension of the pattern matching problem is to find multiple patterns simultaneously instead of a single pattern. That is, given a set $D =
\{p_1, p_2, \ldots, p_r\}$ of patterns called *dictionary* and a text, we find all occurrences
of the patterns in the text. This problem is called the *dictionary matching problem* [1,2]. The dictionary matching problem can be solved by the Aho-Corasick
algorithm [1] or the Commentz-Walter algorithm [7]. Both of the algorithms
preprocess the dictionary first in $O(d \log \sigma)$ time, and then find the occurrences
of patterns in the text in $O(n \log \sigma)$ time for the AC-algorithm and $O(nd \log \sigma)$
time for the Commentz-Walter algorithm in the worst case, where d is the total
length of the patterns, n is the length of the text, and σ is the alphabet size.

Meyer [15] introduced incremental string matching, which is also known as
semi-dynamic dictionary matching problem, as a variant of dictionary matching

© Springer International Publishing AG 2016
S. Inenaga et al. (Eds.): SPIRE 2016, LNCS 9954, pp. 110–121, 2016.
DOI: 10.1007/978-3-319-46049-9_11

Table 1. Comparison of the algorithms for (semi-)dynamic dictionary matching. n is the length of text, m is the length of pattern, and d is the total length of dictionary. $k \geq 2$ is any constant. $lmax$ is the length of the longest pattern, and s is the size of the AC-automaton before insertion.

Algorithm	Update	Query
Idury and Schäffer [12]	$O(m(kd^{1/k} + \log \sigma))$	$O(n(k + \log \sigma))$
Amir et al. [3]	$O(m(\log d/\log \log d + \log \sigma))$	$O(n(\log d/\log \log d + \log \sigma))$
Chan et al. [6]	$O(m \log^2 d)$	$O(n \log^2 d)$
Hon et al. [11]	$O(m \log \sigma + \log d)$	$O(n \log d)$
Feigenblat et al. [10]	$O(m \log d \log \log d)$	$O(n(\log \log d) + \log \sigma)$
Meyer [15]	$O(lmax \cdot s \cdot \sigma)$	$O(n \log \sigma)$
Tsuda et al. [16]	$O(s \log \sigma)$	$O(n \log \sigma)$
Proposed	$O(m \log \sigma + s)$	$O(n \log \sigma)$

that allows insertion of a pattern into the dictionary. He proposed an algorithm for semi-dynamic dictionary matching, which updates the AC-automaton when a new pattern is inserted into the dictionary. Amir et al. [2] introduced *dynamic dictionary matching problem* by allowing deletion of a pattern from dictionary. Dynamic dictionary matching can be solved by constructing sophisticated data structures from the dictionary [2,3,6,12]. Moreover, succinct data structures are also introduced to deal with dynamic dictionary matching, where the main concern is to save the memory usage to store the dictionary [10,11]. Remark that in all these approaches except Meyer's, the query time to search a text for patterns is more or less sacrificed.

Tsuda et al. [16] also considered dynamic dictionary matching by directly following and extending Meyer's method [15], and showed how to update the dictionary, while keeping the query time in $O(n \log \sigma)$. They also performed some computational experiments. Ishizaki and Toyama [13] introduced a data structure called an *expect tree* to efficiently update the dictionary for inserting patterns, and showed some experimental results, but with no theoretical analysis. Along this line, in this paper, we propose a simple and more efficient algorithm for the semi-dynamic dictionary matching problem, where the query time is still in $O(n \log \sigma)$. The algorithm updates an AC-automaton when a new pattern is inserted into the dictionary by using a directed acyclic word graph (DAWG) [4,5] of the dictionary. For each new pattern p, the algorithm updates the AC-automaton in $O(m \log \sigma + u_f + u_o)$ time, where m is the length of p, u_o is the number of states whose output function needs to be updated, and u_f is the number of states whose failure link needs to be updated. Table 1 shows a summary of the proposed algorithm and existing algorithms. Note that $u_f + u_o \leq 2s$. In this sense, our algorithm is more efficient than any other existing ones, under the constraint of the query time in $O(n \log \sigma)$.

2 Preliminaries

Let Σ denote an *alphabet* of size σ. An element of Σ^* is called a *string*. For a string w, the length of w is denoted by $|w|$. The *empty string*, denoted by ε, is a string of length 0. For a string $w = xyz$, strings x, y, and z are called *prefix*, *substring*, and *suffix* of w, respectively. For a string w, let $Substr(w)$ denote the set of all substrings of w, and for a set $D = \{w_1, w_2, \ldots, w_r\}$ of strings, let $Substr(D) = \cup_{i=1}^{r} Substr(w_i)$. Similarly, let $Pref(D)$ be the set of all prefixes of strings in D.

Let $D = \{p_1, p_2, \ldots, p_r\}$ be a set of *patterns* over Σ. D is often called a *dictionary*. Let $d = \sum_{i=1}^{r} |p_i|$, the total length of the patterns in D. An *Aho-Corasick Automaton* of D, denoted by $AC(D)$, is a *trie* of all patterns in D, consisting of *goto*, *failure* and *output* functions. We often identify a state s with the string obtained by concatenating all the labels found on the path from the root to the state s. The state transition function *goto* is defined so that for any two states $s, s' \in Substr(D)$ and any character $c \in \Sigma$, if $s' = sc$ then $s' = goto(s, c)$. The failure function is defined by $flink(s) = s'$ where s' is the longest proper suffix of s such that $s' \in Substr(D)$. Finally, $output(s)$ is the set of all patterns that are suffixes of s. $AC(D)$ is used to find occurrences of any pattern in D on a text. We omit to explain the basic construction algorithm of $AC(D)$ and how to use it for text search (see [1,9]).

For any $x, y \in Substr(D)$, we define $x \equiv_D y$ iff $endPos_D(x) = endPos_D(y)$, where $endPos_D(x)$ is the set of all positions in the patterns in D where an occurrence of x ends. We denote by $[x]_D$ the equivalence class of x with respect to $\equiv_D$. The *Directed Acyclic Word Graph* for D, denoted by $DAWG(D)$, is a directed acyclic graph with the set of nodes[1] $\{[x]_D \mid x \in Substr(D)\}$ and the set of edges $\{([x]_D, [xc]_D) \mid xc \in Substr(D), c \in \Sigma\}$. Each edge $([x]_D, [xc]_D)$ is labeled by c. The node $[\varepsilon]_D$ is called the *source* of $DAWG(D)$. Figure 1 shows an example of a DAWG. For each node except the root, *suffix link* is defined as $slink([x]_D) = [y]_D$, where y is the longest suffix of x satisfying $[x]_D \neq [y]_D$. For convenience, we define $slink^i([x]_D) = slink^{i-1}(slink([x]_D))$ and $slink^1([x]_D) = slink([x]_D)$. We call a node v a *trunk node* if there exists $u \in Pref(D)$ such that u is obtained by concatenating all the labels found on the path from the *source* to the node v. Other nodes are called *branch nodes*. For example, in Fig. 2(b), trunk nodes are numbered, while branch nodes are blank.

From the properties of AC-automaton and DAWG, for each state s in $AC(D)$, there exists a unique node v in $DAWG(D)$ that corresponds to s, so that $AC(D)$ can be consistently embedded into $DAWG(D)$. Because $s \in Pref(D)$, the corresponding node v is a trunk node and each trunk node has its corresponding state. Therefore, there exists a one-to-one mapping from the set of trunk nodes in $DAWG(D)$ to the states of $AC(D)$. We denote this mapping by $s = \pi(v)$. Figure 3(a) and (b) show the AC-automaton and DAWG of $D = \{abba, aca, cbb\}$, respectively, where each number expresses the correspondence.

---

[1] To avoid confusion, we refer to vertex in DAWG as *node*, and vertex in AC-automaton as *state* in this paper.

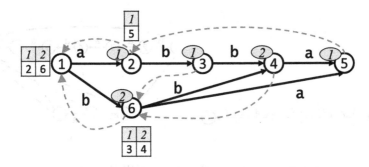

Fig. 1. $DAWG(\{\text{abba}\})$. The table at each node v represents the inverse suffix links v_a, and the number on top of each node v' shows its index v'_{index} in the table

We can verify the following property, which will be utilized in our algorithms.

Lemma 1. *Let s and s' be any states in $AC(D)$, and let $v = \pi^{-1}(s)$ and $v' = \pi^{-1}(s')$ be corresponding trunk nodes in $DAWG(D)$. Then, $s' = flink(s)$ if and only if there exists an integer $k > 0$ such that $v' = slink^k(v)$ and each $slink^i(v)$ is a branch node for $0 < i < k$.*

3   Maintenance of Inverse Suffix Links of DAWG

Meyer [15] and Tsuda *et al.* [16] used the inverse of the failure function of the AC-automaton to update it. Although the inverse failure function can be stored in $O(d)$ space in total, it is not trivial that the access and update time of inverse failure function can be done in $O(1)$ time, because the number of inverse failure links of each state may change dynamically and can be as large as the number of states in the AC-automaton. For instance, let us consider $AC(D)$ for $D = \{\text{baaaac}\}$ over $\Sigma = \{\text{a}, \text{b}, \text{c}\}$ in Fig. 2(a). Its root is pointed by 6 failure links. When adding a new pattern c to D, these algorithms first create a new state s, a new transition from the root to s, and a failure link from s to the root. The real difficulty arises when they try to find which suffix links should be updated to point at s; they must follow all the 6 inverse failure links from the root and get 6 states numbered $2, 3, \ldots, 7$, and check whether there is an edge labeled c from each of them, although only one state 7 should be updated. Ishizaki and Toyama [13] introduced an auxiliary tree structure called an *expect tree* to reduce the number of the candidates and showed some experimental results, but no theoretical analysis is provided. Unfortunately, their algorithm behaves the same for the above example. Therefore, maintaining the inverse failure links to update the AC-automaton might be inefficient.

In order to deal with this difficulty, we pay our attention to the suffix links of $DAWG(D)$, instead of the failure links of $AC(D)$. It is known (see, e.g. [9]) that for any node v in $DAWG(D)$, the number of suffix links that point at v is at most σ. Therefore, accessing and updating the inverse suffix links can be

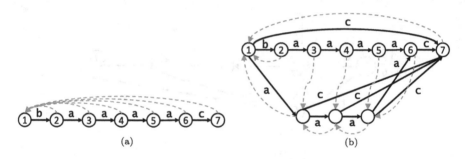

Fig. 2. (a) $AC(\{\texttt{baaaac}\})$ and (b) $DAWG(\{\texttt{baaaac}\})$.

done in $O(1)$ time, for a fixed alphabet Σ. However, this method is inefficient in space if the alphabet is large, such as Chinese and Japanese ones. We need a more space efficient method to maintain the inverse suffix links.

We store the inverse suffix links of each node v in a dynamic array v_a, which is a resizable array [17]. We grow the array when a new suffix link pointed at the node is added and the size exceeds the capacity. The ith suffix link that points at v is represented as $v_a[i]$. Moreover, in order to make the array accessible and updatable in $O(1)$ time, we associate the index i to u as $u_{index} = i$ if $u = v_a[i]$, i.e., u is the ith node whose suffix link points $v = slink(u)$. Figure 1 shows an example of a DAWG with its inverse suffix links.

The inverse suffix links can be maintained in $O(d)$ time overall when we construct $DAWG(D)$. First, from a property of the DAWG constructing algorithm, the number of suffix links that point at each node is never reduced, thus the size of array storing inverse suffix links from the node is also never reduced. Therefore, we do not have to worry about deleting any elements from the array. Next, each inverse suffix link is updated only when the node is split [4]. In this case, we can assign the space in the inverse suffix link array which has been occupied by the old node to the new one. Suppose that there is a node x whose suffix link points at z, i.e., $slink(x) = z$, $z_a[i] = x$ and $x_{index} = i$ for some i. When x is split into x and y, we update the inverse suffix link as $slink(y) = z$, $z_a[i] = y$, $y_{index} = i$, $slink(x) = y$, $y_a[1] = x$, and $x_{index} = 1$. This operation can be performed in $O(1)$ time, because the inverse suffix link can be accessed randomly.

One may think that a DAWG can be used for dictionary matching directly. This is indeed possible for a single pattern [8]. However, it is difficult to maintain the output function efficiently for multiple patterns, as is mentioned in [14]. Therefore, we efficiently update $AC(D)$ with the guide of $DAWG(D)$, and then update $DAWG(D)$ too, as we will see in the next section.

4   AC-Automaton Update Algorithm

We consider inserting a new pattern p of length m into the dictionary D, and we denote the new dictionary by $D' = D \cup \{p\} = \{p_1, p_2, \ldots, p_r, p\}$. It is known

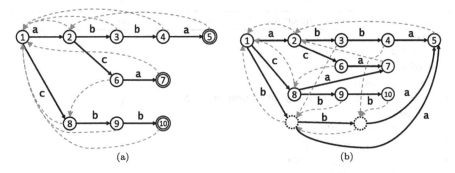

Fig. 3. For a dictionary $D = \{\texttt{abba}, \texttt{aca}, \texttt{cbb}\}$ (a) $AC(D)$, and (b) $DAWG(D)$.

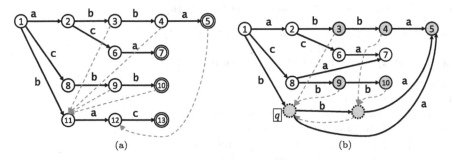

Fig. 4. Illustration of updating process when inserting a pattern $p = \texttt{bac}$ into the dictionary $D = \{\texttt{abba}, \texttt{aca}, \texttt{cbb}\}$. Compare them with Fig. 3. (a) The updated automaton $AC(D')$, where the only updated failure links are shown. (b) In $DAWG(D)$, the only suffix links that are used for the update are shown, and the visited nodes are colored. (Color figure online)

that $DAWG(D)$ can be constructed in $O(d \log \sigma)$ time, and can be updated to $DAWG(D')$ online in $O(m \log \sigma)$ time [4]. We update $AC(D)$ to $AC(D')$ by using $DAWG(D)$, and then update $DAWG(D)$ to $DAWG(D')$. The key point of our algorithm is to update the output and failure functions of $AC(D)$ in linear time with respect to the number of states that should be modified. The *goto* function can be updated easily by adding a new transition for a new state in the same way as in the AC-automaton construction algorithm. We then update the output and failure functions efficiently by using inverse suffix links of $DAWG(D)$. Algorithm 1 updates $AC(D)$ when a new pattern is inserted to D, and Algorithms 2 and 3 find the states whose output function and failure link should be updated, respectively.

For any node v in $DAWG(D)$, let $isuf(v) = \{x \mid slink(x) = v\}$ be the set of its inverse suffix links. The set $isuf(v)$ of inverse suffix links for each v is stored in the dynamic array v_a as described in Sect. 3. For the new pattern p, we can divide p to $p = xyz$ and categorize the prefixes of p into three categories, so that for any i, j, k with $1 \leq i \leq |x| < j \leq |x| + |y| < k \leq m$;

Algorithm 1. Pattern insertion algorithm of AC-automaton

Input: new pattern p

1 $activeState = rootState$;
2 $newStatesSet = empty$;
3 **for** $1 \leq i \leq m$ **do**
4   **if** $goto(activeState, p[i]) \neq$ **fail then** $activeState = goto(activeState, p[i])$;
5   **else**
6   **create** $newState$;
7          $goto(activeState, p[i]) = newState$;
8          $activeState = newState$;
9          $newStatesSet = newStatesSet \cup \{newState\}$;
10   **if** $i = m$ **then**
11         $output(newState) = output(newState) \cup \{p\}$;

12 $failStates = getFailStates(p, m - |newStatesSet| + 1)$;
13 **for** $(s, i) \in failStates$ **do**
14     $flink(s) = newStatesSet[i - |newStatesSet| + 1]$;

15 $activeState = rootState$;
16 **for** $1 \leq i \leq m$ **do**
17   **if** $goto(activeState, p[i]) \in newStatesSet$ **then**
18         $failureState = flink(activeState)$;
19   **while** $goto(failureState, p[i]) =$ **fail do**
20             $failureState = flink(failureState)$;
21         $activeState = goto(activeState, p[i])$;
22         $flink(activeState) = failureState$;
23         $output(activeState) = output(activeState) \cup output(failureState)$;
24   **else**
25         $activeState = goto(activeState, p[i])$;

26 $outStates = getOutStates(p)$;
27 **for** $, \in outStates$ **do** $output(s) = output(s) \cup \{p\}$;

---

1. $p[1 : i]$ exists both in $AC(D)$ and $DAWG(D)$,
2. $p[1 : j]$ does not exist in $AC(D)$ but exists in $DAWG(D)$, and
3. $p[1 : k]$ exists in neither $AC(D)$ nor $DAWG(D)$.

To update both output and failure functions of $AC(D)$ to $AC(D')$ we only use nodes in $DAWG(D)$ that represent prefixes in the second category. Algorithm 2 follows inverse suffix links of a node representing p recursively in $DAWG(D)$, in order to find the states in $AC(D)$ whose output functions need to be updated. On the other hand, Algorithm 3 follows inverse suffix link of nodes that represent $p[i : j]$ for $|x| < j \leq |x| + |y|$ (category 2) recursively, until it finds a trunk node and then saves the state that corresponds to the trunk node to update its failure link later.

Algorithm 2. *getOutStates(p)*

Output: set of states whose output functions should be updated.

1 *activeNode = root*;
2 **for** $1 \leq i \leq m$ **and** *activeNode* $\neq$ **null do**
3 $\quad$ *activeNode = trans(activeNode, p[i])*;
4 $\quad$ **if** $i = m$ **then**
5 $\quad\quad$ push *activeNode* to *stack*;

6 **if** *activeNode* $\neq$ **null then**
7 $\quad$ *queue = * **empty**;
8 $\quad$ push *activeNode* to *queue*;
9 $\quad$ **while** *queue* $\neq$ **empty do**
10 $\quad\quad$ pop *node* from *queue*;
11 $\quad\quad$ **if** *node is a trunk node* **then**
12 $\quad\quad\quad$ *outStates = outStates* $\cup$ $\{\pi(node)\}$
13 $\quad\quad$ **for** *lnode* $\in$ *isuf(node)* **do**
14 $\quad\quad\quad$ push *lnode* to *queue*;

15 **return** *outStates*;

Figure 4 illustrates an example, where we insert a pattern $p =$ bac into the dictionary $D = \{$abba, aca, cbb$\}$. First, we create new states 11, 12, and 13. The string b is represented by node x in $DAWG(D)$, and by the new state 11 in $AC(D')$, thus there is at least one state whose failure link should be updated to point at the state 11. We will explain how to find these states below. Similarly, we know that at least one failure link should be updated to point at the state 12, because the string ba represented by the state 12 in $AC(D')$ is also represented by node 5 in $DAWG(D)$. However, the string bac, which is represented by the new state 13, is not represented in $DAWG(D)$, thus we know that there is no state whose failure link should be updated to state 13. As a result, we have the set $\{11, 12\}$ of states. (Lines 3–6 in Algorithm 3)

We now explain how to find states whose failure links should be updated. We begin by the deepest state in $\{11, 12\}$, that is, state 12. We search the states from node 5 in $DAWG(D)$, which represents the same string ba as state 12 in $AC(D')$. When searching from node 5, we do not search further because node 5 is a trunk node. Therefore, we update the failure link of state 5 to state 12. Next, to find states whose failure links should be updated to state 11, we search the states from node q in $DAWG(D)$, which represents the same string b as state 11 in $AC(D')$. By following the inverse suffix links recursively from node q until reaching a trunk node, we get the set $\{3, 4, 9, 10\}$ of trunk nodes (see Fig. 4(d)). Therefore, we update the failure link of states 3, 4, 9, and 10 to state 11. (Lines 7–19)

We now show the correctness and running time of our Algorithms 2 and 3.

Algorithm 3. *getFailStates(p, start)*

Output: set of states whose failure link should be updated.

1 *stack* = **empty**;
2 *activeNode* = *root*;
3 **for** $1 \leq i \leq m$ **and** *activeNode* $\neq$ **null do**
4   *activeNode* = *trans(activeNode, p[i])*;
5   **if** $i \geq start$ **and** *activeNode* $\neq$ **null then**
6   push *(activeNode, i)* to *stack*;

7 **while** *stack* $\neq$ **empty do**
8   pop *(activeNode, i)* from *stack*;
9   *queue* = **empty**;
10   push *activeNode* to *queue* ;
11   **while** *queue* $\neq$ **empty do**
12   pop *node* **from** *queue*;
13   **if** *node is not marked* **then**
14   mark *node*;
15   **if** *node is a trunk node* **then**
16   *failStates* = *failStates* $\cup \{(\pi(node), i)\}$
17   **else**
18   **for** *lnode* $\in$ *isuf(node)* **do**
19   push *lnode* to *queue*;

20 **return** *failStates*;

Lemma 2 ([5]). *A string $x \in Substr(D)$ is the longest member of $[x]_D$ if and only if either $x \in Pref(D)$, or $ax, bx \in Substr(D)$ for some distinct $a, b \in \Sigma$.*

Lemma 3. *For any branch node in DAWG, there exist at least two suffix links that point at it.*

Proof. Let $[x]_D$ be any branch node in $DAWG(D)$, and $x \in Substr(D)$ be the longest member of $[x]_D$. Then $x \notin Pref(D)$ because $[x]_D$ is a branch node. By Lemma 2, there exist two distinct $a, b \in \Sigma$ such that $ax, bx \in Substr(D)$. Because x is the longest member of $[x]_D$, we have $[ax]_D \neq [x]_D$. Thus, $slink([ax]_D) = [x]_D$ because x is a suffix of ax. Similarly, $slink([bx]_D) = [x]_D$. Because $[ax]_D \neq [bx]_D$, the branch node $[x]_D$ is pointed by at least two suffix links.   □

Lemma 4. *Algorithm 2 correctly returns the set of states whose output functions should be updated.*

Proof. When a new pattern p is inserted to a dictionary D, we have to update the output function of every state s in $AC(D)$ such that p is a suffix of the string s. If there is no node in $DAWG(D)$ representing p, we know that no such a string s exists in D, done. Let s_p be a new state created in $AC(D')$ to represent the pattern p. The output function of some state s should be updated if and only if s_p is reachable from s via a chain of failure links. From Lemma 1, for nodes $u = \pi^{-1}(s)$ and $v_p = [p]_D$, we have $v_p = slink^i(u)$ for some i. Therefore, $s = \pi(u)$ can be found by following inverse suffix links from v_p recursively. $\square$

Lemma 5. *Algorithm 2 runs in $O(m \log \sigma + u_o)$ time, where u_o is the number of states whose output function should be updated.*

Proof. At first, Algorithm 2 finds the node v representing the pattern p, by traversing the nodes from the root, in Lines 3–6. It takes $O(m \log \sigma)$ time. If it failed, done. Then we analyze the runnning time consumed in Lines 6–14 by counting the number ℓ of visited nodes in $DAWG(D)$. These nodes form a tree, rooted at v and connected by inverse suffix links chains. Let b (resp. t) be the number of branch (resp. trunk) nodes in this tree, and let q be the number of nodes (either branch or trunk) that are child nodes of some branch node. Because every branch node has at least two child nodes by Lemma 3, we have $2b \leq q$, and obviously $q \leq b + t$. Thus, $b \leq t$, which yields that $\ell = b + t \leq 2t = 2|outStates| = 2u_o$. Therefore, Algorithm 2 runs in $O(m \log \sigma + u_o)$. $\square$

Lemma 6. *Algorithm 3 correctly returns the set of states whose failure links should be be updated.*

Proof. By argument similar to the proof of Lemma 4, all the states that should be updated is reachable via chains of inverse suffix links from the nodes in $DAWG(D)$ that correspond to the new states in $AC(D')$. Next, we will show that Algorithm 3 only returns the set S of the states that should be updated. Let x be a new state, and $t = [x]_D$ be a node that represents the string x. Assume that S contains a state s that can be reached by following inverse failure links from x recursively, but should not be updated. Let $u = \pi^{-1}(s)$ and $v = \pi^{-1}(flink(s))$ be trunk nodes in $DAWG(D)$ corresponding to s and $flink(s)$, respectively. From Lemma 1, $v = slink^i(u)$ and $t = slink^j(v)$ for some i and j. Since Algorithm 3, started from t, stops a recursive search after reaching a trunk node (v in this case), it would not find u. Therefore, $s = \pi(u) \notin S$. $\square$

Lemma 7. *Algorithm 3 runs in $O(m \log \sigma + u_f)$ time, where u_f is the number of states that their failure link be updated.*

Proof. At first, Algorithm 3 finds the set V of nodes representing the pattern $p[1:j]$ for $1 \leq j \leq m$ such that $p[1:j]$ does not exist in $AC(D)$ but does exist in $DAWG(D)$, by traversing the nodes from the root, in Lines 3–6. The algorithm saves the nodes in a stack, because the algorithm will search from the deepest node. It takes $O(m \log \sigma)$ time. Then we analyze the running time consumed in Lines 7–19 by counting the number ℓ of visited nodes in $DAWG(D)$. These nodes form a forest, where each tree is rooted by some node in V and connected by inverse suffix links chains, where some node in V can be an inner node of a tree rooted by another in V. In this case we mark the nodes that have been visited, thus each node is visited at most twice. Let b (resp. t) be the number of branch (resp. trunk) nodes in this forest, and let q be the number of nodes (either branch or trunk) that are child nodes of some branch node. Because every branch node has at least two child nodes by Lemma 3, we have $2b \leq q$, and obviously $q \leq b+t$. Thus, $b \leq t$, which yields that $\ell = b + t \leq 2t = 2|failStates| = 2u_f$.   □

Theorem 1. *AC-automaton can be updated for each pattern in* $O(m \log \sigma + u_f + u_o)$ *time.*

Proof. The goto, failure and output functions of newly created states can be calculated in $O(m \log \sigma)$, similarly to the original AC-automaton construction algorithm. From Lemmas 5 and 7, output and failure functions of existing states can be updated in $O(m \log \sigma + u_o)$ and $O(m \log \sigma + u_f)$, respectively. Therefore, AC-automaton can be updated in $O(m \log \sigma + u_f + u_o)$ time in total.   □

5   Conclusion

We proposed a new algorithm to update an AC-automaton when a new pattern is inserted to the dicitonary. Our algorithm uses a directed acyclic word graph in order to update the AC-automaton. We showed that our algorithm updates the AC-automaton in $O(m \log \sigma + u_f + u_o)$ time, which is faster than existing AC-automaton update algorithms [13,15,16], where m is the length of the inserted pattern, u_o (resp. u_f) is the number of states whose output function (resp. failure links) should be updated. It means that our update process is minimized, compared to the existing algorithms.

Acknowledgments. This work is supported by Tohoku University Division for Interdisciplinary Advance Research and Education, JSPS KAKENHI Grant Numbers JP15H05706, JP24106010, and ImPACT Program of Council for Science, Technology and Innovation (Cabinet Office, Government of Japan).

References

1. Aho, A.V., Corasick, M.J.: Efficient string matching: an aid to bibliographic search. Commun. ACM **18**(6), 333–340 (1975)
2. Amir, A., Farach, M., Galil, Z., Giancarlo, R., Park, K.: Dynamic dictionary matching. J. Comput. Syst. Sci. **49**(2), 208–222 (1994)
3. Amir, A., Farach, M., Idury, R.M., Lapoutre, J.A., Schaffer, A.A.: Improved dynamic dictionary matching. Inf. Comput. **119**(2), 258–282 (1995)
4. Blumer, A., Blumer, J., Haussler, D., McConnell, R., Ehrenfeucht, A.: Complete inverted files for efficient text retrieval and analysis. J. ACM **34**(3), 578–595 (1987)
5. Blumer, A., Blumer, J., Haussler, D., Ehrenfeucht, A., Chen, M.T., Seiferas, J.: The smallest automation recognizing the subwords of a text. Theor. Comput. Sci. **40**, 31–55 (1985)
6. Chan, H.L., Hon, W.K., Lam, T.W., Sadakane, K.: Dynamic dictionary matching and compressed suffix trees. In: Proceedings of the Sixteenth Annual ACM-SIAM Symposium on Discrete Algorithms. Society for Industrial and Applied Mathematics, pp. 13–22 (2005)
7. Commentz-Walter, B.: A string matching algorithm fast on the average. In: Maurer, H.A. (ed.) Automata, Languages and Programming. LNCS, vol. 71, pp. 118–132. Springer, Heidelberg (1979)

8. Crochemore, M.: String matching with constraints. In: Chytil, M.P., Koubek, V., Janiga, L. (eds.) Mathematical Foundations of Computer Science 1988. LNCS, vol. 324, pp. 44–58. Springer, Heidelberg (1988)
9. Crochemore, M., Rytter, W.: Jewels of Stringology. World Scientific Publishing Co. Pte. Ltd., Singapore (2002)
10. Feigenblat, G., Porat, E., Shiftan, A.: An improved query time for succinct dynamic dictionary matching. In: Kulikov, A.S., Kuznetsov, S.O., Pevzner, P. (eds.) CPM 2014. LNCS, vol. 8486, pp. 120–129. Springer, Heidelberg (2014)
11. Hon, W.-K., Lam, T.-W., Shah, R., Tam, S.-L., Vitter, J.S.: Succinct index for dynamic dictionary matching. In: Dong, Y., Du, D.-Z., Ibarra, O. (eds.) ISAAC 2009. LNCS, vol. 5878, pp. 1034–1043. Springer, Heidelberg (2009)
12. Idury, R.M., Schäffer, A.A.: Dynamic dictionary matching with failure functions. Theor. Comput. Sci. 131(2), 295–310 (1994)
13. Ishizaki, F., Toyama, M.: An incremental update algorithm for large Aho-Corasick automaton. In: Proceedings of the 4th Forum on Data Engineering and Information Management, F11–5, pp. 1–6 (2012). (In Japanese)
14. Kucherov, G., Rusinowitch, M.: Matching a set of strings with variable length don't cares. Theor. Comput. Sci. 178(12), 129–154 (1997)
15. Meyer, B.: Incremental string matching. Inf. Process. Lett. 21(5), 219–227 (1985)
16. Tsuda, K., Fuketa, M., Aoe, J.I.: An incremental algorithm for string pattern matching machines. Int. J. Comput. Math. 58, 33–42 (1995)
17. Wikipedia: Dynamic array – Wikipedia, the free encyclopedia (2016). Accessed 15 May 2016

Parallel Computation for the All-Pairs Suffix-Prefix Problem

Felipe A. Louza[1]([⊠]), Simon Gog[2], Leandro Zanotto[1], Guido Araujo[1], and Guilherme P. Telles[1]

[1] Institute of Computing, University of Campinas, São Paulo, Brazil
{louza,guido,gpt}@ic.unicamp.br, leandro.zanotto@reitoria.unicamp.br
[2] Institute of Theoretical Informatics, Karlsruhe Institute of Technology, Karlsruhe, Germany
gog@kit.edu

Abstract. We show how to parallelize the optimal algorithm proposed by Tustumi *et al.* [19] to solve the all-pairs suffix-prefix matching problem for general alphabets. We compared our parallel algorithm with SOF [17], a practical solution for DNA sequences that exhibits good time and space performance in multithreading environments. The experimental results showed that our parallel algorithm achieves a consistent speedup when compared with the sequential algorithm, and it is competitive with SOF when the minimum overlap length is small.

Keywords: Suffix-prefix matching · Parallel algorithm · Multithreading · Suffix array · LCP array

1 Introduction

Given a collection of strings the all-pairs suffix-prefix matching problem (APSP) is to find all longest overlaps among string ends [7]. This problem is well know in stringology [14] and appears often as a bottleneck part of DNA assembly, where the number of strings ranges from thousands to billions [3]. Other applications include EST clustering [9] and approximating the shortest common superstring [7].

The APSP has been approached in different ways. In practice with DNA sequences, filtering strategies have been used and rendered very efficient algorithms that are able to cope with the huge scale of DNA assembly [5,18]. On the theoretical side, optimal algorithms exist [8,15,19] that are able to handle strings from general alphabets, and, while not beating specialized algorithms for DNA, still perform fairly.

In this article we introduce a parallel algorithm for the APSP. Our solution builds on a previous optimal sequential algorithm by Tustumi *et al.* [19]. We were able to obtain a consistent speedup with a small memory footprint with this approach. Experimental results showed that our algorithm is competitive with the practical solution SOF [17] when the minimum overlap length is small.

© Springer International Publishing AG 2016
S. Inenaga et al. (Eds.): SPIRE 2016, LNCS 9954, pp. 122–132, 2016.
DOI: 10.1007/978-3-319-46049-9_12

The next sections are organized as follows. In Sect. 2 we introduce notation and discuss the practical and the optimal algorithms to solve the APSP. In Sect. 3 we present the optimal algorithm by Tustumi *et al.*In Sect. 4 we present our parallel algorithm and in Sect. 5 we show experimental results. In Sect. 6 we conclude the article.

2   Preliminaries

2.1   Notation

Let S be a string of length n over an ordered alphabet Σ. The i-th symbol of S is denoted by $S[i]$ and the substring including symbols in the interval $[i, j]$, $1 \leq i \leq j \leq n$, is denoted by $S[i, j]$. A prefix of S is a substring of the form $S[1, i]$ and a suffix is a substring of the form $S[i, n]$, which will be denoted by S_i. We use the symbol $<$ for the lexicographic order relation between strings.

The suffix array of $S[1, n]$, SA, is an array of integers in the range $[1, n]$ that gives the lexicographic order of all suffixes of S, such that $S_{\mathsf{SA}[1]} < S_{\mathsf{SA}[2]} < \ldots < S_{\mathsf{SA}[n]}$ [6,13]. We denote the position of suffix S_i in SA as $\mathsf{pos}(S_i)$. The LCP-array is an array of integers that stores the length of the longest common prefix (lcp) of two consecutive suffixes in SA, such that $\mathsf{LCP}[1] = 0$ and $\mathsf{LCP}[i] = lcp(S_{\mathsf{SA}[i]}, S_{\mathsf{SA}[i-1]})$ for $1 < i \leq n$. Both SA and the LCP-array can be constructed in linear time [10,16].

The range minimum query (rmq) with respect to the LCP gives the smallest lcp value in an interval of SA. We define $rmq(i, j) = \min_{i < k \leq j}\{\mathsf{LCP}[k]\}$. Given a string $S[1, n]$ and its LCP-array, it is easy to see that $lcp(S_{\mathsf{SA}[i]}, S_{\mathsf{SA}[j]}) = rmq(i, j)$, with $1 \leq i < j \leq n$.

Let $\mathcal{S} = S^1, S^2, \ldots, S^m$ be a collection of strings of lengths $n_i = |S^i|$, $\forall i \in [1, n]$. The generalized suffix array of $\mathcal{S}$ is the suffix array of the concatenated string $S^{cat} = S^1\$_1 S^2\$_2 \ldots S^m\$_m$ of length $N = m + \Sigma_{i=1}^m n_i$, where each symbol $\$_i$ is a distinct separator that does not occur in Σ, precedes every symbol in Σ, and $\$_i < \$_j$ if $i < j$. For a suffix $S^{cat}_{\mathsf{SA}[i]}$, we denote the prefix of $S^{cat}_{\mathsf{SA}[i]}$ that ends at the first separator $\$_j$ by $S^{\$}_{\mathsf{SA}[i]}$. The generalized suffix array can also be constructed in linear time [11].

For a clearer notation, we introduce the arrays STR and SA'. STR indicates which string in $\mathcal{S}$ a suffix came from, that is, $\mathsf{STR}[i] = j$ if the suffix $S^{\$}_{\mathsf{SA}[i]}$ ends with symbol $\$_j$. SA' holds the position of a suffix with respect to the string it came from (up to the separator), defined as $\mathsf{SA}'[i] = k$ if $S^{\$}_{\mathsf{SA}[i]} = S^j_k\$_j$. In other words, STR and SA' specify the order of all suffixes in the collection. We will denote the generalized suffix array enhanced with the arrays STR, SA' and LCP as GESA. The GESA of the collection $\mathcal{S} = \{\mathsf{aac}, \mathsf{aca}, \mathsf{aa}, \mathsf{caa}\}$ is illustrated in Fig. 1.

Let S^k be the j-th (lexicographically) smallest string in $\mathcal{S}$. P is an array of $m + 1$ integers that stores in $\mathsf{P}[j]$ the position of the complete suffix $S^k[1, n_k]$ in GESA. We define $\mathsf{P}[0] = m + 1$. Let the interval $B^j = (\mathsf{P}[j - 1], \mathsf{P}[j]]$ be a block of GESA corresponding to S^k. GESA can be partitioned into m blocks

	i	SA	LCP	STR	SA'	$S^\$_{SA[i]}$
	1	4	0	1	4	$\$_1$
	2	8	0	2	4	$\$_2$
	3	11	0	3	3	$\$_3$
	4	15	0	4	4	$\$_4$
P[0] →	5	7	0	2	3	a$\$_2$
	6	10	1	3	2	a$\$_3$
	7	14	1	4	3	a$\$_4$
P[1] →	8	9	1	3	1	aa$\$_3$
	9	13	2	4	2	aa$\$_4$
P[2] → 10	10	1	2	1	1	aac$\$_1$
	11	2	1	1	2	ac$\$_1$
P[3] → 12	12	5	2	2	1	aca$\$_2$
	13	3	0	1	3	c$\$_1$
	14	6	1	2	2	ca$\$_2$
P[4] → 15	15	12	2	4	1	caa$\$_4$

Fig. 1. The GESA of $\mathcal{S} = \{\text{aac}, \text{aca}, \text{aa}, \text{caa}\}$. Suffixes in block 1 are highlighted.

$B^1, B^2, \ldots, B^m$, one for each string S^k in $\mathcal{S}$. In Fig. 1, block $B^1 = (5, 8]$ of the corresponding string $S^3 = \text{aa}$ is shown by a gray rectangle.

2.2 APSP

The all-pairs suffix-prefix matching problem (APSP) is to find, for all pairs of strings S^i and S^j in $\mathcal{S}$, the longest suffix of S^i that is a prefix of S^j [7]. In other words, S^i overlaps S^j. The solution of the APSP can be stored in an "overlap" squared matrix Ov of size m^2, where $\text{Ov}[i, j]$ represents the length of the longest suffix of S^i that overlaps S^j. Ov can also be stored using a compact representation [2].

Practical Algorithms: The most demanding application for the APSP currently is overlap detection for DNA assembly, which has been solved much faster in practice by non-optimal algorithms. SGA [18] and Readjoiner [5] are genome assemblers that have a very fast and isolated overlap detection stage with very low space consumption, which find all suffix-prefix overlaps (not only the longest overlaps). Recently, Rachid and Malluhi [17] presented SOF, a practical algorithm to solve the APSP for DNA sequences that is competitive with the genome assemblers. SGA, Readjoiner and SOF may be executed in multithreading environments. Previous experiments [17] have shown that SOF has a better performance with multiple threads.

Optimal Algorithms: The APSP has been solved in optimal time by Gusfield et al. [8] in 1992 through generalized suffix trees [20] and stacks. In 2010, Ohlebusch and Gog [15] improved memory usage through enhanced suffix arrays [1,13] and stacks, reducing the practical running time. Recently,

Tustumi *et al.* [19] proposed a different traversal of the enhanced suffix array and replaced stacks by linked lists to achieve an even better practical running time. We stress that all these algorithms are theoretically optimal and are able to deal with strings from general alphabets.

3   Related Work

The algorithm by Tustumi *et al.* [19] solves the APSP in optimal $O(N + m^2)$ time, based on the following remarks.

All suffixes that are a prefix of S^k are either in positions prior to $\mathsf{pos}(S^k)$ or are identical to S^k and directly succeed $\mathsf{pos}(S^k)$ in the GESA [15, Lemma 3.1]. Given a prefix $S^t < S^k$, if a suffix of S^r, of length ℓ, is a prefix of S^t and $\ell > lcp(S^t, S^k)$, then such suffix of S^r is not a prefix of S^k [19, Lemma 2]. Furthermore, if two different suffixes of S^r are a prefix of S^k, the longest is closer to $\mathsf{pos}(S^k)$ in GESA.

The blocks are processed in order. For each block B^j, with $j = 1, 2, \ldots, m$, a local solution is found scanning B^j backwards and a global solution is obtained reusing the local solutions of the blocks processed previously.

For block B^j, GESA is scanned backwards, from $i = \mathsf{P}[j]$ to $\mathsf{P}[j-1] + 1$. Suppose that $\mathsf{pos}(S^k) = \mathsf{P}[j]$ and $\mathsf{pos}(S^t) = \mathsf{P}[j-1]$. The algorithm uses m local lists and m global lists to track all overlaps seen so far. The value of $\ell = rmq(i, \mathsf{P}[j])$ is computed in $O(1)$ time as the minimum lcp value between the entries processed during the scanning of B^j. If the length of the current suffix is equal to $lcp(S^{\mathsf{STR}[i]}_{\mathsf{SA}'[i]}, S^k) = rmq(i, \mathsf{P}[j])$, then it is a prefix of S^k and its length is inserted at the end of its local list $L_{local}[\mathsf{STR}[i]]$. At the end, the longest overlaps in B^j are at the front of each local list, and ℓ is equal to $lcp(S^t, S^k)$.

The longest suffix of S^r that overlaps S^k may be positioned in a previous block $B^{j-1}, B^{j-2}, \ldots, B^1$ processed so far. The global lists store these overlaps. Each global list $L_{global}[r]$ is composed by the elements inserted in $L_{local}[r]$ and has to be updated as each block is processed. To do this, at the end of each local solution the algorithm removes the suffixes that have length larger than $lcp(S^t, S^k)$ from all local lists. These suffixes no longer overlap S^k or any the following complete strings that will be processed by the algorithm. Each local list $L_{local}[r]$ is prepended in the global list $L_{global}[r]$. Finally, the first element of each $L_{global}[r]$ corresponds to the length of the longest suffix of S^r that overlaps S^k. This value is inserted in $\mathsf{Ov}[k, r]$. Note that, to improve the memory access reference, the algorithm stores in $\mathsf{Ov}[k, r]$ the length of the longest suffix of S^r that overlaps S^k. If $L_{global}[r]$ is empty, there is no overlap of S^r with S^k.

The suffixes identical to S^k that directly succeed $\mathsf{pos}(S^k)$ are found scanning GESA forward from $i = \mathsf{P}[j] + 1$ to q, while $\mathsf{LCP}[q] = n_k$ [15, Lemma 3.1]. The length of these suffixes are inserted in Ov, possibly overwriting the results in $L_{global}[r]$, which is correct as such overlaps are larger.

4   Parallel Algorithm

In this section we show how to split the computation of all overlaps performed by Tustumi *et al.*'s algorithm to solve the APSP in parallel in a shared-memory multithreading environment.

At a glance, our algorithm is composed by three phases. First, all local solutions are computed scanning the blocks B^j concurrently. Then, the local lists are accessed in parallel to obtain the global solutions. Finally, the identical suffixes of all strings are identified scanning the GESA in parallel.

Algorithm: Suppose that for each block B^j, $pos(S^k) = $ P$[j]$ and pos(S^t) = P$[j-1]$. Algorithm 1 works as follows.

In Phase 1, the m blocks of the GESA are processed in parallel (Lines 1 to 12) to compute the local solutions of each string S^k. The algorithm scans each block B^j backwards (Lines 4 to 10), and whenever a suffix of S^r (Line 5) that overlaps S^k is found (Line 7), its length is inserted at the end of a local list, which is pointed by an entry in a two dimensional array (matrix) of lists, namely $L_{local}[r][j]$ (Line 8). Note that the list in $L_{local}[r][j]$ is ordered decreasingly by the overlap lengths due to the backward scan. An array of integers Min of length m is used to store in Min$[j]$ the value of $rmq($P$[j-1],$P$[j])$ (Line 11). At the end of Phase 1 (Line 12), all local solutions have been computed and are stored into the local lists, which will be used together with Min in Phase 2 to obtain the global solutions.

In Phase 2, the m arrays of local lists, that is, the m lines of matrix L_{local} that correspond to each S^r, are processed in parallel (Lines 13 to 23) to find the suffixes of S^r that overlap the other strings (Line 20). Let S^k be such overlapped string (Line 16). A global list for S^r is used to keep track of all valid overlaps in $L_{local}[r][j]$, according to Min$[j]$, as blocks B^j are processed in the inner for-loop (Lines 15 to 22). L_{global} is initially empty (Line 14). For each B^j, with $j = 1, 2, \ldots, m$ (Lines 15 to 22), the algorithm updates L_{global}, first removing all suffixes larger than Min$[j]$ (Lines 17 to 19), since these overlaps are no longer valid for S^k [19, Lemma 2]. Recall that Min$[j] = rmq($P$[j-1],$P$[j]) = lcp(S^t, S^k)$. In the sequel, $L_{local}[r][j]$ is prepended to L_{global} (Line 20) and the longest suffix of S^r that overlaps S^k will be the first element of L_{global} that is inserted at Ov$[r,k]$ (Line 21).

In Phase 3, the suffixes identical to each strings S^k are found in parallel (Lines 24 to 33). For each block B^j and its corresponding string S^k (Line 25), all suffixes of S^r (Line 27) identical to S^k that appear directly after pos(S^k) in positions $i = $ P$[j]+1$ to q are found (Lines 28 to 32). As in Tustumi *et al.*'s algorithm, the length of these suffixes are inserted in Ov (Line 29).

Theoretical Costs: In Phase 1, the parallel for loop in Line 1 is executed m/t times, where t is the number of threads. Then, Phase 1 is $O(\max_{1 \leq j \leq m} |B^j|)$. In Phase 2 the parallel for loop in Line 13 is executed m/t times, whereas Lines 15 to 22 are executed m^2/t times. The parallel for loop of Phase 3 is executed m/t times and each execution reads at most N elements of GESA. Thus our

---

Algorithm 1. Parallel APSP (p-apsp)

Data: GESA of the collection $\mathcal{S} = \{S^1, S^2 \ldots, S^m\}$
Result: result matrix Ov
// Local solutions
1 **for** $j \leftarrow 1$ *to* m **do in parallel**
2 | $k \leftarrow$ STR[P[j]];
3 | $\ell \leftarrow \infty$;
4 | **for** $i \leftarrow$ P[j] *to* P[$j-1$] $+ 1$ **do**
5 | | $r \leftarrow$ STR[i]
6 | | $\ell \leftarrow min(\ell, \mathsf{LCP}[i+1])$;
7 | | **if** $|S^r_{\mathsf{SA}'[i]}| = \ell$ **then**
8 | | | $insert\_at\_end(L_{local}[r][j], \ell)$;   // S^r overlaps S^k
9 | | **end**
10 | **end**
11 | Min[k] $\leftarrow \ell$;
12 **end**
// Global solutions
13 **for** $r \leftarrow 1$ *to* m **do in parallel**
14 | $L_{global} \leftarrow null$;
15 | **for** $j \leftarrow 1$ *to* m **do**
16 | | $k \leftarrow$ STR[P[j]];
17 | | **while** $first(L_{global}) >$ Min[j] **do**
18 | | | $remove\_first(L_{global})$
19 | | **end**
20 | | $L_{global} \leftarrow insert\_at\_front(L_{local}[r][j])$;
21 | | Ov[r, k] $\leftarrow first(L_{global})$;   // S^r overlaps S^k
22 | **end**
23 **end**
// Identical suffixes
24 **for** $j \leftarrow 1$ *to* m **do in parallel**
25 | $k \leftarrow$ STR[P[j]];
26 | $i \leftarrow$ P[j] $+ 1$;
27 | $r \leftarrow$ STR[i];
28 | **while** $|S^r_{\mathsf{SA}'[i]}| = \mathsf{LCP}[i]$ *and* $i < N$ **do**
29 | | Ov[r, k] $\leftarrow \mathsf{LCP}[i]$;   // $S^r_{\mathsf{SA}'[i]}$ is identical to S^k
30 | | $i \leftarrow i + 1$;
31 | | $r \leftarrow$ STR[i];
32 | **end**
33 **end**

---

algorithm runs in $O(N + m^2/t)$ time, which only occurs in bad cases (when the string lengths are very unbalanced). In realistic cases ($m \gg t$ and all strings with about the same size) the parallel time is close to $N/t + m^2/t$.

Overall, all threads insert at most N suffixes into the local lists. Then, the space complexity is given by the $O(N)$ space of the GESA, and by the $O(m^2 + N)$ space of the matrix of local lists, where each list stores at most N overlaps.

Thus the space complexity of our parallel algorithm is $O(N + m^2)$, which is equal to the sequential algorithm by Tustumi *et al.*

5 Experiments

We implemented Algorithm 1 in C++ using OpenMP directives for the parallelization. We used the SDSL [4] version 2.0[1] to construct the GESA. Our source code is publicly available at https://github.com/felipelouza/p-apsp.

We have compared the performance of our parallel algorithm, called p-apsp, with the sequential optimal algorithm by Tustumi *et al.*[2], called apsp, and with the practical solution SOF [17][3], which has good time and space performance in multithreading environments. We used different number of threads $t = \{1, 2, 4, 8, 16, 32\}$ set by the directive omp_set_num_threads() for p-apsp and SOF. All programs were compiled with g++ (v. 4.9.2) with the same optimization flags.

The experiments were executed in a 64 bits Debian GNU/Linux 8 (kernel 3.16.0-4) system with an Intel Xeon Processor E5-2630 v3 20M Cache 2.40-GHz, with 384 GB of internal memory and a 13 TB SATA storage. We used the EST database from *C. elegans*[4]. The number of strings used in the experiments varied from 10.000 to 300.000. We used different minimum overlap length values $\tau = \{5, 10, 15, 20\}$ to limit the number of overlaps found by the algorithms.

We also tested the algorithms with 200.000 ESTs from *Citrus sinensis*[5] and obtained very close results, which are not shown.

Table 1 shows the number of overlaps found solving the APSP for the *C. elegans* dataset. Notice that as the value of τ increases the number of overlaps decreases. In particular, the number of overlaps for 300.000 strings when $\tau = 5$ is 23 times larger than when $\tau = 10$. We shall see that such variation impacts the performance of all algorithms, both in time and space.

Table 1. Number of overlaps found in the experiments with 100.000, 200.000 and 300.000 ESTs of the *C. elegans* varying τ.

τ	5	10	15	20
100.000	18,853,491	206,154	88,725	82,427
200.000	71,451,170	2,675,759	2,139,431	2,077,125
300.000	162,135,112	7,044,274	5,800,397	5,617,779

---

[1] *sdsl-lite* library is available at https://github.com/simongog/sdsl-lite.

[2] https://github.com/felipelouza/apsp.

[3] http://confluence.qu.edu.qa/download/attachments/9240580/Prefix.tgz.

[4] http://www.uni-ulm.de/in/theo/research/seqana.html.

[5] ftp://ftp.bioinfo.wsu.edu/www.citrusgenomedb.org/Citrus_sinensis/C.sinesis_unige ne_v1.0/.

5.1   Running Time

Figure 2 shows the running time of each algorithm not accounting for the time to build the auxiliary data structures, that is, the GESA for apsp and p-apsp, and the compact prefix tree for SOF. The elapsed time was taken by the directive omp_get_wtime().

SOF was the fastest algorithm in all experiments. However, p-apsp has shown a good performance when the minimum overlap length is small, a situation where the specialized strategy used by SOF is not so efficient. This result is coherent with the theoretical optimality of Tustumi *et al.*'s algorithm. Note that the parallel implementation has some overhead with $\tau = 5$ when comparing p-apsp with a single thread (p-apsp$_1$) to the sequential apsp. Note also that p-apsp and SOF improve their running time as the number of threads increases.

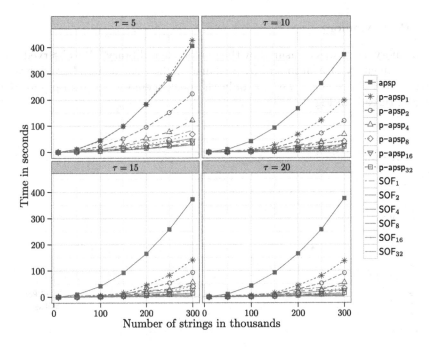

Fig. 2. Running time of p-apsp, apsp and SOF for varying values of τ.

Table 2 shows the running time and the speedup of each algorithm for 300.000 ESTs with $\tau = 5$. However, p-apsp achieved a much better speedup with the increasing number of threads, indicating that this parallel algorithm may be a practical solution for large instances of the APSP on strings coming from a general alphabet.

Table 2. Experiments with 300.000 ESTs of the *C. elegans* dataset with $\tau = 5$. The table shows the running time (in seconds) and the speedup of the parallel algorithms over its serial versions, when the numbers of threads is 1

n. threads	apsp	p-apsp		SOF	
	Time	Time	Speedup	Time	Speedup
1	397.17	463.33		**82.80**	
2		222.78	**1.91**	**52.49**	1.58
4		121.35	**3.51**	**29.50**	2.81
8		68.11	**6.26**	**28.30**	2.93
16		43.41	**9.82**	**28.64**	2.89
32		34.65	**12.31**	**23.62**	3.50

5.2   Peak Memory

The memory usage was measured by the malloc_count library[6]. We observed that the peak memory of p-apsp and SOF change slightly as the number of threads varies. For a clearer view we plot only the peak memory for p-apsp and SOF using 32 threads in Fig. 3.

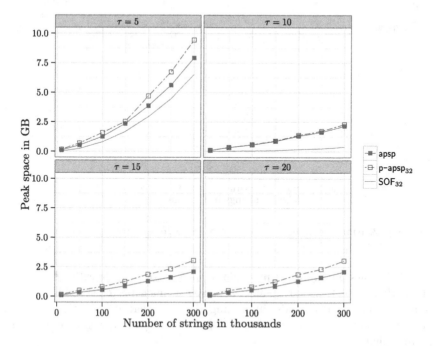

Fig. 3. Peak memory of p-apsp, apsp and SOF for varying values of τ

---

[6] *malloc_count* library is available at http://panthema.net/2013/malloc_count.

As expected, memory usage varies according to τ. SOF uses less memory in all experiments and p-apsp memory usage is very similar to the sequential version apsp. As demonstrated by the theoretical analysis the experiments confirm that practical memory usage differs only by a constant factor when comparing apsp and p-apsp.

6   Conclusion

We showed how to parallelize the optimal algorithm by Tustumi *et al.*[19] to solve the APSP. We separated the computation of the local solution followed by the global solution and by the identical suffixes searching into independent phases. We compared our parallel algorithms with SOF [17], a practical solution with the best parallel performance, as shown in previous work. Our experimental results showed that the parallel algorithm achieves a 12-fold speedup, and that it is competitive with practical algorithm, such as SOF, when the minimum overlap length is small and offers the ability to deal with larger alphabets.

As the algorithm by Tustumi *et al.*, our algorithm can be also improved to work in semi-external memory, since the GESA can be constructed in external memory [12] and its blocks can be accessed as necessary, reducing the peak memory. Our parallel implementation is general enough that it can be executed on a different architecture model, such as cloud distributed computing, possibly enabling the usage of hundreds of threads.

Acknowledgments. FAL acknowledges the financial support CAPES and CNPq (grant No. 162338/2015-5). GPT acknowledges the support of CNPq. The authors thank Prof. Nalvo Almeida for granting access to the machine used for the experiments.

References

1. Abouelhoda, M.I., Kurtz, S., Ohlebusch, E.: Replacing suffix trees with enhanced suffix arrays. J. Discrete Algorithms **2**(1), 53–86 (2004)
2. Dinh, H., Rajasekaran, S.: A memory-efficient data structure representing exact-match overlap graphs with application for next-generation DNA assembly. Bioinformatics **27**(14), 1901–1907 (2011)
3. El-Metwally, S., Hamza, T., Zakaria, M., Helmy, M.: Next-generation sequence assembly: four stages of data processing and computational challenges. PLoS Comput. Biol. **9**(12), e1003345 (2013)
4. Gog, S., Beller, T., Moffat, A., Petri, M.: From theory to practice: plug and play with succinct data structures. In: Gudmundsson, J., Katajainen, J. (eds.) SEA 2014. LNCS, vol. 8504, pp. 326–337. Springer, Heidelberg (2014)
5. Gonnella, G., Kurtz, S.: Readjoiner: a fast and memory efficient string graph-based sequence assembler. BMC Bioinform. **13**(1), 82 (2012)
6. Gonnet, G.H., Baeza-Yates, R.A., Snider, T.: New indices for text: pat trees and pat arrays. In: Information Retrieval, pp. 66–82. Prentice-Hall Inc, Upper Saddle River (1992)

7. Gusfield, D.: Algorithms on Strings, Trees, and Sequences: Computer Science and Computational Biology. Cambridge University Press, New York (1997)

8. Gusfield, D., Landau, G.M., Schieber, B.: An efficient algorithm for the all pairs suffix-prefix problem. Inf. Process. Lett. **41**(4), 181–185 (1992)

9. Kalyanaraman, A., Aluru, S.: Handbook of computational molecular biology, chap. In: Expressed Sequence Tags: Clustering and applications. CRC Press, Boca Raton (2005)

10. Kasai, T., Lee, G.H., Arimura, H., Arikawa, S., Park, K.: Linear-time longest-common-prefix computation in suffix arrays and its applications. In: Amir, A., Landau, G.M. (eds.) CPM 2001. LNCS, vol. 2089, pp. 181–192. Springer, Heidelberg (2001)

11. Louza, F.A., Gog, S., Telles, G.P.: Induced suffix sorting for string collections. In: Proceeding DCC, pp. 43–52. IEEE, Snowbird (2016)

12. Louza, F.A., Telles, G.P., Ciferri, C.D.D.A.: External memory generalized suffix and LCP arrays construction. In: Fischer, J., Sanders, P. (eds.) CPM 2013. LNCS, vol. 7922, pp. 201–210. Springer, Heidelberg (2013)

13. Manber, U., Myers, E.W.: Suffix arrays: a new method for on-line string searches. SIAM J. Comput. **22**(5), 935–948 (1993)

14. Ohlebusch, E.: Bioinformatics Algorithms: Sequence Analysis, Genome Rearrangements, and Phylogenetic Reconstruction. Verlag, Oldenbusch (2013)

15. Ohlebusch, E., Gog, S.: Efficient algorithms for the all-pairs suffix-prefix problem and the all-pairs substring-prefix problem. Inf. Process. Lett. **110**(3), 123–128 (2010)

16. Puglisi, S.J., Smyth, W.F., Turpin, A.H.: A taxonomy of suffix array construction algorithms. ACM Comp. Surv. **39**(2), 1–31 (2007)

17. Rachid, M.H., Malluhi, Q.: A practical and scalable tool to find overlaps between sequences. BioMed Res. Int. **2015**, 1–12 (2015)

18. Simpson, J.T., Durbin, R.: Efficient construction of an assembly string graph using the FM-index. Bioinformatics **26**(12), i367–i373 (2010)

19. Tustumi, W.H., Gog, S., Telles, G.P., Louza, F.A.: An improved algorithm for the all-pairs suffix-prefix problem. J. Discrete Algorithms **47**, 34–43 (2016)

20. Weiner, P.: Linear pattern matching algorithms. In: Proceeding Annual Symposium on Switching and Automata Theory, pp. 1–11. IEEE Computer Society, Washington, DC (1973)

Dynamic and Approximate Pattern Matching in 2D

Raphaël Clifford[1]($\boxtimes$), Allyx Fontaine[1], Tatiana Starikovskaya[1],
and Hjalte Wedel Vildhøj[2]

[1] Department of Computer Science, University of Bristol, Bristol, UK
Raphael.Clifford@bristol.ac.uk
[2] Technical University of Denmark, DTU Compute, Kongens Lyngby, Denmark

Abstract. We consider dynamic and online variants of 2D pattern matching between an $m \times m$ pattern and an $n \times n$ text. All the algorithms we give are randomised and give correct outputs with at least constant probability.

- For dynamic 2D exact matching where updates change individual symbols in the text, we show updates can be performed in $\mathcal{O}(\log^2 n)$ time and queries in $\mathcal{O}(\log^2 m)$ time.
- We then consider a model where an update is a new 2D pattern and a query is a location in the text. For this setting we show that Hamming distance queries can be answered in $\mathcal{O}(\log m + H)$ time, where H is the relevant Hamming distance.
- Extending this work to allow approximation, we give an efficient algorithm which returns a $(1+\varepsilon)$ approximation of the Hamming distance at a given location in $\mathcal{O}(\varepsilon^{-2} \log^2 m \log \log n)$ time.

Finally, we consider a different setting inspired by previous work on locality sensitive hashing (LSH). Given a threshold k and after building the 2D text index and receiving a 2D query pattern, we must output a location where the Hamming distance is at most $(1 + \varepsilon)k$ as long as there exists a location where the Hamming distance is at most k.

- For our LSH inspired 2D indexing problem, the text can be preprocessed in $\mathcal{O}(n^{2(4/3+1/(1+\varepsilon))} \log^3 n)$ time into a data structure of size $\mathcal{O}(n^{2(1+1/(1+\varepsilon))})$ with query time $\mathcal{O}(n^{2(1/(1+\varepsilon))}m^2)$.

1 Introduction

Two dimensional pattern matching has been a topic of study and great interest for many years. The original motivation comes from image processing and recognition where one is attempting to find possibly approximate occurrences of a 2D-pattern inside a larger 2D-text. For exact matching offline, linear time solutions are known [11,12,15] and the indexing problem is solved efficiently with the help of 2D-suffix trees [16]. A number of other variants have also been studied including 2D-compressed pattern matching, matching with rotations, pattern matching with non-rectangular patterns as well as others [2–7,9,14].

© Springer International Publishing AG 2016
S. Inenaga et al. (Eds.): SPIRE 2016, LNCS 9954, pp. 133–144, 2016.
DOI: 10.1007/978-3-319-46049-9_13

We will consider a number of variants of 2D-pattern matching which have to date received little attention. These can broadly be described under the headings of online and dynamic pattern matching. Our focus will be both on exact matching as well as exact and approximate Hamming distance computation. We will also tackle a problem formulation inspired by the locality sensitive hashing work of Andoni and Indyk [10]. Here we are given a pattern as a query and we must report a location in the text where the Hamming distance is not too large as long as one exists. We will now formalise the problems we tackle. All the algorithms we develop will be randomised giving correct answers with at least constant probability. For each problem our input text will be a square matrix $\mathbf{T}$ (the text) of size $n \times n$ and the pattern $\mathbf{P}$ will be of size $m \times m$.

To start we consider a dynamic version of the classic 2D-pattern matching problem. The problem can be seen as a generalisation of the 1D problem considered in [8], where updates are only allowed in the text and the pattern remains static. Our solution relies heavily on Karp-Rabin fingerprinting [18]. The main technical hurdle we overcome is the difficulty in combining fingerprints of adjacent rectangular matrices. We circumvent this problem by only ever combining the fingerprints of two matrices if they are placed horizontally next to each other.

Problem 1 (Dynamic Text Static Pattern Matching in 2D). Given a text $\mathbf{T}$ and a pattern $\mathbf{P}$, build a dynamic index that supports an update $(\sigma, (i,j))$ which sets $T[i,j] \leftarrow \sigma$ and query (i,j) which returns True if there is an exact match at location (i,j) in the text and False otherwise.

Our solution to Problem 1 will in fact support the arrival of entire new patterns efficiently as well. For our next two problems we consider online pattern matching problems where the only update is the arrival of a new pattern and a query will return the exact or approximate Hamming distance at some position in the text. Our aim is to perform all three steps, preprocessing, updates and queries as quickly as possible. We denote by $Ham(\mathbf{P}, \mathbf{T})(i,j)$ the Hamming distance between the 2D-pattern $\mathbf{P}$ and the $m \times m$ submatrix of $\mathbf{T}$ with top left corner (i,j).

Problem 2 (Online Exact Hamming Distance in 2D). Given a text $\mathbf{T}$, build a dynamic index that supports updates with a pattern $\mathbf{P}$ and queries which return the value $Ham(\mathbf{P}, \mathbf{T})(i,j)$.

Our solution uses as a preliminary step linearisation of the input by encoding carefully selected substrings of the 2D-text with their Karp-Rabin fingerprints. This will allow us to search efficiently first for mismatches within columns and then rows using dynamic lower common ancestor queries in suitably constructed suffix trees.

To provide faster solutions we then extend this online Hamming distance problem to allow a $(1+\varepsilon)$ approximation. We show that we can find the approximate value considerably faster than the exact value. To achieve this we use the technique known as sketching [1]. This technique was originally developed for

1D strings but can be transferred to our case by storing sketches of selected substrings of the text **T**.

Problem 3 (Online Approximate Hamming Distance in 2D). Given a binary text **T**, construct a dynamic index that supports updates with a binary pattern **P** and queries which return a $(1 + \varepsilon)$ approximation of the Hamming distance $Ham(\mathbf{P}, \mathbf{T})(i, j)$.

Finally we turn to a closely related indexing problem. Here we may preprocess the 2D-text and we receive a 2D-pattern as a query along with a threshold k and a constant ε. We must output a location in the text where the Hamming distance is no more than $(1 + \varepsilon)k$ as long as there exists a location where the Hamming distance is no more than k.

Problem 4 (Submatrix Near Neighbour Problem). We are initially given a text **T**, an integer k and a constant $\varepsilon > 0$. Construct an index that supports the following query. Given a pattern **P**, output a position (i, j) such that $Ham(\mathbf{T}, \mathbf{P})(i, j) \leq (1 + \varepsilon) \cdot k$ if there exists a submatrix of **T** with Hamming distance at most k from **P**. Otherwise if there is not, the query may either report a location with true Hamming distance up to $(1 + \varepsilon)k$ or no location at all.

In the 1D case Andoni and Indyk [10] solved the same problem we study by developing an index on suffixes of a 1D string. To construct their index Andoni and Indyk [10] heavily relied on relationships between suffixes of a 1D string. These relationships do not exist in the 2D case and so we have introduced new techniques and ideas to construct the index. These are our main contribution for Problem 4.

Definitions and Notation. We will use two kinds of partitioning of the text and pattern which we term *belts* and *canonical submatrices*. Let S be an $s \times t$ matrix. A *belt* of height $h \leq s$ for the matrix S is a submatrix of S with size $h \times t$. A *canonical submatrix* of S is a submatrix of S with size $2^i \times 2^j$ where $i \leq \log s$ and $j \leq \log t$ are both integers. We will also write $\mathbf{T}[i, i + x - 1; j, j + y - 1]$ to denote the $x \times y$ submatrix of **T** with top left corner at some position (i, j) in the text. We assume throughout that all logarithms are taken base two and for convenience of presentation that both m and n are an exact power of two.

2   Dynamic Text Static Pattern Matching in 2D

As our first contribution we describe a dynamic randomised index that supports efficient exact pattern matching queries as well as updates to **T** and hence solves Problem 1.

Theorem 1. *The text* **T** *can be preprocessed in* $\mathcal{O}(n^2 \log n)$ *time into a data structure of size* $\mathcal{O}(n^2)$ *so that after processing the pattern* **P** *in* $\mathcal{O}(m^2 \log m)$ *time, we can support single character updates in* $\mathcal{O}(\log^2 n)$ *time and query if* **P** *occurs at a position* (i, j) *of* **T** *w.h.p. in* $\mathcal{O}(\log^2 m)$ *time.*

The main idea of our dynamic index is to compute the Karp-Rabin finger-prints of submatrices of $\mathbf{T}$ of power of two size in order to be able to compute the fingerprint of the $m \times m$ submatrix with the top left corner at the position (i, j) of $\mathbf{T}$ efficiently. A straightforward partitioning will not suffice however due to the difficulty in computing fingerprints of the concatenation of rectangular matrices.

We start by giving the definition of Karp-Rabin fingerprints for matrices.

Definition 1. *Let S be an $s \times t$ matrix for some $s, t \leq n$. Let $p \geq n^4$ be a prime and r be a random integer in $\mathbb{F}_p$. We define the Karp-Rabin fingerprint φ for S as:*

$$\varphi(S) = \sum_{i=1}^{s} \sum_{j=1}^{t} S[i, j] r^{i+(j-1)s} \pmod{p}$$

Lemma 1. *The Karp-Rabin fingerprints of any two $s \times t$ matrices S, S', where $s, t \leq n$, have the following properties:*

1. *If $S = S'$, then $\varphi(S) = \varphi(S')$;*
2. *If $S \neq S'$, then the probability $\varphi(S) = \varphi(S')$ is at most $1/n^2$.*

Proof. The first claim of the lemma is trivial. To prove the second claim notice that since $\varphi(S) - \varphi(S')$ is a non-trivial polynomial of degree $s \cdot t$, the number of its roots $\in \mathbb{F}_p$ is at most $s \cdot t$. The probability we choose a root randomly from $\mathbb{F}_p$ is at most $\mathcal{O}(s \cdot t/n^4)$. The result holds since $s \cdot t \leq n^2$. $\qquad\square$

Moreover, from the definition of Karp-Rabin fingerprints we immediately obtain the following observation. We say that two submatrices are adjacent on the vertical side if they are placed horizontally next to each other. That is $S = \mathbf{T}[i : i + s - 1, j : j + t - 1]$ and $S' = \mathbf{T}[i : i + s - 1, j + t : j + t + t' - 1]$

Lemma 2. *Let S, S' be two submatrices of $\mathbf{T}$ adjacent on the vertical side. We can compute the Karp-Rabin fingerprint of $S'' = \mathbf{T}[i : i + s - 1, j : j + t + t' - 1]$ as*

$$\varphi(S'') = \varphi(S) + r^{st} \cdot \varphi(S') \pmod{p}$$

Proof. The proof follows immediately from the definition. $\qquad\square$

We now present our dynamic index. For each $i = 0, 1, \ldots, \log n$ we divide $\mathbf{T}$ into $n/2^i$ non-overlapping belts of height 2^i. For each $j = 0, 1, \ldots, \log n$ we then partition each belt into $n/2^j$ canonical submatrices of width 2^j. For each of the canonical submatrices we store its Karp-Rabin fingerprint in a lookup table. It follows from the fingerprint definition that an individual fingerprint can be updated in constant time if a letter at a particular position in the text is changed. When we change one letter in $\mathbf{T}$, we need to update only $\mathcal{O}(\log^2 n)$ fingerprints, which can be therefore be done in $\mathcal{O}(\log^2 n)$ time in total. The partitioning into belts and canonical submatrices is illustrated in Fig. 1.

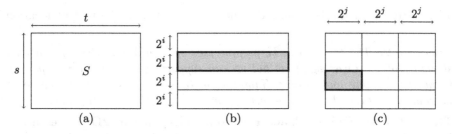

Fig. 1. (a) A matrix S of size $s \times t$. (b) Partition of S into non-overlapping belts of height 2^i. In gray is represented one such belt. (c) Partition of S into canonical submatrices of height 2^i and width 2^j. In gray is represented one such canonical submatrix

When a pattern arrives, we process it in the following way. For each $i = 0, 1, \ldots, \log m$ we compute and store the Karp-Rabin fingerprints of all $m - 2^i + 1$ belts of height 2^i. For a fixed value of i we compute the 2D-fingerprints of all $m - 2^i + 1$ belts of height 2^i in two steps. The first step is computing the 2D-fingerprints of all submatrices of size $2^i \times 1$, which we do column by column. For each column j, we first compute the fingerprint φ_1 of the string $\mathbf{P}[1 : 2^i, j]$, and then for each $\ell \geq 1$ we compute the fingerprint $\varphi_{\ell+1}$ of $\mathbf{P}[\ell + 1 : \ell + 2^i, j]$ from the fingerprint $\varphi_{\ell-1}$ of $\mathbf{P}[\ell : \ell + 2^i - 1, j]$ in constant time. As there are m columns of length m each, this step requires $\mathcal{O}(m^2)$ time. The second step consists in computing for each belt its 2D-fingerprint from its columns' fingerprints as described in Lemma 2 in time $\mathcal{O}(m)$.

Suppose now that we are asked if $\mathbf{T}[i, i+m-1; j, j+m-1]$ matches pattern $\mathbf{P}$. We can divide $\mathbf{T}[i : i+m-1, j : j+m-1]$ into $\mathcal{O}(\log m)$ non-overlapping belts with heights that are powers of two. Each belt can then be divided vertically into $\mathcal{O}(\log m)$ canonical submatrices for which we already know their Karp-Rabin fingerprints. With the help of Lemma 2 we compute Karp-Rabin fingerprints of the belts in $\mathcal{O}(\log^2 m)$ time and compare them to those of the pattern.

Construction of the Index. We now explain how we construct the text index. We iteratively compute Karp-Rabin fingerprints of canonical matrices of height 2^i, $i = 0, 1, \ldots, \log n$. When the height is fixed, we iteratively compute Karp-Rabin fingerprints of canonical matrices of width 2^j, for $j = 0, 1, \ldots, \log n$.

For each i we start by computing Karp-Rabin fingerprints of all $2^i \times 1$ submatrices in $\mathcal{O}(n^2)$ time in a straightforward manner. When Karp-Rabin fingerprints of $2^i \times 2^j$ submatrices are computed, we can compute Karp-Rabin fingerprints of $2^i \times 2^{j+1}$ submatrices in $\mathcal{O}(n^2/2^j)$ time using Lemma 2. In total, to compute the fingerprints of all submatrices of height 2^i, we need $\mathcal{O}(n^2)$ time. In total, we will need $\mathcal{O}(n^2 \log n)$ time for all submatrices.

3 Online Exact 2D Hamming Distance

In this section we consider Problem 2. We are given an $n \times n$ text $\mathbf{T}$ which we process first. Updates come in the form of new $m \times m$ patterns $\mathbf{P}$ and a query asks us to return the Hamming distance between $\mathbf{P}$ and the text at location (i, j).

Theorem 2. *The text $\mathbf{T}$ can be preprocessed in $\mathcal{O}(n^2 \log n)$ time into a data structure of size $\mathcal{O}(n^2 \log n)$ so that we can support updates with a new pattern $\mathbf{P}$ in $\mathcal{O}(m^2)$ time and process Hamming distance queries to return up to H mismatches between $\mathbf{P}$ and $\mathbf{T}$ at a position (i, j) in $\mathcal{O}(\log m + H)$ time.*

The Index for Online Exact Hamming Distance in 2D. For each $i = 1, 2, \ldots, \log n$ we consider $n - 2^i + 1$ belts of height 2^i. We define a linearisation of a belt as a string of length n, where the j-th supercharacter is the Karp-Rabin fingerprint of the j-th column of the belt.

Lemma 3. *The linearisations for all belts of height 2^i for a fixed i can be computed in $\mathcal{O}(n^2)$ time.*

Proof. It suffices to note that for a fixed j the Karp-Rabin fingerprints of j-th columns of all $n - 2^i$ belts can be computed in $\mathcal{O}(n)$ time [18]. $\qquad\square$

The main idea will be to first find columns within the pattern that mismatch and then to look within those columns to find individual mismatches. In order to do this efficiently, we compute all linearisations for all belts in $\mathcal{O}(n^2 \log n)$ time and then build a suffix tree for them. We also augment the suffix tree with an efficient dynamic lower common ancestor (LCA) data structure [13]. The suffix tree and the data structure can be built in $\mathcal{O}(n^2 \log n)$ time. We then build a suffix tree for all columns of $\mathbf{T}$ and augment it with the dynamic LCA data structure as well.

When the pattern arrives, we partition it into $\mathcal{O}(\log m)$ non-overlapping belts of power of two heights. We linearise the belts in the way described above and add the linearisations to the generalised suffix tree for the text belts. We also add columns of the pattern to the generalised suffix tree for the columns. This takes time $\mathcal{O}(m^2)$, see [17]. Finally, we update the LCA data structures. In total, this takes $\mathcal{O}(m^2)$ time.

We then work with each of the pattern belts independently. We will use the technique known as *kangaroo jumping* [17, Chap. 9.4]. To find the first H mismatches between the pattern belt of height 2^i and the text, we find the leaf in the suffix tree for the text belt of height 2^i containing the pattern belt and the leaf for the pattern belt and use an LCA query to find the first column of the pattern belt that does not match the corresponding column of the text belt. We then use the generalised suffix tree for the columns and kangaroo jump using LCA queries to report all mismatches in the column in constant time per mismatch. We then go back to the suffix tree for the belts and proceed. When a new pattern update arrives we need first to delete the previous pattern which was added to the two trees.

4   Online Approximate Hamming Distance in 2D

In this section we consider Problem 3. Assume that we are given an $n \times n$ matrix $\mathbf{T}$ and a constant $\varepsilon > 0$. We assume that we are also given an $m \times m$

pattern matrix $\mathbf{P}$ and that we can process it before answering queries. We will give a text index for $\mathbf{T}$ that will support the following queries: Given a position (i, j) return a $(1 + \varepsilon)$-approximation of the Hamming distance between $\mathbf{P}$ and $\mathbf{T}[i : i + m - 1, j : j + m - 1]$.

Theorem 3. *The text $\mathbf{T}$ can be preprocessed in $\mathcal{O}(\varepsilon^{-2}n^2 \log^3 n \log \log n)$ time into a data structure of size $\mathcal{O}(\varepsilon^{-2}n^2 \log^2 n \log \log n)$. After processing a new pattern $\mathbf{P}$ in $\mathcal{O}(\varepsilon^{-2}m^2 \log \log n)$ time, we can compute a $(1 + \varepsilon)$-approximation of the Hamming distance for any position (i, j) in $\mathbf{T}$ in $\mathcal{O}(\varepsilon^{-2} \log^2 m \log \log n)$ time. The answer is correct with constant probability.*

The Index for Online Approximate Hamming Distance in 2D. Consider all $\mathcal{O}(n^2 \log^2 n)$ canonical submatrices of $\mathbf{T}$ of sizes $2^i \times 2^j$ for $i = 1, 2, \ldots, \log n$ and $j = 1, 2, \ldots, \log n$. Let C be a constant to be defined later. For each canonical submatrix we create and store $\gamma = C \log \log n$ vectors (sketches) of length $1/\varepsilon^2$ as follows.

For each pair (i, j) and for each $k = 1, 2, \ldots, \gamma$ we create and store $1/\varepsilon^2$ sign matrices $S_\ell^{i,j,k}$ of size $2^i \times 2^j$. Each entry of a sign matrix is an i.u.d. ± 1 random variable. We now define the k-th sketch of a $2^i \times 2^j$ matrix M as:

$$(\langle M, S_1^{i,j,k} \rangle, \langle M, S_2^{i,j,k} \rangle, \ldots, \langle M, S_{1/\varepsilon^2}^{i,j,k} \rangle)$$

where $\langle M, S_\ell^{i,j,k} \rangle = tr(M^T, S_\ell^{i,j,k})$ is also known as the Hilbert-Schmidt inner product of matrices M and $S_\ell^{i,j,k}$. This sketching technique is a simple variant of the second moment sketches of Alon et al. [1].

Suppose we have two $2^i \times 2^j$ matrices A and B. For each k we approximate the Hamming distance between A and B using the sketches obtained with the help of the sign matrices $S_1^{i,j,k}, S_2^{i,j,k}, \ldots, S_{1/\varepsilon^2}^{i,j,k}$. In particular, the Hamming distance approximation we derive from the k-th sketches is $h_k = \varepsilon^2 \| \langle S_1^{i,j,k}, (A - B) \rangle, \ldots, \langle S_{1/\varepsilon^2}^{i,j,k}, (A - B) \rangle \|_2^2$. It follows from standard techniques that:

Lemma 4. *We can choose a constant C so that the median of the Hamming distance approximations over all $\gamma = C \log \log n$ sketches for the matrices A and B will belong to the interval $[H, (1 + \varepsilon)H]$ with probability at least $1 - \frac{1}{2 \log^2 n}$, where H is the Hamming distance between A and B.*

We process the queries in the following way. For each arriving pattern, we partition $\mathbf{P}$ into $\mathcal{O}(\log^2 m)$ non-overlapping submatrices of sizes $2^i \times 2^j$. Next, we compute sketches of all submatrices in the partition with the help of sign matrices, which takes $\mathcal{O}(\varepsilon^{-2}m^2 \log \log n)$ time, but we only need to do this once. When a query arrives, that is when we receive a position (i, j), we consider the same partitioning of $\mathbf{T}[i : i + m - 1, j : j + m - 1]$. For each corresponding pair of submatrices in the partitioning of $\mathbf{P}$ and $\mathbf{T}[i : i + m - 1, j : j + m - 1]$ we compute the $(1 + \varepsilon)$-approximations of Hamming distances with the help of the sketches. By Lemma 4 and the union bound, the sum of these values will be a $(1 + \varepsilon)$-approximation between $\mathbf{P}$ and $\mathbf{T}[i : i + m - 1, j : j + m - 1]$ with constant probability. Processing a query takes $\mathcal{O}(\varepsilon^{-2} \log^2 m \log \log n)$ time.

4.1   Construction of the Index

We finally explain how to compute the sketches of the canonical matrices. To compute the sketches for one canonical matrix of size $2^i \times 2^j$ we need only perform a sequence of 2D convolutions. In total, computing the sketches of all canonical submatrices of size $2^i \times 2^j$ takes $\mathcal{O}(\varepsilon^{-2} n^2 \log n \log \log n)$ time. Therefore, computing all sketches of all canonical submatrices over all sizes takes $\mathcal{O}(\varepsilon^{-2} n^2 \log^3 n \log \log n)$ time.

5   Submatrix Near Neighbour Problem

In this section we consider Problem 4. Assume that we are given an $n \times n$ matrix $\mathbf{T}$, an integer k, and a constant $\varepsilon > 0$. We will give a text index for $\mathbf{T}$ which will support the following queries: Given an $m \times m$ pattern matrix $\mathbf{P}$ such that there is a k-mismatch occurrence of $\mathbf{P}$ in $\mathbf{T}$, return an occurrence where the Hamming distance is at most $(1 + \varepsilon) \cdot k$. Let $N = n^2$ and $M = m^2$. We will show that

Theorem 4. $\mathbf{T}$ *can be preprocessed in* $\mathcal{O}(N^{4/3+1/(1+\varepsilon)} \log^3 N)$ *time into a data structure of size* $\mathcal{O}(N^{1+1/(1+\varepsilon)})$ *with query time* $\mathcal{O}(N^{1/(1+\varepsilon)} M)$. *If* $\mathbf{T}$ *contains a k-mismatch occurrence of* $\mathbf{P}$, *then the data structure w.h.p. retrieves a $(1+\varepsilon) \cdot k$-mismatch occurrence of* $\mathbf{P}$ *in* $\mathbf{T}$.

The Index for Submatrix Nearest Neighbour Search. We will start by recalling the notion of the L-encoding of a matrix.

Definition 2 ([16]). *The L-encoding of an $n \times n$ matrix* $\mathbf{T}$ *is a string* $s_1 s_2 \ldots s_n$ *of length* n^2, *where* $s_i = T[i : i, 1 : i - 1]T[1 : i, i : i]$. *(See Fig. 2)*

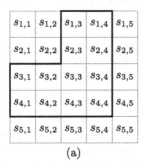   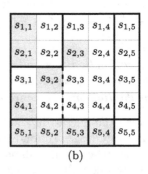

(a)   (b)

Fig. 2. A submatrix S of the text matrix $\mathbf{T}$. (a) The L-encoding of the submatrix S is $s_{1,1} s_{2,1} s_{1,2} s_{2,2} s_{3,1} s_{3,2} s_{1,3} s_{2,3} s_{3,3} \ldots s_{5,5}$. $L_{3:4}$ is the L-shape formed by the 3-rd and the 4-th rows and the 3-rd and the 4-th columns (shown in bold). (b) Let g be a projection onto a set of $\ell = 9$ positions $\{1, 2, 6, 8, 10, 17, 18, 19, 20\}$ (highlighted in gray), i.e. $g(S) = s_{1,1} s_{2,1} s_{3,2} s_{2,3} s_{4,1} s_{5,1} s_{5,2} s_{5,3} s_{5,4}$. The blocks will be $\{1, 2\}$, $\{6, 8, 10\}$, $\{17, 18, 19\}$, $\{20\}$. The corresponding partitioning of S into L-shapes and rectangles is shown on the figure by bold lines

Note that if **P** occurs in the top left corner of **T** with k mismatches, then the L-encoding of **T** starts with a k-mismatch occurrence of the L-encoding of **P**. A suffix of **T** is the L-encoding of a square submatrix with bottom right hand corner in the last row or in the last column of **T**. Let $S_1, S_2, \ldots, S_N$ be the suffixes of **T**. A k-mismatch occurrence of **P** in **T** guarantees that at least one of the L-encodings $S_1, S_2, \ldots, S_N$ starts with a k-mismatch occurrence of the L-encoding of **P**, and vice versa. We will make use of data structure by Andoni and Indyk which we call *sketch forest*. The following corollary follows directly from the work of [10, Sect. 2].

Corollary 1. *A sketch forest on a set of strings* $S = \{S_1, S_2, \ldots, S_N\}$ *occupies* $\mathcal{O}(N^{1+1/(1+\varepsilon)})$ *space. If at least one of the strings starts with a k-mismatch of the L-encoding of* **P***, then the data structure will identify in* $\mathcal{O}(N^{1/(1+\varepsilon)}M)$ *time a subset of* $\mathcal{O}(N^{1/(1+\varepsilon)})$ *suffixes of* **T** *that w.h.p. contains at least one suffix starting with a* $(1 + \varepsilon) \cdot k$*-mismatch occurrence of the L-encoding of* **P***.*

After having identified the subset of $\mathcal{O}(N^{1/(1+\varepsilon)})$ suffixes of **T**, we check for each of them if it starts with a $(1+\varepsilon) \cdot k$-mismatch occurrence of the L-encoding of **P** in a straightforward manner, comparing the letters of the suffix and the L-encoding of **P** one by one. In total, this takes $\mathcal{O}(N^{1/(1+\varepsilon)}M)$ more time.

The work of Andoni and Indyk heavily relied for its efficiency on the fact that different suffixes of a single string are suffixes of each other. However, in our linearisation of the text **T** this is no longer true. This requires us to devise a new method to construct the sketch forest efficiently which we now describe.

5.1 Construction

In this section we explain how we build the sketch forest. We start by describing its main elements.

Let $p_1 = 1 - k/N$, and $p_2 = 1 - (1 + \varepsilon) \cdot k/N$. The intuition behind these values is as follows: If S_1, S_2 are two strings of length N, then p_1 is a lower bound for the probability of two letters $S_1[i], S_2[i]$ to be equal if the Hamming distance between S_1 and S_2 is at most k. On the other hand, p_2 is an upper bound for the probability of two letters $S_1[i], S_2[i]$ to be equal if the Hamming distance between S_1 and S_2 is at least $(1 + \varepsilon) \cdot k$.

Let $\mathcal{H}$ be a set of projections of a string along a fixed coordinate, i.e. the j-th projection maps a string onto its j-th letter. A sketch forest is defined by a family of $N^\rho = \mathcal{O}(N^{1/(1+\varepsilon)})$ random functions $g_i \in \mathcal{H}^\ell$, where $\rho = \frac{\log p_1}{\log p_2}$ and $\ell = \frac{\log N}{\log 1/p_2}$. The choice of ρ and ℓ guarantees low error probability and space complexity. Each of the functions g_i can be considered as a projection along a randomly chosen set of coordinates of size $\ell \leq N$. The sketch forest contains exactly one trie for each projection function in the family. A trie $\mathcal{T}_{g_i}$ contains sketches $g_i(S_1), g_i(S_2), \ldots, g_i(S_N)$ of all strings in the set.

Fix a projection function $g \in \{g_1, g_2, \ldots, g_{N^\rho}\}$. We will show that the trie $\mathcal{T}_g$ can be built in $\mathcal{O}(N^{4/3} \log^2 N)$ time. As an immediate corollary, all tries in the sketch forest can be built in $\mathcal{O}(N^{4/3+1/(1+\varepsilon)} \log^2 N)$ time.

We start building the trie $\mathcal{T}_g$ by sorting the strings $g(S_1), g(S_2), \ldots, g(S_N)$ lexicographically and computing the longest common prefixes of all adjacent strings in that order. Below we show that this can be done in $\mathcal{O}(N^{4/3} \log^2 N)$ time. After having sorted the strings we build $\mathcal{T}_g$ in $\mathcal{O}(N)$ time by using this longest common prefix information.

We now explain how we sort $g(S_1), g(S_2), \ldots, g(S_N)$. Our algorithm will follow the lines of that of [10], but because $S_1, S_2, \ldots, S_N$ are suffixes of a 2D string and not a 1D string as in [10], we will have to introduce some new techniques.

String Sorting in $\mathcal{O}(N^{4/3} log^2 N)$ Time. We will give two methods for sorting strings $g(S_1), g(S_2), \ldots, g(S_N)$. *Sort A* will run in $\mathcal{O}(N\sqrt{\ell} \log^2 N)$ time and *Sort B* will run in $\mathcal{O}(N \log^2 N/\ell)$ time. We will use Sort A if $\ell \le N^{2/3}$ and Sort B if $\ell > N^{2/3}$.

Both Sort A and Sort B need to make at most $N \log N$ string comparisons. Note that in fact all we need to compare two strings is to find the first mismatch between them. For Sort A, we will show that after $\mathcal{O}(N\sqrt{\ell} \log^2 N)$-time preprocessing it is possible to find the first mismatch between any two strings in $\mathcal{O}(\sqrt{\ell})$ time. As a result, the total running time of sort A is $\mathcal{O}(N\sqrt{\ell} \log^2 N)$. For Sort B, we will show that the first mismatch between any two strings can be found in $\mathcal{O}(N \log N/\ell)$ time, which will give $\mathcal{O}(N^2 \log^2 N/\ell)$ time in total.

Sort A. Let g be a projection function onto positions $p_1 < p_2 < \ldots < p_\ell$. We will divide this set into $\mathcal{O}(\sqrt{\ell})$ blocks of consecutive positions of length at most $\sqrt{\ell}$ each. The method will consist of two steps. We will start by finding the first block containing a mismatch. After having found the block, we will iterate over all positions in it to find the desired mismatch. The second step can be implemented in a straightforward manner and requires $\mathcal{O}(\sqrt{\ell})$ time.

We will now explain how we implement the first step. Let us start by explaining how we divide the sequence $p_1 < p_2 < \ldots < p_\ell$ into blocks. Remember that these are positions in the L-encoding of an $n \times n$ matrix. Let $L_{i:j}$ be the L-shape formed by the i-th to j-th rows and the i-th to j-th columns (see Fig. 2 for an example).

We start by greedily dividing the matrix into L-shapes, where each L-shape either contains at most $\sqrt{\ell}$ sampled positions (type I L-shapes) or is of form $L_{i:i}$ (type II L-shapes). We first find the largest i_1 such that $L_{1:i_1}$ contains at most $\sqrt{\ell}$ sampled positions. We then try to find the largest i_2 such that $L_{i_1+1:i_2}$ contains at most $\sqrt{\ell}$ sampled positions. If such i_2 does not exist, we let $i_2 = i_1 + 1$, and continue in the same fashion. We further divide each type-II L-shape into the smallest number of horizontal and vertical rectangles containing at most $\sqrt{\ell}$ sampled positions each. The corner element forms a separate 1×1 rectangle.

This partitioning of the matrix into L-shapes and rectangles defines a partitioning of $p_1 < p_2 < \ldots < p_\ell$ into $\mathcal{O}(\sqrt{\ell})$ blocks, containing at most $\sqrt{\ell}$ of the sampled positions each. Note that positions in each block are consecutive, that is they form a single range of the sequence $p_1 < p_2 < \ldots < p_\ell$. Each block defines a projection of a matrix onto at most $\sqrt{\ell}$ positions, and we will now define and compute a hash function of these projections.

For a rectangular block, we define the hash function to be the Karp-Rabin fingerprint of the projection. We can compute the values of this hash function for all suffixes $S_1, S_2, \ldots, S_N$ in $\mathcal{O}(N \log N)$ time as a convolution of rows or columns of **T** with a suitable vector.

Example 1. Consider Fig. 2. The hash function for the block $\{17, 18, 19\}$ is the Karp-Rabin fingerprint of $s_{5,1}s_{5,2}s_{5,3}$.

For an L-shaped block we define the hash function differently. First, we divide the L-shape into two halves, a horizontal one and a vertical one. The hash function will be defined as a pair of fingerprints. The first fingerprint will be defined to be the Karp-Rabin fingerprint of a permutation of the projection on the sampled positions in the horizontal half obtained by reading the positions by columns, and the second fingerprint as the Karp-Rabin fingerprint of a permutation of the projection on the sampled positions in the vertical half obtained by reading the positions by columns.

Example 2. Consider Fig. 2. The L-shape $L_{3:4}$ is divided into two halves by a dashed line. The hash function of the horizontal half is the Karp-Rabin fingerprint of $s_{4,1}s_{3,2}$. The hash function of the vertical half is the Karp-Rabin fingerprint of $s_{2,3}$.

The Karp-Rabin fingerprints of the horizontal and vertical parts for a fixed L-shape and all suffixes $S_1, S_2, \ldots, S_N$ can be computed in $\mathcal{O}(N \log N)$ time as a sequence of 2D convolutions. In total, computing the hash functions for all L-shaped blocks takes $\mathcal{O}(N\sqrt{\ell} \log N)$ time.

Sort B. Similarly to Sect. 3, we consider $n - 2^i$ belts of **T** of height 2^i for each $i = 1, 2, \ldots, \log n$. We then linearise them, build a suffix tree and augment it with the LCA data structure. The tree can be constructed in $\mathcal{O}(N \log N)$ time and occupies $\mathcal{O}(N \log N)$ space. With the help of the suffix tree and kangaroo jumps we can report up to t mismatches between any two $2^i \times j$ submatrices S_1, S_2 of **T** in $\mathcal{O}(t)$ time.

We also build a generalised suffix tree for all columns and rows of **T**, which occupies $\mathcal{O}(N)$ space and augment it with the LCA data structure as well.

As it was shown in [10], w.h.p. the first mismatch between $g(S_i)$ and $g(S_j)$ is contained in the first $3N \log N/\ell$ mismatches between S_i and S_j. We will use binary search and the suffix trees for the belts to extract these mismatches. When a mismatch is extracted, we check if it belongs to $\{p_1, p_2, \ldots, p_\ell\}$ in constant time and stop if it does.

We start by finding the smallest t such that there are at least $3N \log N/\ell$ mismatches between the $t \times t$ top left submatrices of S_i and S_j. We do so by binary search on t. For each value of t we divide the $t \times t$ top left submatrices into a logarithmic number of even smaller submatrices of size power of two by t. For any pair of such submatrices of S_i and S_j we can use the suffix trees for the belts and for the columns to list the mismatches between them in constant time per mismatch using the kangaroo method. We stop when we have found $3N \log N/\ell$ mismatches, so we never spend more than $\mathcal{O}(3N \log N/\ell)$ time.

We guarantee that there are at least $3N \log N/\ell$ mismatches between the $t \times t$ submatrices of S_i and S_j. Unfortunately, there can be much more mismatches if the L-shapes $L_{t:t}$ of these submatrices contain many mismatches. However, using the suffix trees for columns and for rows, we can list the mismatches between these two L-shapes in order in constant time per mismatch.

References

1. Alon, N., Matias, Y., Szegedy, M.: The space complexity of approximating the frequency moments. In: STOC 1996, pp. 20–29. ACM (1996)
2. Amir, A., Benson, G.: Efficient two-dimensional compressed matching. In: Data Compression Conference, DCC 1992, pp. 279–288. IEEE (1992)
3. Amir, A., Benson, G.: Two-dimensional periodicity in rectangular arrays. SIAM J. Comp. **27**(1), 90–106 (1998)
4. Amir, A., Benson, G., Farach, M.: Optimal two-dimensional compressed matching. J. Algorithms **24**(2), 354–379 (1997)
5. Amir, A., Butman, A., Crochemore, M., Landau, G.M., Schaps, M.: Two-dimensional pattern matching with rotations. Theor. Comput. Sci. **314**(1), 173–187 (2004)
6. Amir, A., Farach, M.: Efficient 2-dimensional approximate matching of non-rectangular figures. In: SODA, pp. 212–223 (1991)
7. Amir, A., Farach, M.: Efficient 2-dimensional approximate matching of half-rectangular figures. Inf. Comput. **118**(1), 1–11 (1995)
8. Amir, A., Landau, G.M., Lewenstein, M., Sokol, D.: Dynamic text and static pattern matching. ACM Trans. Algorithms (TALG), **3**(2) (2007)
9. Amir, A., Landau, G.M., Sokol, D.: Inplace run-length 2D compressed search. Theor. Comput. Sci. **290**(3), 1361–1383 (2003)
10. Andoni, A., Indyk, P.: Efficient algorithms for substring near neighbor problem. In: SODA 2006, pp. 1203–1212 (2006)
11. Baker, T.P.: A technique for extending rapid exact-match string matching to arrays of more than one dimension. SIAM J. Comp. **7**(4), 533–541 (1978)
12. Bird, R.S.: Two dimensional pattern matching. IPL **6**(5), 168–170 (1977)
13. Cole, R., Hariharan, R.: Dynamic LCA queries on trees. SIAM J. Comput. **34**(4), 894–923 (2005)
14. Fredriksson, K., Navarro, G., Ukkonen, E.: Optimal exact and fast approximate two dimensional pattern matching allowing rotations. In: Apostolico, A., Takeda, M. (eds.) CPM 2002. LNCS, vol. 2373, pp. 235–248. Springer, Heidelberg (2002)
15. Galil, Z., Park, K.: Truly alphabet-independent two-dimensional pattern matching. In: FOCS 1992, pp. 247–256 (1992)
16. Giancarlo, R.: A generalization of the suffix tree to square matrices, with applications. SIAM J. Comp. **24**(3), 520–562 (1995)
17. Gusfield, D.: Algorithms on Strings, Trees and Sequences. Computer Science and Computational Biology. Cambridge University Press, Cambridge (1997)
18. Karp, R.M., Rabin, M.O.: Efficient randomized pattern-matching algorithms. IBM J. Res. Dev. **31**(2), 249–260 (1987)

Fully Dynamic de Bruijn Graphs

Djamal Belazzougui[1], Travis Gagie[2,3(✉)], Veli Mäkinen[2,3],
and Marco Previtali[4]

[1] CERIST, Ben Aknoun, Algiers, Algeria
[2] Helsinki Institute for Information Technology, Helsinki, Finland
[3] University of Helsinki, Helsinki, Finland
travis.gagie@gmail.com
[4] University of Milano-Bicocca, Milan, Italy

Abstract. We present a space- and time-efficient fully dynamic implementation of de Bruijn graphs, which can also support fixed-length jumbled pattern matching.

1 Introduction

Bioinformaticians define the kth-order de Bruijn graph for a string or set of strings to be the directed graph whose nodes are the distinct k-tuples in those strings and in which there is an edge from u to v if there is a $(k+1)$-tuple somewhere in those strings whose prefix of length k is u and whose suffix of length k is v.[1] These graphs have many uses in bioinformatics, including *de novo* assembly [17], read correction [15] and pan-genomics [16]. The datasets in these applications are massive and the graphs can be even larger, however, so pointer-based implementations are impractical. Researchers have suggested several approaches to representing de Bruijn graphs compactly, the two most popular of which are based on Bloom filters [9,14] and the Burrows-Wheeler Transform [5–7], respectively. In this paper we describe a new approach, based on minimal perfect hash functions [13], that is similar to that using Bloom filters but has better theoretical bounds when the number of connected components in the graph is small, and is fully dynamic: i.e., we can both insert and delete nodes and edges efficiently, whereas implementations based on Bloom filters are usually semi-dynamic and support only insertions. We also show how to modify our implementation to support, e.g., jumbled pattern matching [8] with fixed-length patterns.

Our data structure is based on a combination of Karp-Rabin hashing [11] and minimal perfect hashing, which we will describe in the full version of this paper and which we summarize for now with the following technical lemmas:

Lemma 1. *Given a static set N of n k-tuples over an alphabet Σ of size σ, with high probability in $O(kn)$ expected time we can build a function $f : \Sigma^k \to \{0, \ldots, n-1\}$ with the following properties:*

---

[1] An alternative definition, which our data structure can be made to handle but which we do not consider in this paper, has an edge from u to v whenever both nodes are in the graph.

© Springer International Publishing AG 2016
S. Inenaga et al. (Eds.): SPIRE 2016, LNCS 9954, pp. 145–152, 2016.
DOI: 10.1007/978-3-319-46049-9_14

- *when its domain is restricted to N, f is bijective;*
- *we can store f in $O(n + \log k + \log \sigma)$ bits;*
- *given a k-tuple v, we can compute $f(v)$ in $\mathcal{O}(k)$ time;*
- *given u and v such that the suffix of u of length $k-1$ is the prefix of v of length $k-1$, or vice versa, if we have already computed $f(u)$ then we can compute $f(v)$ in $\mathcal{O}(1)$ time.*

Lemma 2. *If N is dynamic then we can maintain a function f as described in Lemma 1 except that:*

- *the range of f becomes $\{0, \dots, 3n-1\}$;*
- *when its domain is restricted to N, f is injective;*
- *our space bound for f is $\mathcal{O}(n(\log \log n + \log \log \sigma))$ bits with high probability;*
- *insertions and deletions take $\mathcal{O}(k)$ amortized expected time.*
- *the data structure may work incorrectly with very low probability (inversely polynomial in n).*

Suppose N is the node-set of a de Bruijn graph. In Sect. 2 we show how we can store $\mathcal{O}(n\sigma)$ more bits than Lemma 1 such that, given a pair of k-tuples u and v of which at least one is in N, we can check whether the edge (u, v) is in the graph. This means that, if we start with a k-tuple in N, then we can explore the entire connected component containing that k-tuple in the underlying undirected graph. On the other hand, if we start with a k-tuple not in N, then we will learn that fact as soon as we try to cross an edge to a k-tuple that is in N. To deal with the possibility that we never try to cross such an edge, however — i.e., that our encoding as described so far is consistent with a graph containing a connected component disjoint from N — we cover the vertices with a forest of shallow rooted trees. We store each root as a k-tuple, and for each other node we store $1 + \lg \sigma$ bits indicating which of its incident edges leads to its parent. To verify that a k-tuple we are considering is indeed in the graph, we ascend to the root of the tree that contains it and check if that k-tuple is what we expect. The main challenge for making our representation dynamic with Lemma 2 is updating the covering forest. In Sect. 3 how we can do this efficiently while maintaining our depth and size invariants. Finally, in Sect. 4 we observe that our representation can be easily modified for other applications by replacing the Karp-Rabin hash function by other kinds of hash functions. To support jumbled pattern matching with fixed-length patterns, for example, we hash the histograms indicating the characters' frequencies in the k-tuples.

2   Static de Bruijn Graphs

Let G be a de Bruijn graph of order k, let $N = \{v_0, \dots, v_{n-1}\}$ be the set of its nodes, and let $E = \{a_0, \dots, a_{e-1}\}$ be the set of its edges. We call each v_i either a node or a k-tuple, using interchangeably the two terms since there is a one-to-one correspondence between nodes and labels.

We maintain the structure of G by storing two binary matrices, IN and OUT, of size $n \times \sigma$. For each node, the former represents its incoming edges whereas the latter represents its outgoing edges. In particular, for each k-tuple $v_x = c_1c_2 \ldots c_{k-1}a$, the former stores a row of length σ such that, if there exists another k-tuple $v_y = bc_1c_2 \ldots c_{k-1}$ and an edge from v_y to v_x, then the position indexed by b of such row is set to 1. Similarly, OUT contains a row for v_y and the position indexed by a is set to 1. As previously stated, each k-tuple is uniquely mapped to a value between 0 and $n-1$ by f, where f is as defined in Lemma 1, and therefore we can use these values as indices for the rows of the matrices IN and OUT, i.e., in the previous example the values of $\text{IN}[f(v_x)][b]$ and $\text{OUT}[f(v_y)][a]$ are set to 1. We note that, e.g., the SPAdes assembler [2] also uses such matrices.

Suppose we want to check whether there is an edge from bX to Xa. Letting $f(bX) = i$ and $f(Xa) = j$, we first assume bX is in G and check the values of $\text{OUT}[i][a]$ and $\text{IN}[j][b]$. If both values are 1, we report that the edge is present and we say that the edge is *confirmed* by IN and OUT; otherwise, if any of the two values is 0, we report that the edge is absent. Moreover, note that if bX is in G and $\text{OUT}[i][a] = 1$, then Xa is in G as well. Symmetrically, if Xa is in G and $\text{IN}[j][b] = 1$, then bX is in G as well. Therefore, if $\text{OUT}[i][a] = \text{IN}[j][b] = 1$, then bX is in G if and only if Xa is. This means that, if we have a path P and if all the edges in P are confirmed by IN and OUT, then either all the nodes touched by P are in G or none of them is.

We now focus on detecting false positives in our data structure maintaining a reasonable memory usage. Our strategy is to sample a subset of nodes for which we store the plain-text k-tuple and connect all the unsampled nodes to the sampled ones. More precisely, we partition nodes in the undirected graph G' underlying G into a forest of rooted trees of height at least $k \lg \sigma$ and at most $3k \lg \sigma$. For each node we store a pointer to its parent in the tree, which takes $1 + \lg \sigma$ bits per node, and we sample the k-mer at the root of such tree. We allow a tree to have height smaller than $k \lg \sigma$ when necessary, e.g., if it covers a connected component. Figure 1 shows an illustration of this idea.

We can therefore check whether a given node v_x is in G by first computing $f(v_x)$ and then checking and ascending at most $3k \lg \sigma$ edges, updating v_x and $f(v_x)$ as we go. Once we reach the root of the tree we can compare the resulting k-tuple with the one sampled to check if v_x is in the graph. This procedure requires $\mathcal{O}(k \lg \sigma)$ time since computing the first value of $f(v_x)$ requires $\mathcal{O}(k)$, ascending the tree requires constant time per edge, and comparing the k-tuples requires $\mathcal{O}(k)$.

We now describe a Las Vegas algorithm for the construction of this data structure that requires, with high probability, $\mathcal{O}(kn + n\sigma)$ expected time. We recall that N is the set of input nodes of size n. We first select a function f and construct bitvector B of size n initialized with all its elements set to 0. For each elements v_x of N we compute $f(v_x) = i$ and check the value of $B[i]$. If this value is 0 we set it to 1 and proceed with the next element in N, if it is already set to 1, we reset B, select a different function f, and restart the procedure from the first element in N. Once we finish this procedure — i.e., we found that f do not produces collisions when applied to N — we store f and proceed to initialize IN and OUT correctly. This procedure requires with high probability $\mathcal{O}(kn)$ expected

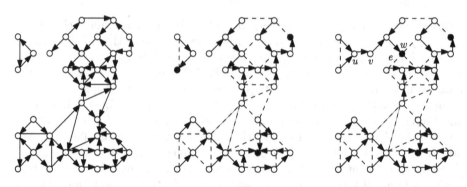

Fig. 1. Given a de Bruijn graph (left), we cover the underlying undirected graph with a forest of rooted trees of height at most $3k \lg \sigma$ (center). The roots are shown as filled nodes, and parent pointers are shown as arrows; notice that the directions of the arrows in our forest are not related to the edges' directions in the original de Bruijn graph. We sample the k-tuples at the roots so that, starting at a node we think is in the graph, we can verify its presence by finding the root of its tree and checking its label in $\mathcal{O}(k \log \sigma)$ time. The most complicated kind of update (right) is adding an edge between a node u in a small connected component to a node v in a large one, v's depth is more than $2k \lg \sigma$ in its tree. We re-orient the parent pointers in u's tree to make u the temporary root, then make u point to v. We ascend $k \lg \sigma$ steps from v, then delete the parent pointer e of the node w we reach, making w a new root. (To keep this figure reasonably small, some distances in this example are smaller than prescribed by our formulas.)

time for constructing f and $\mathcal{O}(n\sigma)$ time for computing IN and OUT. Notice that if N is the set of k-tuples of a single text sorted by their starting position in the text, each $f(v_x)$ can be computed in constant time from $f(v_{x-1})$ except for $f(v_0)$ that still requires $\mathcal{O}(k)$. More generally, if N is the set of k-tuples of t texts sorted by their initial position, we can compute $n-t$ values of the function $f(v_x)$ in constant time from $f(v_{x-1})$ and the remaining in $\mathcal{O}(k)$. We will explain how to build the forest in the full version of this paper. In this case the construction requires, with high probability, $\mathcal{O}(kt + n + n\sigma) = \mathcal{O}(kt + n\sigma)$ expected time.

Combining our forest with Lemma 1, we can summarize our static data structure in the following theorem:

Theorem 1. *Given a static σ-ary kth-order de Bruijn graph G with n nodes, with high probability in $\mathcal{O}(kn + n\sigma)$ expected time we can store G in $\mathcal{O}(\sigma n)$ bits plus $\mathcal{O}(k \log \sigma)$ bits for each connected component in the underlying undirected graph, such that checking whether a node is in G takes $\mathcal{O}(k \log \sigma)$ time, listing the edges incident to a node we are visiting takes $\mathcal{O}(\sigma)$ time, and crossing an edge takes $\mathcal{O}(1)$ time.*

In the full version we will show how to use monotone minimal perfect hashing [3] to reduce the space to $(2 + \epsilon)n\sigma$ bits of space (for any constant $\epsilon > 0$). We will also show how to reduce the time to list the edges incident to a node of degree d to $O(d)$, and the time to check whether a node is in G to $\mathcal{O}(k)$. We note that the obtained space and query times are both optimal up to constant factors, which is unlike previous methods which have additional factor(s) depending on k and/or σ in space and/or time.

3   Dynamic de Bruijn Graphs

In the previous section we presented a static representation of de Buijn graphs, we now present how we can make this data structure dynamic. In particular, we will show how we can insert and remove edges and nodes and that updating the graph reduces to managing the covering forest over G. In this section, when we refer to f we mean the function defined in Lemma 2. We first show how to add or remove an edge in the graph and will later describe how to add or remove a node in it. The updates must maintain the following invariant: any tree must have size at least $k \log \sigma$ and height at most $3k \log \sigma$ except when the tree covers (all nodes in) a connected component of size at most $k \log \sigma$.

Let v_x and v_y be two nodes in G, $e = (v_x, v_y)$ be an edge in G, and let $f(v_x) = i$ and $f(v_y) = j$.

Suppose we want to add e to G. First, we set to 1 the values of $\mathtt{OUT}[i][a]$ and $\mathtt{IN}[j][b]$ in constant time. We then check whether v_x or v_y are in different components of size less than $k \lg \sigma$ in $\mathcal{O}(k \lg \sigma)$ time for each node. If both components have size greater than $k \lg \sigma$ we do not have to proceed further since the trees will not change. If both connected components have size less than $k \lg \sigma$ we merge their trees in $\mathcal{O}(k \lg \sigma)$ time by traversing both trees and switching the orientation of the edges in them, discarding the samples at the roots of the old trees and sampling the new root in $\mathcal{O}(k)$ time.

If only one of the two connected components has size greater than $k \lg \sigma$ we select it and perform a tree traversal to check whether the depth of the node is less than $2k \lg \sigma$. If it is, we connect the two trees as in the previous case. If it is not, we traverse the tree in the bigger component upwards for $k \lg \sigma$ steps, we delete the edge pointing to the parent of the node we reached creating a new tree, and merge it with the smaller one. This procedure requires $\mathcal{O}(k \lg \sigma)$ time since deleting the edge pointing to the parent in the tree requires $\mathcal{O}(1)$ time, i.e., we have to reset the pointer to the parent in only one node.

Suppose now that we want to remove e from G. First we set to 0 the values of $\mathtt{OUT}[i][a]$ and $\mathtt{IN}[j][b]$ in constant time. Then, we check in $\mathcal{O}(k)$ time whether e is an edge in some tree by computing $f(v_x)$ and $f(v_y)$ checking for each node if that edge is the one that points to their parent. If e is not in any tree we do not have to proceed further whereas if it is we check the size of each tree in which v_x and v_y are. If any of the two trees is small (i.e., if it has fewer than $k \lg \sigma$ elements) we search any outgoing edge from the tree that connects it to some other tree. If such an edge is not found we conclude that we are in a small connected component that is covered by the current tree and we sample a node in the tree as a root and switch directions of some edges if necessary. If such an edge is found, we merge the small tree with the bigger one by adding the edge and switch the direction of some edges originating from the small tree if necessary. Finally if the height of the new tree exceeds $3k \log \sigma$, we traverse the tree upwards from the deepest node in the tree (which was necessarily a node in the smaller tree before the merger) for $2k \lg \sigma$ steps, delete the edge pointing to the parent of the reached node, creating a new tree. This procedure requires $\mathcal{O}(k \lg \sigma)$ since the number of nodes traversed is at most $O(k \lg \sigma)$ and the number of changes to the data structures is also at most $O(k \lg \sigma)$ with each change taking expected constant time.

It is clear that the insertion and deletion algorithms will maintain the invariant on the tree sizes and heights. It is also clear that the invariant implies that the number of sampled nodes is $O(n/(k \log \sigma))$ plus the number of connected components.

We now show how to add and remove a node from the graph. Adding a node is trivial since it will not have any edge connecting it to any other node. Therefore adding a node reduces to modify the function f and requires $\mathcal{O}(k)$ amortized expected time. When we want to remove a node, we first remove all its edges one by one and, once the node is isolated from the graph, we remove it by updating the function f. Since a node will have at most σ edges and updating f requires $\mathcal{O}(k)$ amortized expected time, the amortized expected time complexity of this procedure is $\mathcal{O}(\sigma k \lg \sigma + k)$.

Combining these techniques for updating our forest with Lemma 2, we can summarize our dynamic data structure in the following theorem:

Theorem 2. *We can maintain a σ-ary kth-order de Bruijn graph G with n nodes that is fully dynamic (i.e., supporting node and edge insertions and deletions) in $\mathcal{O}(n(\log \log n + \sigma))$ bits (plus $\mathcal{O}(k \log \sigma)$ bits for each connected component) with high probability, such that we can add or remove an edge in expected $\mathcal{O}(k \lg \sigma)$ time, add a node in expected $\mathcal{O}(k + \sigma)$ time, and remove a node in expected $\mathcal{O}(\sigma k \lg \sigma)$ time, and queries have the same time bounds as in Theorem 1. The data structure may work incorrectly with very low probability (inversely polynomial in n).*

4   Jumbled Pattern Matching

Karp-Rabin hash functions implicitly divide their domain into equivalence classes — i.e., subsets in which the elements hash to the same value. In this paper we have chosen Karp-Rabin hash functions such that each equivalence class contains only one k-tuple in the graph. Most of our efforts have gone into being able, given a k-tuple and a hash value, to determine whether that k-tuple is the unique element of its equivalence class in the graph. In some sense, therefore, we have treated the equivalence relation induced by our hash functions as a necessary evil, useful for space-efficiency but otherwise an obstacle to be overcome. For some applications, however — e.g., parameterized pattern matching, circular pattern matching or jumbled pattern matching — we are given an interesting equivalence relation on strings and asked to preprocess a text such that later, given a pattern, we can determine whether any substrings of the text are in the same equivalence class as the pattern. We can modify our data structure for some of these applications by replacing the Karp-Rabin hash function by other kinds of hash functions.

For indexed jumbled pattern matching [1,8,12] we are asked to pre-process a text such that later, given a pattern, we can determine quickly whether any substring of the text consists of exactly the same multiset of characters in the pattern. Consider fixed-length jumbled pattern matching, when the length of the patterns is fixed at pre-processing time. If we modify Lemmas 1 and 2 so that, instead of using Karp-Rabin hashes in the definition of the function f, we use a hash function on the histograms of characters' frequencies in k-tuples, our function f will map all permutations of a k-tuple to the same value. The rest of our implementation stays the same, but now the nodes of our graph are multisets of characters of size k and there is an edge between two nodes u and v if it is possible to replace an element of u and obtain v. If we build our graph for the multisets of characters in k-tuples in a string S, then our process for checking whether a node is in the graph tells us whether there is a jumbled match in S for a pattern of length k. If we build a tree in which the root is a graph for all of S, the left and right children of the root are graphs for the first and second halves of S, etc., as described by Gagie et al. [10], then we increase the space by a logarithmic factor but we can return the locations of all matches quickly.

Theorem 3. *Given a string $S[1..n]$ over an alphabet of size σ and a length $k \ll n$, with high probability in $\mathcal{O}(kn + n\sigma)$ expected time we can store $(2n \log \sigma)(1 + o(1))$ bits such that later we can determine in $\mathcal{O}(k \log \sigma)$ time if a pattern of length k has a jumbled match in S.*

Acknowledgements. Many thanks to Rayan Chikhi and the anonymous reviewers for their comments.

References

1. Amir, A., Chan, T.M., Lewenstein, M., Lewenstein, N.: On hardness of jumbled indexing. In: Esparza, J., Fraigniaud, P., Husfeldt, T., Koutsoupias, E. (eds.) ICALP 2014. LNCS, vol. 8572, pp. 114–125. Springer, Heidelberg (2014)
2. Bankevich, A., et al.: SPAdes: a new genome assembly algorithm and its applications to single-cell sequencing. J. Comput. Biol. **19**, 455–477 (2012)
3. Belazzougui, D., Boldi, P., Pagh, R., Vigna, S.: Monotone minimal perfect hashing: searching a sorted table with o(1) accesses. In: Proceedings of the Twentieth Annual ACM-SIAM Symposium on Discrete Algorithms, pp. 785–794. Society for Industrial and Applied Mathematics (2009)
4. Belazzougui, D., Gagie, T., Mäkinen, V., Previtali, M.: Fully dynamic de bruijn graphs. arXiv preprint (2016). arXiv:1607.04909
5. Belazzougui, D., Gagie, T., Mäkinen, V., Previtali, M., Puglisi, S.J.: Bidirectional variable-order de Bruijn graphs. In: Kranakis, E., et al. (eds.) LATIN 2016. LNCS, vol. 9644, pp. 164–178. Springer, Heidelberg (2016). doi:10.1007/978-3-662-49529-2_13
6. Boucher, C., Bowe, A., Gagie, T., Puglisi, S.J., Sadakane, K.: Variable-order de Bruijn graphs. In: Data Compression Conference (DCC), pp. 383–392. IEEE (2015)

7. Bowe, A., Onodera, T., Sadakane, K., Shibuya, T.: Succinct de Bruijn graphs. In: Raphael, B., Tang, J. (eds.) WABI 2012. LNCS, vol. 7534, pp. 225–235. Springer, Heidelberg (2012)

8. Burcsi, P., Cicalese, F., Fici, G., Lipták, Z.: Algorithms for jumbled pattern matching in strings. Int. J. Found. Comput. Sci. **23**(02), 357–374 (2012)

9. Chikhi, R., Rizk, G.: Space-efficient and exact de Bruijn graph representation based on a Bloom filter. Algorithm Mol. Biol. **8**(22), 1–9 (2012)

10. Gagie, T., Hermelin, D., Landau, G.M., Weimann, O.: Binary jumbled pattern matching on trees and tree-like structures. Algorithmica **73**(3), 571–588 (2015)

11. Karp, R.M., Rabin, M.O.: Efficient randomized pattern-matching algorithms. IBM J. Res. Dev. **31**(2), 249–260 (1987)

12. Kociumaka, T., Radoszewski, J., Rytter, W.: Efficient indexes for jumbled pattern matching with constant-sized alphabet. In: Bodlaender, H.L., Italiano, G.F. (eds.) ESA 2013. LNCS, vol. 8125, pp. 625–636. Springer, Heidelberg (2013)

13. Mehlhorn, K.: On the program size of perfect and universal hash functions. In: 23rd Annual Symposium on Foundations of Computer Science, SFCS'08, pp. 170–175. IEEE (1982)

14. Salikhov, K., Sacomoto, G., Kucherov, G.: Using cascading Bloom filters to improve the memory usage for de Brujin graphs. In: Darling, A., Stoye, J. (eds.) WABI 2013. LNCS, vol. 8126, pp. 364–376. Springer, Heidelberg (2013)

15. Salmela, L., Rivals, E.: Lordec: accurate and efficient long read error correction. Bioinformatics **30**(24), 3506–3514 (2014). http://dx.doi.org/10.1093/bioinformatics/btu538

16. Sirén, J., Välimäki, N., Mäkinen, V.: Indexing graphs for path queries with applications in genome research. IEEE/ACM Trans. Comput. Biol. Bioinform. (TCBB) **11**(2), 375–388 (2014)

17. Zerbino, D.R., Birney, E.: Velvet: algorithms for de novo short read assembly using de Bruijn graphs. Genome Res. **18**(5), 821–829 (2008)

Bookmarks in Grammar-Compressed Strings

Patrick Hagge Cording[1]([✉]), Pawel Gawrychowski[2], and Oren Weimann[2]

[1] Technical University of Denmark, DTU Compute, Kongens Lyngby, Denmark
phaco@dtu.dk
[2] University of Haifa, Haifa, Israel

Abstract. We consider the problem of storing a grammar of size n compressing a string of size N, and a set of positions $\{i_1, \ldots, i_b\}$ (*bookmarks*) such that any substring of length l crossing one of the positions can be decompressed in $O(l)$ time. Our solution uses space $O((n+b)\max\{1, \log^* n - \log^*(\frac{n}{b} + \frac{b}{n})\})$. Existing solutions for the bookmarking problem either require more space or a super-constant "kick-off" time to start the decompression.

1 Introduction

Textual databases for e.g. biological or web-data are growing rapidly, and it is often only feasible to store the data in compressed form. However, compressing the data comes at a price: it may be necessary to decompress the entire file in order to retrieve just a small portion of it. Inserting *bookmarks* in the compressed file can accommodate this problem. A bookmark in a compressed string is a position i from which any substring of length l crossing position i can be decompressed in $O(l)$ time.

A popular technique for compressing a string is to instead store a small grammar that generates the string (and only the string). The idea dates back far and has received much attention in the theory community while also being widely used in practice. In particular, popular compression schemes such as LZ78 [15], LZW [13], Re-pair [9], and Sequitur [11] produce grammars. Even the LZ77 [14] compression scheme that does not produce a grammar, can be converted to a grammar with only a logarithmic overhead in the space [5,12]. For our purposes, we consider Straight Line Programs (SLPs). These are context-free grammar in Chomsky Normal Form that generate exactly one string. SLPs capture any grammar-based compression scheme.

For the *bookmarking problem*, we are given an SLP $\mathcal{S}$ of size n compressing a string S of size N and a set of positions $\{i_1, \ldots, i_b\}$, and we want to construct a data structure that supports linear-time decompression of substrings crossing any of the b positions.

P.H. Cording—Supported by the Danish Research Council under the Sapere Aude Program (DFF 4005-00267).

S. Inenaga et al. (Eds.): SPIRE 2016, LNCS 9954, pp. 153–159, 2016.
DOI: 10.1007/978-3-319-46049-9_15

Related Work. Gagie et al. [6] presented a bookmarking data structure that uses $O(n + b\log^* N)$ space[1] for *balanced* SLPs (i.e., SLPs whose parse tree is balanced). When the SLP is unbalanced, we may use an algorithm to balance it at the cost of adding nodes [5,12], and as a result the space usage of their data structure increases to $O(n\log\frac{N}{n} + b\log^* N)$.

A more general problem is to support random access to the compressed string (i.e., access to a single character of S without decompression). This does not require any bookmarks to be predefined, but in turn incurs a "kick-off" time when decompressing a substring. If we allow the kick-off time to be $O(\log N)$ (i.e., $O(l + \log N)$ time to decompress a substring of length l), we may use the $O(n)$-space data structure of Bille et al. [4]. For a faster kick-off time of $O(\log_\tau N)$, for any $2 < \tau \le \log N$, we may instead apply the data structure of Belazzougui et al. [1] at the cost of increasing the space to $O(n\tau\log_\tau\frac{N}{n})$. The data structure of Belazzougui et al. [2] supports random access to any character of the compressed string in $O(1)$ time and thereby allows decompression of any substring in time linear in the substring's length. However, this data structure uses space $O(n^{1-\epsilon}N^\epsilon)$ for some constant $0 < \epsilon \le 1$.

The compressed finger search problem is somehow a hybrid of the bookmarking problem and the random access problem. For this problem, we place a set of *fingers*, and now we may answer random access queries in $O(\log D)$ time, where D is the distance from a given finger to the index we query for [3]. Using this data structure, we get a bookmarking data structure of $O(n)$ space that can decompress any substring of length l in $O(l\log l)$ time.

Our Results. In this paper we present a bookmarking data structure for SLP-compressed strings that uses space $O((n + b)\max\{1, \log^* n - \log^*(\frac{n}{b} + \frac{b}{n})\})$ and supports decompression of length-l substrings crossing bookmarks in $O(l)$ time. The space is measured in words and we assume the standard RAM model of computation.

The general idea is to make τ copies of the SLP for some parameter τ. Each copy is modified so that the decompression kick-off takes less time but only supports decompression of substrings up to a certain length and from certain positions. At query time, we then select the copy of the SLP that provides a kick-off time of $O(l)$.

2   Preliminaries

Let S be a string of length $|S|$ consisting of characters from an alphabet of size σ. We use $S[i, j]$, $1 \le i \le j \le |S|$, to denote the substring starting in position i and ending in position j of S.

A Straight Line Program (SLP) $\mathcal{S}$ is a context-free grammar in Chomsky normal form with n production rules that derives a single string S of size N. We

---

[1] The bound is in fact $O(z + b\log^* N)$, where z is the size of the LZ77 parse of S. Since it is known that $z \le n' \le n$ [12], where n' is the size of the smallest SLP generating S, we replace z by n for clarity.

represent the SLP as a rooted, ordered, and node-labelled directed acyclic graph (DAG) with outdegree 2 and we will refer to production rules as nodes where it is appropriate. We denote by $v = uw$ that node v in the DAG has left-child u and right-child w. A depth-first left-to-right traversal starting from a node v in the DAG produces the string $S(v)$. As a shorthand we sometimes use $|v|$ instead of $|S(v)|$.

All logarithms in this paper are base 2. As a shorthand to denote the logarithm applied i times to a number n we write $\log^{(i)} n$, e.g., $\log^{(3)} n = \log \log \log n$. The iterated logarithm $\log^* n$ is equal to the number of times the logarithm can be applied to n before the result is less than 1, i.e., $\log^* n = \arg\min_i \{\log^{(i)} n \leq 1\}$. We also need the up-arrow notation of Knuth [8] defined as follows: $2 \uparrow\uparrow 0 = 1$ and $2 \uparrow\uparrow (k+1) = 2^{2\uparrow\uparrow k}$. Observe that $k = \log^* n$ if and only if $2 \uparrow\uparrow (k-1) < n \leq 2 \uparrow\uparrow k$.

3   A Simple Solution

In this section we give a simple data structure to the bookmarking problem with the following bounds.

Theorem 1. *Given an SLP for $S[1, N]$ with n rules and positions $i_1, \ldots, i_b$ in S, we can store S in $O(n + b + \min\{n, b\} \log N)$ space such that later, given $i \in \{i_1, \ldots, i_b\}$ we can extract $S[i, i+l]$ in $O(l)$ time.*

Our solution builds on the following data structure by Bille et al. [4].

Lemma 1 ([4]). *Let S be a string of length N compressed by an SLP S of size n. There is data structure of size $O(n)$ that, given a node v in S, supports decompression of a substring $S(v)[i, i+l]$ in $O(\log |v| + l)$ time.*

Notice that when $l \geq \log N$ the decompression time in Lemma 1 is dominated by the $O(l)$ term. This means we only need to focus on the case where $l < \log N$.

To obtain a $O(n + b \log N)$-space solution, since $l < \log N$, we can simply store the substring $S[i - \log N, i + \log N]$ for each bookmark $i \in \{i_1, \ldots, i_b\}$ along with the data structure of Lemma 1.

In the case where $n < b$, to obtain a $O(n \log N + b)$-space solution, we show that it is sufficient to store n substrings each of length $O(\log N)$. For this we use the following lemma, stating that any substring of S is the concatenation of a suffix of $S(u)$ and a prefix of $S(w)$ for some node v whose left child is u and right child is w. The observation was first used for compressed pattern matching [10]. For the sake of completeness, we will give a proof using our terminology.

Lemma 2 ([10]). *Let S be a string of length N compressed by an SLP S of size n. Let $r(v) = S(u)[\max\{1, |u| - k\}, |u|]S(w)[1, \min\{k, k-1\}]$ be the relevant substring with respect to k of a node $v = uw$ in S. Then any substring of S of length at most k is also a substring of some string in $\{r(v) \mid v \in S \wedge |v| \geq k\}$.*

Proof. The proof is by induction. For the base case, consider a node $v = uw$ where $|v| \leq 2k - 2$ and $|u| < k$ and $|w| < k$. Since $r(v) = S(v)$ this obviously contains every substring of length k. For the inductive step we again consider some node $v = uw$ and we know that $S(v) = S(u) \circ S(w)$. Assume that $|u| \geq k$ and $|w| \geq k$, then by the induction hypothesis it holds that the set of strings $\{r(u') \mid u' \in \mathcal{S}(u) \wedge |u'| \geq k\} \cup \{r(w') \mid w' \in \mathcal{S}(w) \wedge |w'| \geq k\}$ contains all substrings of length k in $S(u)$ and $S(w)$. The substrings of length k starting in $S(u)$ and $S(w)$ are not guaranteed to be in this set, but since $r(v)$ contains exactly all these, they will be after adding $r(v)$ to the set. For the cases when $|u| < k$ or $|w| < k$ the same argument holds.   □

For our data structure, we set $k = 2 \log N$ and store the strings $r(v)$ for all $v \in \mathcal{S}$. For each bookmark i we store the deepest node that generates the string $S[i - \log N, i + \log N]$. Furthermore, we build the data structure of Lemma 1 for use for the case where $l \geq \log N$.

Since $|r(v)| \leq 4 \log N - 2$ and we store $O(1)$ words (pointers) for each bookmark, and the data structure of Lemma 1 uses $O(n)$ space, our data structure uses $O(n \log N + b)$ space in total. This concludes the proof of Theorem 1.

4   A Leveled Solution

We now describe a data structure that seeks to reduce the $\log N$ factor of the space usage in Theorem 1. The time to decompress a substring of length l crossing some bookmark is still $O(l)$. The key to our solution is a technique due to Gawrychowski [7] captured by the following lemma.

Lemma 3 ([7]). *Let S be a string of length N compressed into an SLP $\mathcal{S}$ of size n. We can choose an arbitrary ℓ and modify $\mathcal{S}$ in $O(N)$ time by adding $O(n)$ new variables such that we can write S as $S = S(v_1)S(v_2) \ldots S(v_m)$ with $m = O(N/\ell)$ and $|S(v_i)| \leq 2\ell - 2$. Furthermore, for any substring $S[i, i + \ell]$ there are a constant number of nodes $v_1, \ldots, v_c$ such that $S[i, i + \ell]$ is a substring of $S(v_1) \ldots S(v_c)$.*

The lemma says that we can restructure the given SLP such that for any substring $S[i, i + \ell]$ we can find $O(1)$ nodes whose concatenation has total length $O(\ell)$ and contains $S[i, i + \ell]$ as a substring. We now describe how to apply this restructuring procedure to get a bookmarking data structure using almost linear space. In the description we use the parameter τ which is later to be minimized subject to n and b.

Construction. First we make τ copies of $\mathcal{S}$, denoted by $\mathcal{S}_1, \ldots, \mathcal{S}_\tau$. We then apply the restructuring procedure for $\ell = \log N, \log^{(2)} N, \ldots, \log^{(\tau)} N$ to the τ copies of $\mathcal{S}$ to get $\mathcal{S}'_1, \ldots, \mathcal{S}'_\tau$. Next, we build the data structure of Lemma 1 for each SLP $\mathcal{S}'_1, \ldots, \mathcal{S}'_\tau$. For each $\mathcal{S}'_j$, let a *block node* be a node v for which $|S(v)| = \Theta(\log^{(j)} N)$. For each SLP $\mathcal{S}'_j$ and for each bookmark i we store the $O(1)$ block nodes generating the string containing $S[i - \log^{(j)} N, i + \log^{(j)} N]$.

We also store the relative index of position i in the string generated by the first block node. On the lowest level (i.e., for $\mathcal{S}'_\tau$) we apply the technique from the previous section. I.e., if $b \leq n$ we use $O(n + b \log^{(\tau)} N)$ space and if $b > n$ we use $O(n \log^{(\tau)} N + b)$ space.

Decompression. To decompress a substring of length l from a bookmark position $i \in \{i_1, \ldots, i_b\}$ we do the following.

If $\log^{(j+1)} N < l \leq \log^{(j)} N$ for some $j < \tau$. We locate the block node that contains i in $\mathcal{S}'_j$ and decompress the string starting in the relative position stored for the current bookmark using the data structure of Lemma 1. If we reach the end of the string generated by the current block node, we move on to the next node that we stored and repeat the process from relative position 1. When we decompress from a block node v in $\mathcal{S}'_j$, the query time of Lemma 1 becomes $O(\log \log^{(j)} N + l) = O(l)$ since $\log^{(j+1)} N < l$. We visit $O(1)$ block nodes so the total time to decompress $S[i, i+l]$ becomes $O(l)$.

If on the other hand $l < \log^{(\tau)} N$, then we use the solution chosen for the bottom level, which according to Theorem 1 yields a decompression time of $O(l)$.

Analysis. Our data structure creates τ copies of $\mathcal{S}$. Each has size $O(n)$ after the restructuring of Lemma 3 and the application of Lemma 1, i.e., this requires $O(\tau n)$ space. For each bookmark, we store references to $O(1)$ nodes in each copy for a total of $O(\tau b)$ space. For $\mathcal{S}'_\tau$ we need $O(\min\{n, b\} \log^{(\tau)} N)$ space as stated in Theorem 1. Hence, the total space usage is $O(\tau(n + b) + \min\{n, b\} \log^{(\tau)} N)$, which is equal to $O(\tau(n + b) + \min\{n, b\} \log^{(\tau)} n)$, because $n \leq N \leq 2^n$ in any SLP. It remains to choose τ as to minimize this expression.

We define $x = \frac{n+b}{\min\{n, b\}} \geq 1$. Then, the goal is to minimize $\min\{n, b\} f(\tau)$, where $f(\tau) = x \cdot \tau + \log^{(\tau)} n$, over all $\tau \geq 1$. We claim that $f(\tau)$ is minimized (up to a constant multiplicative factor) for $\tau = \max\{1, \log^* n - \log^* x + 1\}$, when $f(\tau) = O(x \max\{1, \log^* n - \log^* x\})$. If $\log^* n - \log^* x + 1 < 2$ then $x > \log n$, so the expression is minimized for $\tau = 1$. Otherwise, define $p = \log^* n$ and $q = \log^* x$, where $t = p - q + 1 \geq 2$. By the properties of iterated log, $2 \uparrow\uparrow (p-1) < n \leq 2 \uparrow\uparrow p$ and $2 \uparrow\uparrow \leq (q - 1) < x \leq 2 \uparrow\uparrow q$. Hence $\log^{(p-q+1)} n \leq 2 \uparrow\uparrow (q-1) < x$ and $f(t) \leq x(t+1) \leq 2x \cdot t$. We claim that, for any $\tau \geq 1$, $f(\tau) \geq \frac{1}{4} x \cdot t$, that is, $\tau = t$ is the (asymptotically) best choice.

If $\tau \geq \frac{1}{4} t$ then clearly $f(\tau) \geq x \cdot \tau \geq \frac{1}{4} x \cdot \tau$. It remains to analyze the case $\tau < \frac{1}{4} t$. We will prove that, for any $\tau < \frac{1}{4} t$, $\log^{(\tau)} n \geq \frac{1}{4} x \cdot t$. Because $\log^{(\tau)} n$ is monotone in τ, it is enough to prove that $\log^{(\frac{1}{4} t - 1)} n \geq \frac{1}{4} x \cdot t$, or by the properties of iterated log $2 \uparrow\uparrow (p - \frac{1}{4} t) \geq 2 \uparrow\uparrow q \cdot \frac{1}{4} t$. Because $p - \frac{1}{4} t \geq q + \frac{1}{4} t$ by the assumption that $p - q \geq 1$, this reduces to showing that $2 \uparrow\uparrow (q + \frac{1}{4} t) \geq 2 \uparrow\uparrow q \cdot \frac{1}{4} t$.

Lemma 4 *For any $x, y \geq 0$, $2 \uparrow\uparrow (x + y) \geq 2 \uparrow\uparrow x \cdot y$.*

Proof. If $x = 0$, we need to show that $2 \uparrow\uparrow y \geq y$, which holds for any $y \geq 0$. From now on, we assume that $x \geq 1$ and apply induction on $y \geq 0$.

For $y = 0$, the left side is positive and the right side is zero. For $y = 1$, $2^{2\uparrow\uparrow x} \geq 2 \uparrow\uparrow x$ holds for all $x \geq 0$. For $y = 2$, $2 \uparrow\uparrow (x+2) > 2^{2\uparrow\uparrow x} \geq 2 \uparrow\uparrow x \cdot 2$ holds for all $x \geq 0$.

Now assume that $2 \uparrow\uparrow (x+y) \geq 2 \uparrow\uparrow x \cdot y$ for some $y \geq 2$. Then

$$2 \uparrow\uparrow (x+y+1) = 2^{2\uparrow\uparrow(x+y)} \geq 2^{2\uparrow\uparrow x \cdot y} \geq (2 \uparrow\uparrow x)^y.$$

So now it is enough to show that $(2 \uparrow\uparrow x)^{y-1} \geq y+1$. But $x \geq 1$, so this reduces to $2^{y-1} \geq y+1$, which holds for any $y \geq 3$.   □

In conclusion, choosing $\tau = \max\{1, \log^* n - \log^* x + 1\}$ gives us the total space usage of $O((n + b) \max\{1, \log^* n - \log^* x\})$. By rewriting this expression to remove x we get the following Theorem.

Theorem 2. *Given an SLP for $S[1, N]$ with n rules and positions $i_1, \ldots, i_b$ in S, we can store S in space $O((n + b) \max\{1, \log^* n - \log^*(\frac{n}{b} + \frac{b}{n})\})$ such that later, given $i \in \{i_1, \ldots, i_b\}$ we can extract $S[i, i + l]$ in $O(l)$ time.*

5   Conclusion

We have shown a bookmarking data structure that uses a little more than linear space. If $b \leq \frac{n}{\log^{(c)} N}$ or $n \log^{(c)} N \leq b$ the space becomes $O(n + b)$. Furthermore, $O(n + b)$ space can be achieved for any n and b if we are willing to pay a $O(\log^{(c)} N)$ kick-off time for decompression. It remains open whether there exists a bookmarking data structure that uses $O(n+b)$ space and supports linear time decompression, regardless of the relationship between n and b.

References

1. Belazzougui, D., Cording, P.H., Puglisi, S.J., Tabei, Y.: Access, rank, and select in grammar-compressed strings. In: Bansal, N., Finocchi, I. (eds.) ESA 2015. LNCS, vol. 9294, pp. 142–154. Springer, Heidelberg (2015)
2. Belazzougui, D., Gagie, T., Gawrychowski, P., Kärkkäinen, J., Ordónez, A., Puglisi, S.J., Tabei, Y.: Queries on LZ-bounded encodings. In: DCC, pp. 83–92 (2014)
3. Bille, P., Christiansen, A.R., Cording, P.H., Gørtz, I.L.: Finger search in grammar-compressed strings (2015). CoRR arXiv:1507.02853
4. Bille, P., Landau, G.M., Raman, R., Sadakane, K., Satti, S.R., Weimann, O.: Random access to grammar-compressed strings and trees. SIAM J. Comput. 44(3), 513–539 (2015)
5. Charikar, M., Lehman, E., Liu, D., Panigrahy, R., Prabhakaran, M., Sahai, A., Shelat, A.: The smallest grammar problem. IEEE Trans. Inf. Theor. 51(7), 2554–2576 (2005)
6. Gagie, T., Gawrychowski, P., Kärkkäinen, J., Nekrich, Y., Puglisi, S.J.: LZ77-based self-indexing with faster pattern matching. In: Pardo, A., Viola, A. (eds.) LATIN 2014. LNCS, vol. 8392, pp. 731–742. Springer, Heidelberg (2014)

7. Gawrychowski, P.: Faster algorithm for computing the edit distance between SLP-compressed strings. In: Calderón-Benavides, L., González-Caro, C., Chávez, E., Ziviani, N. (eds.) SPIRE 2012. LNCS, vol. 7608, pp. 229–236. Springer, Heidelberg (2012)

8. Knuth, D.E.: Mathematics and computer science: coping with finiteness. Science (New York, NY) **194**(4271), 1235–1242 (1976)

9. Larsson, N.J., Moffat, A.: Off-line dictionary-based compression. In: DCC, vol. 88, no. 11, pp. 1722–1732 (2000)

10. Miyazaki, M., Shinohara, A., Takeda, M.: An improved pattern matching algorithm for strings in terms of straight-line programs. In: Apostolico, A., Hein, J. (eds.) CPM 97. LNCS, vol. 1264, pp. 1–11. Springer, Heidelberg (1997)

11. Nevill-Manning, C.G., Witten, I.H.: Identifying hierarchical structure in sequences: a linear-time algorithm. J. Artif. Intell. Res. (JAIR) **7**, 67–82 (1997)

12. Rytter, W.: Application of Lempel-Ziv factorization to the approximation of grammar-based compression. Theor. Comput. Sci. **302**(1), 211–222 (2003)

13. Welch, T.A.: A technique for high-performance data compression. Computer **6**(17), 8–19 (1984)

14. Ziv, J., Lempel, A.: A universal algorithm for sequential data compression. IEEE Trans. Inf. Theor. **23**(3), 337–343 (1977)

15. Ziv, J., Lempel, A.: Compression of individual sequences via variable-rate coding. IEEE Trans. Inf. Theor. **24**(5), 530–536 (1978)

Analyzing Relative Lempel-Ziv Reference Construction

Travis Gagie[(✉)], Simon J. Puglisi, and Daniel Valenzuela

Department of Computer Science, Helsinki Institute for Information Technology,
University of Helsinki, Helsinki, Finland
{gagie,puglisi,dvalenzu}@cs.helsinki.fi

Abstract. Relative Lempel-Ziv is a popular algorithm designed to compress sets of strings relative to a given reference string, which acts as a kind of dictionary. It can still applied even when there is no obvious natural reference string for a dataset, by sampling substrings from the dataset and concatenating them to obtain an artificial reference. This works well in practice but a theoretical analysis has been lacking. In this paper we provide such an analysis and verify it experimentally.

1 Introduction

Handling massive datasets is one of the most pressing challenges facing computer scientists today. Many of these datasets are highly compressible but their very size prevents us applying most classic compression algorithms in a reasonable amount of time. For example, despite recent advances [5], running Lempel-Ziv '77 [11] (LZ77) with an unbounded window on a dataset that does not fit in internal memory, is often still prohibitively slow; running it with a bounded window, on the other hand, yields poor compression when the distance between repetitions is larger than the window size (e.g., if we try to compress a database of genomes with a window smaller than a genome). Fortunately, new algorithms have been developed that scale well, such as Kuruppu, Puglisi and Zobel's [6] Relative-Lempel Ziv (RLZ). This algorithm was designed to compress sets of strings which are all similar to a given reference: it stores an index for the reference (e.g., a suffix tree) and then greedily parses the rest of the dataset into substrings that either exactly match, or match except for the final character, some substring of the reference; thus, at compression time, it uses internal memory bounded in terms of the size of the reference, not the whole dataset.

What seems at first to be the main drawback of RLZ—i.e., that we can apply it only when we have a natural reference—has turned out not to be a drawback at all in practice. Several authors [3,7,10] have shown how, by sampling substrings from the entire dataset and concatenating them, we can build an *artificial* reference, with which we can usually still obtain excellent compression.

Supported by the Academy of Finland through grants 268324, 284598, and 294143. Part of this work was done while the first author visited the University of A Coruña, Spain. The authors thank Paweł Gawrychowski for his suggestions.

S. Inenaga et al. (Eds.): SPIRE 2016, LNCS 9954, pp. 160–165, 2016.
DOI: 10.1007/978-3-319-46049-9_16

In fact, this approach works so well that RLZ is now considered one of the best general-purpose compression algorithms for repetitive datasets. As far as we know, however, no one has given a theoretical analysis of why the artificial references work so well. In this paper we provide such an analysis and show that, e.g., if we sample $z \lg^{2+\epsilon} n$ blocks each of length $\sqrt{n/(z \lg n)}$, where n is the length of the dataset and z is the number of phrases in its unbounded-window LZ77 parse, then we can expect to use $\mathcal{O}((nz)^{1/2} \log^{2+\epsilon} n)$ bits of space for the encoding. That is, if LZ77 compresses a dataset well, then, given an appropriately sampled reference, so should RLZ. Our experiments show the compression we observe as we vary the number of blocks qualitatively fits our predictions.

2 Theoretical Analysis

It is intuitively clear that as we increase the number and length of the sampled substrings, the size of the reference will increase but, if the dataset is repetitive, then the complexity of the entire dataset with respect to the reference will tend to decrease. To understand this decrease, consider that if a somewhat shorter substring is common, then it is likely to be included somewhere in the sample, in which case all its occurrences are well-compressed; if it is uncommon, then the cost of storing all its occurrences uncompressed is still not large. In this section we formalize this intuition and provide a theoretical analysis showing that when three conditions hold we can expect good compression.

Suppose we want to compress a string S of length n over an alphabet of size σ whose LZ77 parse consists of z phrases. Results by Rytter [9] and Gawrychowski [2] imply that we can divide S into $\mathcal{O}(n/\ell)$ blocks of length at most $\ell/2$ such that $\mathcal{O}(z \log n)$ of them are distinct, for $2 \le \ell \le n$. Suppose we sample k substrings of length ℓ from S and concatenate them, appending the first and last ℓ characters of S, to obtain the reference. This takes $\mathcal{O}(k\ell \log \sigma)$ bits.

If the ith distinct block has frequency f_i, then the probability one of its occurrences is completely included in some sampled substring is at least $1 - p_i$, where $p_i = \left(1 - \frac{f_i \ell}{2n}\right)^k$. In this case, all of its occurrences can be stored in a total of $\mathcal{O}(f_i \log(k\ell))$ bits; otherwise, they are stored in $\mathcal{O}(f_i \ell \log \sigma)$ bits.

Let $b = \mathcal{O}(z \log n)$ be the number of distinct blocks. By the optimality of greedy parsing, the expected size in bits of the RLZ parse is

$$\mathcal{O}\left(\sum_{i=1}^{b} f_i(1 - p_i)\log(k\ell) + \sum_{i=1}^{b} f_i p_i \ell \log \sigma\right) \le \mathcal{O}\left(\frac{n \log(k\ell)}{\ell} + \ell \log(\sigma) \sum_{i=1}^{b} f_i p_i\right).$$

Since $1 - x \le e^{-x}$, we have $p_i = \left(1 - \frac{f_i \ell}{2n}\right)^k \le \frac{1}{e^{f_i k\ell/2n}}$ so $\sum_{i=1}^{b} f_i p_i \le \sum_{i=1}^{b} \frac{f_i}{e^{f_i k\ell/2n}}$, which is concave and, thus, maximum when all the distinct blocks occur equally often. Therefore, calculation shows

$$\ell \log(\sigma) \sum_{i=1}^{b} f_i p_i = \frac{n \log \sigma}{e^{\Omega(k/z \log n)}}.$$

Summing up, we obtain the following theorem:

Theorem 1. *If we randomly sample and concatenate k blocks of length ℓ from the dataset to form an artificial reference, then the expected total size in bits of the RLZ encoding (i.e., the reference and the parse together) is*

$$\mathcal{O}\left(k\ell \log \sigma + \frac{n \log(k\ell)}{\ell}\right) + \frac{n \log \sigma}{e^{\Omega(k/z \log n)}}.$$

This bound guarantees good compression in the expected case for sufficiently large datasets if three conditions hold simultaneously:

1. $k\ell \ll n$,
2. $\log(k\ell) \ll \ell \log \sigma$,
3. $k = \omega(z \log n)$.

One consequence of this observation is that when LZ77 compresses well, so too should RLZ with an artificial reference sequence when sampling is done appropriately. For example, by setting $k = z \lg^{2+\epsilon} n$ and $\ell = \sqrt{n/k}$, we use $\mathcal{O}\left((nk)^{1/2} \log n\right)$ bits of space. We note as an aside that even when a good reference is given, however, either the length of that reference or the number of phrases in the RLZ parse must be at least the square root of the total length of the dataset. Of course, this is not a serious concern in practice.

Corollary 1. *If we randomly sample and concatenate $z \lg^{2+\epsilon} n$ blocks of length $\sqrt{n/k}$, then the expected total size of the RLZ encoding is $\mathcal{O}\left((nk)^{1/2} \log n\right)$.*

The first two conditions listed above are fairly trivial—they mean, essentially, that the reference should be smaller than the dataset and a pointer into the reference should be smaller than the substring it is replacing—but the third is more interesting. For one thing, while choosing k, say, a tenth larger than optimal should not increase the size of the encoding by more than about a tenth, the last term in our space bound suggests that choosing k a tenth *smaller* than optimal could drastically worsen compression. That is, our analysis suggests that if we plot the size of the encoding against the number of blocks we sample, it should start by falling sharply and then rise slowly. For another thing, it is not clear what "optimal" really means here. The $\log n$ factor comes from the approximation ratio of Rytter's algorithm for building a small straight-line program, and it is unknown if that ratio can be improved even using exponential time (see Charikar et al. [1]).

3   Experimental Results

We checked our prediction that if we plot the size of the encoding against the number of blocks we sample, it should start by falling sharply and then rise slowly, by compressing the following datasets with RLZ using various sampling regimes to build the artificial reference:

English: A concatenation of English text files selected from `etext02` to `etext05` collections of Gutenberg Project[1]. Total size is 1 GB.

Gov2$_{1\,GB}$**:** 1GB Prefix of the TREC collection Gov2[2].

Proteins: A sequence of newline-separated protein sequences (without descriptions) obtained from the Swissprot database. Total size is 1.2 GB (see footnote 1).

Rep$_{1\,GB}$**:** Very repetitive text generated by concatenating 400 copies of a random 25 MB string.

We built an artificial reference for each dataset, as described in Sects. 1 and 2, by picking values of k and ℓ and then concatenating k randomly sampled substrings of length ℓ from the dataset, storing the resulting reference in $k\ell \log \sigma$ bits. We stored phrases using $\lg(k\ell)$ bits for pointers and $\lg \ell$ bits for phrase lengths.

We used two approaches to build the references:

Fixed length: we fixed a value of ℓ and tried different values of k;

Corollary 1: we used the values of k and ℓ prescribed by Corollary 1, using varying estimates of z.

Additionally, with the algorithm of [4], we calculated the size of the LZ77 encoding and include it as a baseline, using $2 \lg n$ bits for each phrase.

The results are shown in Fig. 1. The curves we obtained are in accordance with Theorem 1. That is, there is a sharp drop on the left, when increasing the number of samples significantly improves compression. At some point, the curve starts to rise roughly parallel to the reference size, which increases linearly. That is, as noted in Sect. 2, increasing the number of samples by a tenth, for example, never increases the size of the encoding by more than about a tenth. This explains why the reference pruning strategies employed in [7,10] work: it is much safer to oversample in the beginning than to undersample.

4   Discussion

We have given the first theoretical analysis of artificial reference construction for RLZ, and the results of our experiments qualitatively fit our predictions. One possible concern, however, is that our analysis prescribes sampling parameters in terms of z, which depends on the behaviour of LZ77, and if we could run LZ77 then we would not need RLZ in the first place[3]. We note, however, that it is possible to estimate z more efficiently than by computing the LZ77 parse [8]. Another concern is whether our analysis is reasonably tight. It is not difficult to construct instances in which we achieve better compression by sampling fewer blocks than our analysis prescribes: for example, consider a randomly chosen

---

[1] Available at http://pizzachili.dcc.uchile.cl/.

[2] For more details, see http://ir.dcs.gla.ac.uk/test_collections/gov2-summary.htm.

[3] Though RLZ does have attractive properties other than ease of compression; for example, support for random access.

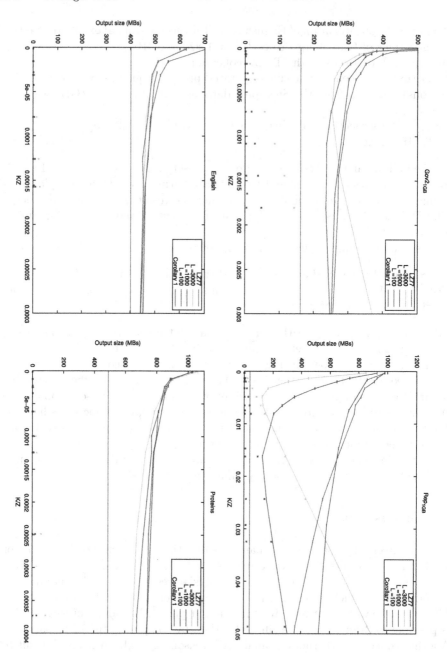

Fig. 1. Size of the RLZ compression using different artificial references on four different collections. For each collection, different regimes for constructing the reference are shown: fixed length of samples of size 3000, 1000, and 100 characters each, and varying amounts (k value) for each of them. Also the strategy proposed by Corollary 1 is used. Our curves show the total size of the encoding as a function of k/z. For each data point in the curves we show a point in the same color below that indicates the size of the reference. In addition, we include the size of the LZ77 encoding as a baseline

binary string B of length $\sqrt{n}$, and let B' be the binary string of length n obtained by replacing each 0 in B by the same randomly-chosen binary string S_0 and replacing each 1 by the same randomly-chosen binary string S_1. Since z is almost certainly around $n^{1/2}/\lg n$, following the sampling regime we described above results in us using about $n^{3/4}\lg^{1+\epsilon} n \lg \sigma$ bits, whereas if we simply sample S_0 and S_1 then RLZ uses about $n^{1/2}\lg n$ bits.

We do not see how to tighten our analysis to address such cases. Even a loose upper bound may still be useful, however: to see why, suppose we want to apply RLZ to a massive dataset in external memory; if our bound tells us that using the prescribed number and lengths of blocks should produce an encoding that fits in internal memory; because of the properties of RLZ, once we have built that encoding we can support fast access to the dataset without reading anything from disk; it follows that we can then try different sampling regimes or pruning strategies [7,10] to find the best, working in internal memory only.

References

1. Charikar, M., Lehman, E., Liu, D., Panigrahy, R., Prabhakaran, M., Sahai, A., Shelat, A.: The smallest grammar problem. IEEE Trans. Inf. Theory **51**(7), 2554–2576 (2005)
2. Gawrychowski, P.: Faster algorithm for computing the edit distance between SLP-compressed strings. In: Calderón-Benavides, L., González-Caro, C., Chávez, E., Ziviani, N. (eds.) SPIRE 2012. LNCS, vol. 7608, pp. 229–236. Springer, Heidelberg (2012)
3. Hoobin, C., Puglisi, S.J., Zobel, J.: Relative Lempel-Ziv factorization for efficient storage and retrieval of web collections. Proc. VLDB **5**, 265–273 (2011)
4. Kärkkäinen, J., Kempa, D., Puglisi, S.J.: Linear time Lempel-Ziv factorization: simple, fast, small. In: Fischer, J., Sanders, P. (eds.) CPM 2013. LNCS, vol. 7922, pp. 189–200. Springer, Heidelberg (2013)
5. Kärkkäinen, J., Kempa, D., Puglisi, S.J.: Lempel-Ziv parsing in external memory. In: Proceedings of the DCC, pp. 153–162 (2014)
6. Kuruppu, S., Puglisi, S.J., Zobel, J.: Relative Lempel-Ziv compression of genomes for large-scale storage and retrieval. In: Chavez, E., Lonardi, S. (eds.) SPIRE 2010. LNCS, vol. 6393, pp. 201–206. Springer, Heidelberg (2010)
7. Liao, K., Petri, M., Moffat, A., Wirth, A.: Effective construction of relative Lempel-Ziv dictionaries. In: Proceedings of the WWW, pp. 807–816 (2016)
8. Raskhodnikova, S., Ron, D., Rubinfeld, R., Smith, A.D.: Sublinear algorithms for approximating string compressibility. Algorithmica **65**(3), 685–709 (2013)
9. Rytter, W.: Application of Lempel-Ziv factorization to the approximation of grammar-based compression. Theor. Comp. Sci. **302**(1–3), 211–222 (2003)
10. Tong, J., Wirth, A., Zobel, J.: Principled dictionary pruning for low-memory corpus compression. In: Proceedings of the SIGIR, pp. 283–292 (2014)
11. Ziv, J., Lempel, A.: A universal algorithm for sequential data compression. IEEE Trans. Inf. Theory **23**, 337–343 (1977)

Inverse Range Selection Queries

M. Oğuzhan Külekci$^{(\boxtimes)}$

Informatics Institute, İstanbul Technical University, Istanbul, Turkey
`kulekci@itu.edu.tr`

Abstract. On a given sequence $X = \langle x_1 x_2 \ldots x_n \rangle$, the range selection queries denoted by $Q(i, j, k)$ return the kth-smallest element on $\langle x_i x_{i+1} \ldots x_j \rangle$. The problem has received significant attention in recent years and many solutions aiming to achieve this task with a cost lower than dynamically sorting the elements on the queried range have been proposed. The reverse problem interestingly has not yet received that much attention, although there exists practical usage scenarios especially in the time–series analysis. This study investigates the inverse range selection query $\bar{Q}(v, k)$ that aims to detect all possible intervals on X such that the kth-smallest element is less than or equal to v. We present the basic solution first and then discuss how that basic solution can be implemented with different data structures previously proposed for regular range selection queries.

1 Introduction

Unprecedented increase in the volume, velocity, and variety of data produced in recent years has led to numerous challenges in algorithmic data management. A massive amount of the data to-be-generated in the near future is expected to be sequences of numbers as a result of the upcoming age of machine-to-machine communication, where many sensors connected to the internet will be continuously transmitting some measurements in the concept of internet-of-things. That fact provides a strong motivation for the studies focusing on management and analysis of number sequences.

One of the fundamental issue in massive number series management is surely the computation of some queries that are particularly important in statistical analysis of those sequences. Range selection queries [1,2,5,7] in that sense has been a good research problem with its possible usage in many practical cases. Given a sequence of integers as $X = \langle x_1, x_2, \ldots, x_n \rangle$, where $x_i \in \{0, 1, 2, \ldots, U\}$ for a large $U >> n$, the range selection query $Q(i, j, k)$ aims to detect the kth-smallest element on $\langle x_i x_{i+1} \ldots x_j \rangle$ for dynamically changing parameters i, j and k. The naive solution for such queries is to sort the numbers in the queried range $\langle x_i x_{i+1} \ldots x_j \rangle$ and then simply return the kth element on that sorted list. However, this would bring a heavy computational load when the sequence as well as the number of queries are huge. Thus, research on this problem targets to provide methods that perform the job without performing a sort operation for each individual query. Many results from different perspectives have appeared

© Springer International Publishing AG 2016
S. Inenaga et al. (Eds.): SPIRE 2016, LNCS 9954, pp. 166–177, 2016.
DOI: 10.1007/978-3-319-46049-9_17

on the range selection queries. An early algorithm, the *quickselect*, developed by Hoare [4] solves the problem in linear time ($O(j - i)$–time) by modifying the *quicksort* algorithm such that once the elements in the queried range are split into two as the ones greater than and smaller than a randomly selected pivot, the piece in which the kth smallest item should appear is recursively split again around a randomly selected pivot in that section until the result is reached. Since only one segment of the array is processed at each step, the $O(n \log n)$ time complexity of quicksort becomes $O(n)$ in the quickselect. The problem has a special interest when $k = \frac{(j-i)}{2}$ as in this case the *median* value, which is a fundamental parameter in statistical analysis of sequences, is returned [6,11].

The inverse of the problem on the other hand has not been mentioned in the studies to the best of our knowledge. Sample practical scenarios to describe the usage of the inverse problem might be given as follows. Assume we have the daily temperature readings from various locations and would like to answer such a query: "What are the intervals in which the temperature did not fall below $10\,°C$ for more than three days?". Yet another exemplary query might be given on the analysis of the sensor data in a large scientific experiment such that one may need to find the intervals where the data read from the sensor has not fallen below a threshold for more than, say 3, times as those 3 cases might be the possible erroneous readings.

Several variants of the problem might also be interesting depending on the practical use cases. For instance, the *inverse range median queries*, where it is required to detect largest intervals whose median value is equal to, greater than, or less than a value might be precious in the analysis of time-series data, e.g., "How many days long is the largest interval where the median temperature had not fallen below $15\,°C$?".

Answering those queries require performing the reverse of the range selection, which we denote with $\bar{Q}(v, k)$ as finding all possible intervals on X such that the k–th smallest element in that interval is less than or equal to k. This definition may be narrowed down to detect the intervals that the k–th smallest is exactly equal to v to precisely reflect the reverse problem of the range selection query definitions. Considering that this would be too restrictive in practice, we prefer to keep it with "less than or equal to". Obviously, all solutions described in this study may support that exact case also.

There is an important distinction between the range selection and its inverse. In range selection, we are given a dynamic range, and the complexity of the operation actually relies on the length of that interval. When that range is small, using the sort based naive solutions might be more efficient in practice. On the other hand, to answer the reverse problem, one needs to investigate the whole sequence regardless of the parameters. Thus, on each query, it is always required to explore X completely to detect the appropriate ranges. When the source sequence X is extremely large, this might turn into a heavy computational load especially when the number of queries is huge. Thus, opportunistic approaches to increase the chance of achieving sub-linear performances on the average would be helpful for the reverse range selection queries.

In this study, we begin by formally describing the problem, and then provide the basic $O(n)$–time solution that requires $O(k)$ words of space. It would not be very meaningful in practice to construct different data structures for the ordinary and the reverse problems. Thus, we consider how to utilize the data structures previously proposed for efficient handling of range selection queries in the reverse problem. Selected previous solutions proposed for the ordinary queries are investigated in this study despite the basic linear solution.

In [2], the authors constructed a wavelet tree [9] over X that returns the answer of a range selection query in $O(\log \sigma)$ time, where σ represents the number of unique integers in X. When all integers are unique in X, the time complexity becomes $O(\log n)$, and the *additional* space requirement reaches $o(n)$ bits. We investigate the opportunity of using the same data structure to answer the reverse problem, and observe that depending on the rank of the v parameter in X, it is possible to achieve sub-linear performance.

Another solution that proposed to use the wavelet trees for the range selection was [7] by which the queries are answered in $O(\log \log U + \log a)$, where a denotes the returned answer, the k-th smallest in the queried range. The data structure introduced in [7] does not only helps efficient handling of the queries, but also keeps the integers in X compact by representing each x_i with its minimal binary description length $\lfloor \log x_i \rfloor$. Despite the space gained by keeping the integers compact, the additional overhead is also smaller as being $n \log \log U + o(n \log \log U)$ bits when compared to $n \log \sigma + o(n \log \sigma)$ bits in the normal wavelet tree solution assuming that usually in practice $\sigma >> \log U$.

In this study, we explore ways of using that data-aware approach [7] for answering the inverse problem, and observe that it may also result a sub-liner performance for some v values.

Besides the wavelet tree based approaches, a sort based solution that first sorts X and stores the rank array in a preprocessing step is also considered.

The outline of the study is as follows. Following the formal definition of the inverse range selection queries in Sect. 2, Sect. 3 is devoted to the possible solutions, where we first analyze the basic linear–time solution, and then investigate the usage of the wavelet–tree based solutions previously proposed for the range selection. The space–efficient and sorting-based approaches are also visited in related subsections of Sect. 3. We conclude with a concise summary of the results including possible future research directions related to the range queries.

2   The Inverse Range Selection Problem

Let $X = \langle x_1, x_2, \ldots x_n \rangle$ be a sequence of integers such that $0 \leq x_i \leq U$ for $1 \leq i \leq n$. The range selection query $v \leftarrow Q(i, j, k)$, for $1 \leq k \leq (j - i + 1)$ and $1 \leq i < j \leq n$, returns the kth smallest value v in the range $\langle x_i, x_{i+1}, \ldots x_j \rangle$.

Definition 1 (Inverse range selection query). *On a given sequence $X = \langle x_1, x_2, \ldots x_n \rangle$, the inverse range selection query $\mathcal{A} \leftarrow \bar{Q}(v, k)$ returns the set $\mathcal{A}$ that includes all possible $\langle i, j \rangle$ tuples, where in $\langle x_i, x_{i+1}, \ldots, x_j \rangle$ the kth smallest value is less than or equal to v.*

On a sample $X = \{16, 17, 3, 6, 2, 11, 5, 2, 3, 15, 16, 9, 13\}$, assume we would like to answer $\bar{Q}(3, 2)$ as detecting all intervals whose 2nd smallest value is less than or equal to 3. The list of valid range pairs for that query is $\langle 1, 5 \rangle$, $\langle 1, 6 \rangle$, $\langle 1, 7 \rangle$, $\langle 2, 5 \rangle$, $\langle 2, 6 \rangle$, $\langle 2, 7 \rangle$, $\langle 3, 5 \rangle$, $\langle 3, 6 \rangle$, $\langle 3, 7 \rangle$, $\langle 4, 8 \rangle$, $\langle 5, 8 \rangle$, $\langle 6, 9 \rangle$, $\langle 7, 9 \rangle$, $\langle 8, 9 \rangle$, $\langle 8, 10 \rangle$, $\langle 8, 11 \rangle$, $\langle 8, 12 \rangle$, $\langle 8, 13 \rangle$.

Definition 2 (Maximal range). *The range $\langle x_i, x_{i+1}, \ldots x_j \rangle$ denoted by the tuple $\langle i, j \rangle$ in the answer set $\mathcal{A}$ of an inverse range selection query $\mathcal{A} \leftarrow \bar{Q}(v, k)$ is a maximal range if there exists no other tuple $\langle m, n \rangle \in \mathcal{A}$ such that $m \leq i < j \leq n$, which means the $\langle i, j \rangle$ interval cannot be expanded either to the right or to the left.*

Following the example above, the intervals $\langle 1, 7 \rangle$, $\langle 4, 8 \rangle$, $\langle 6, 9 \rangle$, and $\langle 8, 13 \rangle$ are the maximal ranges for the $\bar{Q}(3, 2)$ on X.

Lemma 1. *If $\langle x_i, x_{i+1}, \ldots x_j \rangle$ is a maximal range, then $\big[(x_{i-1} \leq v) \vee (i = 1) \big]$ and $\big[(x_{j+1} \leq v) \vee (j = n) \big]$ conditions should hold.*

Proof. According to the Definitions 1 and 2, the $\langle x_i, x_{i+1}, \ldots x_j \rangle$ includes k items that are less than or equal to v, and is not expandable towards right or left. The interval can not be extended to the left when $i = 1$ as this is the leftmost position on X that naturally prohibits moving left. In case $i > 1$, then $x_{i-1} \leq v$ should hold, since the $\langle x_{i-1}, x_i, \ldots x_j \rangle$ interval will otherwise have $k + 1$ items smaller than the queried v parameter. Similarly, the expansion to the right is restricted when $j = n$ or $x_{j+1} \leq v$.

Lemma 2. *If $\langle x_i, x_{i+1}, \ldots x_j \rangle$ is a maximal range, where the $x_{i'}$ and $x_{j'}$ are respectively the leftmost and rightmost values that are less than or equal to v for $i \leq i' \leq j' \leq j$, then all $\langle a, b \rangle$ tuples for $a \in \{i, i+1, \ldots, i'\}$ and $b \in \{j', j' + 1, \ldots, j\}$ are in the set $\mathcal{A}$ of the inverse range selection query $\mathcal{A} \leftarrow \bar{Q}(v, k)$.*

Proof. Following the assumption mentioned, there are k integers that are less than or equal to v in $\langle x_{i'}, x_{i'+1}, \ldots x_{j'} \rangle$. All of the integers in the intervals $\langle x_i, x_{i+1}, \ldots x_{i'-1} \rangle$ and $\langle x_{j'+1}, x_{j'+2}, \ldots x_j \rangle$ should be larger than k, since there will be otherwise more than k items smaller than v, which would violate the query. Then, an interval $\langle x_a, x_{a+1}, \ldots x_b \rangle$, for $a \in \{i, i+1, \ldots, i'\}$ and $b \in \{j', j' + 1, \ldots, j\}$, includes exactly k items less than v, and holds with the $\bar{Q}(v, k)$ query.

We are explicitly interested with the *maximal* ranges since the remaining intervals can be trivially generated from them as proved in Lemma 2. For instance, $\langle 1, 7 \rangle$ is a maximal range as stated in the example of the Definition 2. All $\langle a, b \rangle$ intervals, for $a \in \{1, 2, 3\}$ and $b \in \{5, 6, 7\}$ are in the answer set of the inverse range query for the inverse range selection query $\mathcal{A} \leftarrow \bar{Q}(v, k)$. These subranges as $\langle 1, 5 \rangle$, $\langle 1, 6 \rangle$, $\langle 1, 7 \rangle$, $\langle 2, 5 \rangle$, $\langle 2, 6 \rangle$, $\langle 2, 7 \rangle$, $\langle 3, 5 \rangle$, $\langle 3, 6 \rangle$, $\langle 3, 7 \rangle$ can be computed once it is determined that the leftmost and rightmost positions in the maximal range $\langle 1, 7 \rangle$ having values less than or equal to $v = 3$ are $3(x_3 = 3)$ and $5(x_5 = 2)$. Hence, the primary concern in this study is to detect the maximal ranges for a given inverse range selection query.

Definition 3 (Maximal range set). *The maximal range set $\mathcal{A}' \subseteq \mathcal{A}$ that can be expressed with*

$$\mathcal{A}' = \{\langle i, j \rangle : (Q(i,j,k) \leq v) \wedge [(x_{i-1} \leq v) \vee (i = 1)] \wedge [(x_{j+1} \leq v) \vee (j = n)]\}$$

includes all maximal range $\langle i, j \rangle$ tuples in the answer set $\mathcal{A}$ of the inverse range selection query $\bar{Q}(v, k)$.

3   The Solution

Theorem 1. *The maximal range set of the inverse range selection query $\bar{Q}(v, k)$ on an integer sequence $X = \langle x_1 x_2 \ldots x_n \rangle$ can be detected in $O(n)$-time by using $(k + 1) \cdot \log n$ bits additional space.*

Proof. Following Lemma 1, a maximal range computation requires the knowledge of $(k + 1)$ consecutive positions whose corresponding integers are less than or equal to v on X. A first-in-first-out array $q[1 \ldots (k+1)]$ keeping the positions of the last observed such $(k + 1)$ integers can be maintained while passing over X linearly. The size of that array is $(k+1) \cdot \log n$ bits as each entry is $\log n$ bits long. After detecting the first such $(k + 1)$ integers on X, the initial maximal range is $\langle x_1 \ldots x_{q[k+1]-1} \rangle$, which expands from position 1 to the preceding position of the last item in the array. The next maximal range should begin from the succeeding position of the first element in the q array. After selecting the starting position of the next maximal range as $q[1] + 1$, the next position on X whose corresponding value is less than or equal to v is scanned and the q array is updated. Since q is a FIFO array, this update may be described as shifting all values to the left by one, which disposes $q[1]$, and inserting the newly detected position to the end of the array, $q[k + 1]$. Now the end of the current maximal range is set to the newly computed $(q[k+1] - 1)$. Deciding on the start of the next range, updating the array by keep scanning X linearly, and setting the end according to the latest update is repeated until all elements in X are visited. Since we visit each element of X once during the traversal, and maintain an array of size $(k + 1)$, the procedure detects all maximal ranges in $O(n)$-time and $O(k)$-space.

The pseudo-code shown in Algorithm 1 returns the maximal range set for an inverse range query $\bar{Q}(k, v)$ over X. Notice that we simulate the FIFO array q described in the proof of the Theorem 1 via a circular array in this procedure.

3.1   Solution via Wavelet Tree

Let's assume a hypothetical bitmap $M = m_1 m_2 \ldots m_n$ such that $m_i = 1$ when $x_i \leq v$, and $m_i = 0$ otherwise. Algorithm 1 may be viewed as simulating the search on M for the largest intervals that exactly include k number of 1 bits. While achieving this goal, it does not maintain the bit array M explicitly, but instead computes it on the fly while passing over X linearly with the *"while"* loops appearing at the lines 5 and 17 on Algorithm 1. The efficiency of that

Algorithm 1. InverseRangeSelect(X, k, υ)

Input: $X = \langle x_1, x_2, \ldots x_n \rangle$ is the input integer stream, k and υ are the parameters of the inverse range selection query.
Output: The maximal range set $\mathcal{A}'$ including all possible $\langle i, j \rangle$ pairs such that $\langle x_i, x_{i+1}, \ldots, x_j \rangle$ is a maximal range having k values less than or equal to υ.

```
 1  A' ← {};
 2  z ← 1;
 3  q ← 0;
 4  for i = 1 to k + 1 do   // the positions of the first (k + 1) integers not greater than υ
 5      while ((x_z > υ) ∧ (z ≤ n)) do z ← z + 1;            // find next x_z ≤ υ
 6      arr[q] ← z ;                                 // store the position in arr[q]
 7      z ← z + 1 ;                                  // advance the pointer on X
 8      q ← (q + 1) mod (k + 1) ;        // advance the pointer on the circular array arr
 9      if (z > n) then break;
10  end
11  if (i < (k + 1)) then return A';            // less than (k + 1) items are ≤ υ in X
12  begin ← 1 ;                           // the beginning position of the first interval
13  end ← arr[(q − 1) mod (k + 1)] − 1 ;     // the ending position of the first interval
14  A' ← A' ∪ ⟨begin, end⟩;
15  while (z < n) do
16      begin ← arr[q] + 1;             // save the beginning position of the next interval
17      while ((x_z > υ) ∧ (z ≤ n)) do z ← z + 1;            // find next x_z ≤ υ
18      arr[q] ← z;                      // update the array to store the detected position z
19      end ← arr[q] − 1;                  // compute the ending position of the interval
20      A' ← A' ∪ ⟨begin, end⟩;             // add this interval to the maximal range set
21      q ← (q + 1) mod (k + 1) ;    // manage the circular array to imitate the FIFO list
22  end
23  return A';
```

solution will be improved if a better way of detecting the $x_i \leq \upsilon$ positions can be achieved.

Previously, the study by Gagie *et al.* [2] has proposed to use wavelet trees for the range selection queries. The wavelet tree constructed over a sample X sequence is depicted in Fig. 1. The same structure may also be used for the inverse problem.

Lemma 3. *Given the balanced wavelet tree W constructed over an integer sequence X, detection of the positions of all integers on X that are less than or equal to υ can be computed on the average with*

$$n \cdot \sum_{i=1}^{\lceil \log n \rceil} p_i \cdot \frac{i}{2^i}$$

select *queries, where $p_i \in \{0, 1\}$ indicates whether the top-down traversal for υ on W follows the left ($p_i = 0$) or right child ($p_i = 1$) on the bitmap at level i.*

Proof. We begin searching for the value υ on the balanced wavelet tree W. During this traversal if the pivotal value ($pivot_i$) of the visited node at level i on W is larger than or equal to υ ($\upsilon \leq pivot_i$), we move to the left child that is devoted to the items represented by 0 bits in the current node. Else, if $\upsilon > pivot_i$, then we should follow the right link. However, in this case all the items represented by 0 bits in the current node are known to be less than υ,

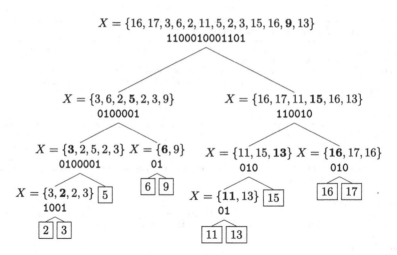

Fig. 1. The wavelet tree constructed on a sample X integer vector. The pivot values at each node are highlighted. The leaf nodes are surrounded with boxes.

and thus, should be traced back on the tree to mark their positions on the M array. For each such 0 bit in the current node at level i, it needs i number of **select** operations until the root is reached and the positions are marked on M. Assuming the number of zeros and ones at each node of the balanced wavelet tree are approximately equal, the number of 0 bits at a node at level i is around $\frac{n}{2^i}$. During the top-down traversal of the W for the value v, no **select** required as we keep going left, and $\frac{n \cdot i}{2^i}$ **select** operations should be performed when we move right at a node at level i.

The pseudocode to detect the integer positions that are less than or equal to a queried value v is given in Algorithm 2. As an example, assume we would like to mark the positions of $x_i \leq 13$ on the given sample X in Fig. 1. Since 13 is larger than the pivotal value 9 of the root node at level 1, the positions of the 7 0-bits in the root bitmap should be marked with 1 on the bitmask M. We move to right child of the root node and compare 13 with the new pivot value 15 of the new node. Since $13 \leq 15$, we follow the left child without any operation on the current node's bitmap. Similarly, as 13 is still less than or equal to the pivotal value of the node at level 3 that is 13, we again take the left link. In the current visited node at 4th level of W, we observe that the pivotal value is 11, which is larger than 13. Thus, we need to traverse the 0-bits in the current node back to the root. We have only one 0 bit here, and the corresponding positions on the 4th, 3rd, 2nd, and 1th level bitmaps are computed as 1, 1, 3, and 6 with a total of 4 **select** queries. After setting $m_6 = 1$ on M, we follow the right child, which takes us to the leaf node devoted to the value 13. As this value is ≤ 13 again, we need to traverse back the 1 bits in the parent node of this leaf. This traversal requires 4 **select** queries as we have only one bit set to 1 in the parental node at level 4. The respective positions computed on the path to the root are

---

Algorithm 2. DetectLEQValues(X, W, v)

Input: $X = \langle x_1, x_2, \ldots x_n \rangle$ is the input integer sequence and the wavelet tree W is the balanced wavelet tree constructed over X according to the scheme proposed by Gaggie *et al.* [2]. The value v is the parameter of the inverse range selection query.

Output: The bitmap $M = m_1 m_2 \ldots m_n$ such that $m_i = 1$ when $x_i \leq v$, else $m_i = 0$.

```
1  B ← W.rootBitmap ;                        // begin with the root bitmap at level i = 1
2  φ ← 0 ;                                    // until the last step we will be selecting 0 bits
3  M[1..n] ← 0 ;                              // initialize bit array M to all 0s
4  continue ← true;                           // iterate until the last step that we reach the leaf
5  while (1) do
6  │   while (v ≤ B.pivotValue) do B ← B.leftChild;      // move left while v ≤ pivot_i
7  │   m ← number of bits set to φ in B ;                //
8  │   for i = 1 to m do     // before moving right do this for each φ bit on current node
9  │   │   B' ← B;
10 │   │   pos ← select_φ(i, B');             // select the i^th φ bit on the current node
11 │   │   while (B' ≠ W.rootBitmap) do       // trace back this position till the root
12 │   │   │   P ← B'.parent;
13 │   │   │   if B' == P.leftChild then
14 │   │   │   │   pos ← select_0(pos, P)
15 │   │   │   else
16 │   │   │   │   pos ← select_1(pos, P)
17 │   │   │   end
18 │   │   │   B' ← P;
19 │   │   end
20 │   │   M[pos] ← 1;                         // mark the detected position on M
21 │   end
22 │   if (continue == false) then break;
23 │   B ← B.rightChild;                       // move to right child
24 │   if ((B.isLeaf)&&(leafValue ≤ v)) then
25 │   │   φ ← 1;              // now we need to trace 1 bits of the parent node back to root
26 │   │   B ← B.parent;
27 │   │   continue ← false ;    // since we reached a leaf, stop after tracing the 1 bits
28 │   else break;
29 end
30 return M;
```

2, 3, 6, and 11. At the end, $M = \{0,0,1,1,1,1,1,1,1,0,0,1,1\}$ is constructed with 15 select queries.

The number of required select queries increase as much as the v value is represented towards the right side of the wavelet tree, which means it resides in the higher quartiles of the sorted X. Thus, the wavelet tree construction over X results sub-linear performance only when the v value resides in the initial quartiles of the sorted X sequence. For instance, on our example, when $v = 2$, 8 select queries are sufficient to create the mask $M = \{0,0,0,0,1,0,0,1,0,0,0,0,0\}$.

Theorem 2. *The inverse range selection query $\bar{Q}(v,k)$ can be answered in $O(\kappa + D)$-time by using $O(n \log \sigma)$ bits space, where κ denotes the number of integers on $X = \langle x_1, x_2, \ldots x_n \rangle$ such that $x_i \leq v$, σ is the number of unique integers in X, and D is the time required to create the mask bitarray M as described in Lemma 3.*

Proof. As a one-time preprocessing operation, the wavelet tree W of size $O(n \log \sigma)$ can be constructed over X, which has σ distinct integers, in $O(n \log \sigma)$ time [3]. Given the $\bar{Q}(v,k)$ query, the mask bit array M with constant-time select support [10] is computed from the wavelet tree W as discussed in

Lemma 3. The time required for this operation is denoted by $O(D)$, where D is the number of **select** queries performed during this computation. Once the M array that includes κ 1 bits is constructed, the *while* loops at lines 5 and 17 can be replaced by constant-time **select** queries to find the next position holding a value less than or equal to v, which reduces the complexity of Algorithm 1 to $O(\kappa)$. The total time complexity for the query becomes $O(\kappa + D)$.

3.2 Solution via Sorting

The number of **select** operations in the wavelet tree based solution may be less than or larger than n depending on the rank of the value of v in the sorted X. When this value is large, the wavelet tree based solution (Theorem 2) is not expected to be competitive with the basic solution (Theorem 1). Yet another approach, which makes use of a space equivalent to the wavelet tree while providing a better time complexity in such a case is described in Theorem 3.

Theorem 3. *When κ denotes the number of integers on $X = \langle x_1, x_2, \ldots x_n \rangle$ such that $x_i \leq v$, the inverse range selection query $\bar{Q}(v, k)$ can be answered in $O(\log n + \kappa)$-time by using $O(n \log \sigma)$ bits additional space.*

Proof. In a preprocessing step that is to be achieved only once, X can be sorted in $O(n \log n)$ time. The positions of the integers in increasing order in this sorted array can be saved by reserving a space of $O(n \log \sigma)$ bits assuming there are σ distinct values in X. For the inverse range selection query, first the position of the largest integer that is less than or equal to v is detected by a binary search that takes $O(\log n)$ time on that indices array. The positions till that point are the positions of the κ integers on X having a value less than or equal to v. As an example, for the sample X depicted in Fig. 1, the sorted values are $\langle 2, 2, 3, 3, 5, 6, 9, 11, 13, 15, 16, 16, 17 \rangle$, and the corresponding indices array is $\langle 5, 8, 3, 9, 7, 4, 12, 6, 13, 10, 1, 11 \rangle$. The positions that hold values less than or equal to 13 are the first 9 items in the indices list since the position of 13 is detected to be position 9 via a binary search incorporating the X and the indices array. The mask array M with the constant-time **select** support is then created by marking those positions on M. The detection of the maximal intervals is achieved according to Algorithm 1 by replacing the *while* loops on lines 5 and 17 with $O(1)$–time select operations on M, which reduces the complexity of the Algorithm 1 from $O(n)$ time to $O(\kappa)$-time.

3.3 Space–Efficient Solution with Compact Integer Representation

Both the wavelet tree based and the sorting based solutions described above require $O(n \log \sigma)$ bits space in addition to the space occupied by the actual integer array X itself. When all integers in X are in between 0 and U, X requires a space of $n \log U$ bits.

Previously, Külekci and Thankachan [7] has proposed to use the compact integer representation introduced in [8] for the range selection problem. Within that

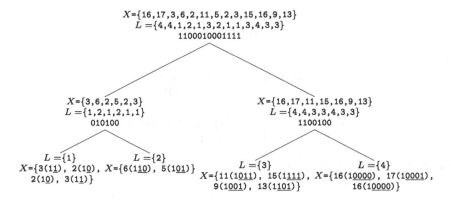

Fig. 2. The wavelet tree constructed over the $\lfloor x_i \rfloor$ values (denoted by L arrays in the tree) of the sample X integer vector as described in [7,8]. The integers at the leaf notes are represented by their minimal binary lengths.

scheme, the integers in X are clustered according to their $\lfloor \log x_i \rfloor$ values by constructing a wavelet tree. Since the $\lfloor \log x_i \rfloor$ values are in $\{0, 1, 2, \ldots \lfloor \log U \rfloor\}$, the height of the balanced wavelet tree is $\lceil \log \log U \rceil$, and thus, the additional space required becomes $O(n \log \log U)$ bits, which is better than $O(n \log \sigma)$ for particularly large n and σ. More than that, since the integers are clustered according to their bit-lengths[1] in the leaves of the wavelet tree, instead of reserving a fixed-length $\log U$ bits per each integer, an individual x_i value is represented by no more than required minimal $\lfloor \log x_i \rfloor$ bits. The reduction caused by the compact representation of the integers compensates the space deposited for the wavelet tree to some extent. Hence, such a representation becomes more space efficient in practice.

Such a wavelet tree constructed over the sample X is depicted in Fig. 2. In [7], it has been proposed to keep the integers in a leaf node again with a secondary wavelet tree that holds the bits of the integers from most significant to least significant one. Figure 3 shows that representation for a sample bunch of integers that all require 3 bits.

Theorem 4. *On a given* $X = \langle x_1 x_2 \ldots x_n \rangle$, *where* $x_i \in \{0, 1, 2, \ldots, U\}$, *the inverse range selection query* $\bar{Q}(v, k)$ *can be answered in* $O(D + \kappa)$-*time with* $O(n \log \log U)$ *bits space by representing* X *with the compact data structure introduced in [7]. Here,* D *denotes the number of* select *operations required on that data structure to detect all* κ *integers that are less than or equal to* v *on* X.

Proof. Since the $\lfloor \log x_i \rfloor$ values are in $\{0, 1, 2, \ldots, \lfloor \log U \rfloor\}$, the balanced wavelet tree constructed over them will require $O(n \log \log U)$ bits. Each leaf node of that tree is devoted to a specific length by which all integers mapped to that node

---

[1] An integer x_i can be represented by $\lfloor \log x_i \rfloor$ bits by omitting the leftmost 1 bit. We assume $\log 0 = 0$ and process the integers 0 and 1 a bit differently on the wavelet tree. See [7,8] for more details.

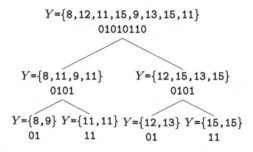

Fig. 3. The wavelet tree representation of a sample Y integer vector from the most-significant to least-significant bits.

can be represented with. The integers belonging a leaf node may be represented by another wavelet tree such that the bits of the integers from the most to least significant positions are collected as depicted in Fig. 1.

To answer a $\bar{Q}(v,k)$ query, first all the items such that $\lfloor \log x_i \rfloor < \lfloor \log v \rfloor$ are detected with the same process described in Algorithm 2 on the top wavelet tree constructed over $\log x_i$ values. This is due to the fact that if $\lfloor \log x_i \rfloor < \lfloor \log v \rfloor$, then $x_i < v$. Notice that we have now at most $\log \log U$ levels to traverse as oppose to $\log \sigma$, which may reduce the number of required select operations significantly when compared to the pure wavelet tree based solution described above.

When we arrive to the leaf node to which the v value should be assigned, we start finding the integers that are less than or equal to v again with Algorithm 2. The number of levels to traverse here is $\lceil \log v \rceil$. At each level if the corresponding bit of the v is 0, we move to the left child without any operation, and if it is 1 we trace back all 0 items to mark their positions.

For large v values the depth of the bottom wavelet tree increases, which also brings an increase in the number of select queries. To avoid this, one may prefer not to keep the integers in the bottom wavelet tree, but instead maintain a rank array as described in Theorem 3. This will bring an additional overhead of $\sum_0^{\lfloor \log U \rfloor} c_i \cdot \log c_i$, where c_i represents the number of integers x_i such that $\log x_i = c_i$ $c_0 + c_1 + \ldots + c_{\lfloor \log U \rfloor} = n$ bits. However, since the efficiency will be improved this trade–off might make sense according to the structure of the X.

4   Conclusions

This study has shed an initial light on the inverse range selection queries that have not been explicitly addressed to date. Unlike the ordinal range selection queries, where the query is focused on a specific region of the data, while answering an individual inverse range selection query, one needs to process the complete sequence at each time.

Although there had been significant theoretical results on range selection queries, studies addressing the practical performances of the proposed schemes is limited. Combined with the inverse range selection queries introduced in this study, evaluating the methods in an experimental framework is believed to make sense in future research.

Acknowledgments. The author thanks to the anonymous reviewers of the paper for their valuable corrections and comments.

References

1. Chan, T.M., Wilkinson, B.T.: Adaptive and approximate orthogonal range counting. In: Proceedings of the Twenty-Fourth Annual ACM-SIAM Symposium on Discrete Algorithms (SODA), pp. 241–251 (2013)
2. Gagie, T., Puglisi, S.J., Turpin, A.: Range quantile queries: another virtue of wavelet trees. In: Karlgren, J., Tarhio, J., Hyyrö, H. (eds.) SPIRE 2009. LNCS, vol. 5721, pp. 1–6. Springer, Heidelberg (2009)
3. Grossi, R., Gupta, A., Vitter, J.S.: High-order entropy-compressed text indexes. In: Proceedings of the 14th Annual ACM-SIAM Symposium on Discrete Algorithms, pp. 841–850. SIAM (2003)
4. Hoare, C.A.R.: Algorithm 65: find. Commun. ACM **4**(7), 321–322 (1961)
5. Jørgensen, A.G., Larsen, K.G.: Range selection and median: tight cell probe lower bounds and adaptive data structures. In: Proceedings of the Twenty-Second Annual ACM-SIAM Symposium on Discrete Algorithms, pp. 805–813 (2011)
6. Krizanc, D., Morin, P., Smid, M.: Range mode and range median queries on lists and trees. Nord. J. Comput. **12**(1), 1–17 (2005)
7. Külekci, M.O., Thankachan, S.V.: Range selection queries in data aware space and time. In: Data Compression Conference (DCC), pp. 73–82. IEEE (2015)
8. Külekci, M.O.: Enhanced variable-length codes: improved compression with efficient random access. In: Data Compression Conference (DCC), pp. 362–371. IEEE (2014)
9. Navarro, G.: Wavelet trees for all. In: Kärkkäinen, J., Stoye, J. (eds.) CPM 2012. LNCS, vol. 7354, pp. 2–26. Springer, Heidelberg (2012)
10. Okanohara, D., Sadakane, K.: Practical entropy-compressed rank/select dictionary. In: Proceedings of the Meeting on Algorithm Engineering & Expermiments, pp. 60–70. Society for Industrial and Applied Mathematics, Philadelphia (2007). http://dl.acm.org/citation.cfm?id=2791188.2791194
11. Petersen, H., Grabowski, S.: Range mode and range median queries in constant time and sub-quadratic space. Inform. Process. Lett. **109**(4), 225–228 (2009)

Low Space External Memory Construction of the Succinct Permuted Longest Common Prefix Array

German Tischler[(✉)]

Max Planck Institute of Molecular Cell Biology and Genetics,
Pfotenhauerstraße 108, 01037 Dresden, Germany
tischler@mpi-cbg.de

Abstract. The longest common prefix (LCP) array is a versatile auxiliary data structure in indexed string matching. It can be used to speed up searching using the suffix array (SA) and provides an implicit representation of the topology of an underlying suffix tree. The LCP array of a string of length n can be represented as an array of length n words, or, in the presence of the SA, as a bit vector of $2n$ bits plus asymptotically negligible support data structures. External memory construction algorithms for the LCP array have been proposed, but those proposed so far have a space requirement of $O(n)$ words (i.e. $O(n \log n)$ bits) in external memory. This space requirement is in some practical cases prohibitively expensive. We present an external memory algorithm for constructing the $2n$ bit version of the LCP array which uses $O(n \log \sigma)$ bits of additional space in external memory when given a (compressed) BWT with alphabet size σ and a sampled inverse suffix array at sampling rate $O(\log n)$. This is often a significant space gain in practice where σ is usually much smaller than n or even constant. The algorithm has average run-time $O(n \log n \log \sigma)$ and worst case run-time $O(n^2 \log \sigma)$. It can be improved to $O(n \log^2 n \log \sigma)$ worst case time while keeping the same space bound in external memory if $O(n/\log n)$ bits of internal memory are available. We also present experimental data showing that our approach is practical.

1   Introduction

The suffix array (SA) and longest common prefix array (LCP) were introduced as a lower memory variant of the suffix tree (cf. [30]) for exact string matching using a precomputed index (cf. [20]). For a text of length n both can be computed in linear time in internal memory (IM) (cf. [4,18,19]) and require n words of memory each. For large texts the space requirements of SA and LCP in IM can be prohibitive. Compressed and succinct variants including compressed suffix arrays (see e.g. [13,14,23]), the FM index and variants (see [7–9]) and succinct LCP arrays (see [24]) use less space, but for practicality it is also crucial to be able to construct these data structures using affordable space requirements. Construction algorithms for compressed suffix arrays and the Burrows Wheeler

© Springer International Publishing AG 2016
S. Inenaga et al. (Eds.): SPIRE 2016, LNCS 9954, pp. 178–190, 2016.
DOI: 10.1007/978-3-319-46049-9_18

transform (BWT, see [3]) using $o(n \log n)$ bits of space in IM (assuming $\sigma \in o(n)$) were introduced (see e.g. [15,22]). However they require an amount of IM which is several times larger than what is needed for the input text. An algorithm for constructing the succinct LCP array in IM using a compressed suffix array was given in [25]. External memory solutions for constructing the suffix array and LCP array have also been presented (see e.g. [2,5,16]). These algorithms require $O(n)$ words ($O(n \log n)$ bits) of external memory (EM). However, as for their IM pendants, this space requirement is large if the algorithms are used as a vehicle to obtain a compressed representation. Recently algorithms for constructing the BWT in EM without explicitly constructing a full suffix array were designed and implemented (see [6,27]). In this paper we present an algorithm for constructing a succinct LCP array in EM based on a BWT and sampled inverse suffix array while using $O(n \log \sigma)$ instead of $O(n \log n)$ bits of space in EM. Both the BWT and sampled inverse suffix array can be produced in space $O(n \log \sigma)$ bits in external memory by the algorithm presented in [27,28]. We also present experimental data showing that our approach is practical.

2   Definitions

Let Σ denote a totally ordered and ranked alphabet, we assume $\Sigma = \{0, 1, \ldots, \sigma - 1\}$ for some $\sigma > 0$. Throughout this paper we consider a string $s = s_0 s_1 \ldots s_{n-1}$ of length $|s| = n > 0$ over Σ s.t. the last symbol of s is the minimal symbol in s and does not appear elsewhere in s. The assumption of $\Sigma = \{0, 1, \ldots, \sigma - 1\}$ for some $\sigma > 0$ is w.l.o.g. if $\sigma \in O(n^k)$ for some finite exponent k. We use $s[i]$ to denote s_i and $s[i \mathinner{.\,.} j]$ for $s_i s_{i+1} \ldots s_j$ for $0 \le i \le j < n$. $s[i \mathinner{.\,.} j]$ is the empty string for $i > j$. The i'th suffix of s denoted by $\tilde{s}_i$ is the string $s[i \mathinner{.\,.} n-1]$. Suffix $\tilde{s}_i$ is smaller than $\tilde{s}_j$ for $i \ne j$ (denoted by $\tilde{s}_i < \tilde{s}_j$) if for the smallest k s.t. $s[i+k] \ne s[j+k]$ we have $s[i+k] < s[j+k]$. The suffix array SA of s is the permutation of $0, 1, \ldots, n-1$ s.t. $\tilde{s}_{\text{SA}[i-1]} < \tilde{s}_{\text{SA}[i]}$ for $i = 1, 2, \ldots, n-1$. For two suffixes $\tilde{s}_i$ and $\tilde{s}_j$ with $i \ne j$ the longest common prefix $\text{lcp}(i, j)$ of the two is $[i \mathinner{.\,.} i + \ell - 1]$ for the smallest ℓ s.t. $s[i + \ell] \ne s[j + \ell]$. The longest common prefix array LCP of s is defined by $\text{LCP}[i] = |\text{lcp}(\text{SA}[i - 1], \text{SA}[i])|$ for $i > 0$ and $\text{LCP}[0] = 0$. The inverse suffix array ISA of s is defined by $\text{ISA}[\text{SA}[i]] = i$ for $0 \le i < n$. The permuted LCP array PLCP of s is given by $\text{PLCP}[i] = \text{LCP}[\text{ISA}[i]]$ for $0 \le i < n$ and $\text{PLCP}[i] = 0$ otherwise. The Burrows Wheeler transform BWT of s is defined by $\text{BWT}[i] = s[(\text{SA}[i] - 1) \bmod n]$ for $0 \le i < n$. Let C be the array of length σ s.t. $C[a] = |\{i \mid s[i] = a\}|$ for $a \in \Sigma$ and let D be an array of length $\sigma + 1$ s.t. $\text{D}[a] = \sum_{i < a} C[i]$ for $0 \le a \le \sigma$. For a sequence $t = t_0, t_1, \ldots, t_{k-1}$ for some $k \ge 0$ let $\text{RANK}_t(a, j) = |\{i | 0 \le i < \min(j, k), t_i = a\}|$, i.e. the number of a elements in t up to but excluding index j and let $\text{SELECT}_t(a, j) = \min\{i \mid \text{RANK}_t(a, i + 1) = j + 1\}$ if $0 \le j < \text{RANK}_t(a, k)$ and undefined otherwise. LF is defined by $\text{LF}(r) = \text{ISA}[(\text{SA}[r] - 1) \bmod n]$. R is defined by $\text{R}(a, i) = \text{D}[a] + \text{RANK}_{\text{BWT}}(a, i)$ for $a \in \sigma, 0 \le i \le n$ and BACKSTEP by $\text{BACKSTEP}(a, (i, j)) = (\text{R}(a, i), \text{R}(a, j))$ for $a \in \Sigma, 0 \le i, j \le n$. A rank r is called reducible (the notion was introduced in [17]) if $\text{BWT}[r] = \text{BWT}[(r - 1) \bmod n]$.

Otherwise it is irreducible. We denote the block transfer size used for EM by B (see [29]) and assume this expresses the number of words per block, i.e. one disk block stores $B \log n$ bits.

3   Previous Work

The first linear time algorithm for computing the LCP array from the suffix array and text appeared in [19]. One of the main combinatorial properties used by this algorithm is the fact that $\texttt{PLCP}[i] \geq \texttt{PLCP}[i-1] - 1$ for $0 < i < n$. This property is also used in [24] to obtain a representation of the PLCP array using $2n + o(n)$ bits while allowing constant time access. Let $\zeta(0) = 1$ and $\zeta(i) = 0\zeta(i-1)$ for $i > 0$ (the concatenation of a zero bit and $\zeta(i-1)$). The $2n$ bits in the data structure are the bit sequence $K = \eta(n-1)$ given by $\eta(0) = \zeta(\texttt{PLCP}[0] + 1)$ and $\eta(i) = \eta(i-1)\zeta(\texttt{PLCP}[i] - \texttt{PLCP}[i-1] + 1)$ for $0 < i < n$. The $o(n)$ additional bits are used for a select index (cf. [21]) on K. K stores the sequence of pairwise differences of adjacent PLCP values shifted by 1 in unary representation (the number i is represented as i zero bits followed by a 1 bit). The value $\texttt{PLCP}[i]$ can be retrieved as $\text{SELECT}_K(1, i) - 2(i+1) - 1$. In [1] Beller et al. present an algorithm for computing the LCP array in IM using a wavelet tree (see [13]). This algorithm runs for $\ell_m + 1$ rounds where ℓ_m is the maximum LCP value produced. Round i for $0 \leq i \leq \ell_m$ sets $\texttt{LCP}[r]$ for exactly those ranks r s.t. $\texttt{LCP}[r] = i$, i.e. the values are produced in increasing order.

4   Computing the Succinct PLCP Array

4.1   Constructing the PLCP Bit Vector by Observing LCP Values in Increasing Order

In this section we modify the algorithm by Beller et al. (cf. [1]) to produce the succinct $2n$ bit PLCP bit vector in EM. The main idea is to use the fact that the algorithm produces the LCP values in increasing order. It starts with a tuple $(\epsilon, (0, n))$ which denotes the empty word and the corresponding rank interval on the suffix array (the lower end 0 is included, the upper n is excluded). Round i takes the tuples from the previous round (or the start tuple for round 0) and considers all possible extensions by one symbol via backward search (cf. [7]), i.e. it produces $(aw, (l', r'))$ from $(w, (l, r))$ for each aw appearing in s. All suffixes considered in round i starting by aw in the rank interval (l', r') have a common prefix of length $i + 1$, while the suffixes at ranks $l' - 1$ and l' (for $l' \neq 0$) as well as at ranks $r' - 1$ and r' (for $r' < n$) have a common prefix of at most length i. Based on this insight we can set $\texttt{LCP}[l']$ and $\texttt{LCP}[r']$ to i, if they have not already been set in a previous round. In the tuples the first (string) component is only provided for the sake of exposition, the algorithm does not require or use it. In addition the algorithm prunes away intervals when a respective LCP value (Beller et al. use the upper bound r' for setting new values in [1], in our implementation we use the lower bound l' as it simplifies the transition to EM) is already set.

The succinct PLCP array K contains n zero and n one bits. The one bits mark positions in the text (remember PLCP is in text order). The zero bits encode the differences between adjacent PLCP values shifted by 1. For computing this bit vector assume that we start off with a vector of n one bits. The information we need in addition is how many 0 bits we have to insert in front of any given 1 bit in the sequence. If PLCP$[i]$ is not smaller than PLCP$[i-1]$, then we have to add PLCP$[i]$ − PLCP$[i-1]$ + 1 zero bits just in front of the $(i+1)$st 1 bit. In the algorithm we can achieve this by starting to add 0 bits for ranks r which did not have their value set in a previous round but which do have the value for the rank of the previous position (obtainable as LF(r)) set in the current round. We call this adding a rank to the active set. We stop adding 0 bits for a rank in the round in which the value for the rank itself gets set, which we call removing a rank from the active set.

4.2   Computing Backward Extensions in External Memory

The first obstacle we need to overcome is to compute backward extensions in external memory. Let

$$I = [I_0 = (\ell_0, r_0), I_1 = (\ell_1, r_1), \ldots, I_{z-1} = (\ell_{z-1}, r_{z-1})]$$

denote a sequence of intervals s.t. $\ell_0 = 0$, $r_{z-1} = n$, $r_{i-1} = \ell_i$ for $1 \leq i < z$ and $\ell_i \neq r_i$ for $0 \leq i < z$. We call any sequence of intervals with these properties a partition. Note that BACKSTEP$(a, (\ell, r))$ for $a \in \Sigma$ and an interval (ℓ, r) can, if the result is non empty, be computed by extracting the smallest rank ℓ' and the largest rank r' in $\ell, \ell + 1, \ldots, r - 1$ s.t. BWT$[\ell'] = $ BWT$[r'] = a$ and subsequently computing $(\text{LF}(\ell'), \text{LF}(r') + 1)$ which by the definition of BACKSTEP is equivalent to $(\text{LF}(\ell'), \text{LF}(\ell') + c)$ for $c = \{i \mid \ell \leq i < r, \text{BWT}[i] = a\}$. Due to the properties of the BWT the LF function for some rank r can be computed by marking rank r in the BWT, sorting the BWT stably by symbol and extracting the index of the marked symbol after sorting (see [3]). Given a sequence of source ranks $\ell_0, \ell_1, \ldots, \ell_{z-1}$ in strictly increasing order we can compute the set LF(ℓ_0), LF(ℓ_1), $\ldots$, LF(ℓ_{z-1}) by marking all of the source ranks in the BWT, sorting it and checking where the marked symbols ended up. If we have BWT$[\ell_0] = $ BWT$[\ell_1] = \ldots = $ BWT$[\ell_{z-1}]$ then LF(ℓ_0), LF(ℓ_1), $\ldots$, LF(ℓ_{z-1}) will also be a strictly increasing sequence as LF$(r) = $ R$(\text{BWT}[r], r)$ and R$(\text{BWT}[r], r)$ is strictly increasing for any strictly increasing sequence of ranks $r_0, r_1, \ldots, r_{x-1}$ s.t. BWT$[r_0] = $ BWT$[r_1] = \ldots$ BWT$[r_{x-1}]$. The algorithm shown in Fig. 1 uses these properties to compute all backward extensions for a given partition I using a bucketing approach. The BWT is scanned sequentially and can thus be in any compressed form allowing efficient sequential scanning. The partition I can be stored as two bit vectors of length n where one bit vector marks the lower interval ends and the other the upper end minus one (see [1]). The K_i sequences are stored in the same way. The algorithm runs in time $O(n\sigma)$. This approach however is not suitable for external memory for all but the smallest alphabets, as it requires σ output streams for storing the output intervals. This is fixed by transforming the algorithm to the one shown in Fig. 2 by turning the pure bucketing approach (use one bucket per alphabet symbol) to a radix sort along the bit representation of the alphabet symbols (use $\lceil \log_2 \sigma \rceil$ rounds of bucket sorting using 2 buckets. The rounds use the bits of the alphabet symbols from the least to most the most significant one for bucketing). This brings the runtime of the algorithm to $O(n \log \sigma)$. However it requires to sort the BWT along the way, which involves decoding and encoding it $\log(\sigma)$ times and thus increases the amount of I/O necessary from a linear scan to $O(\frac{n \log^2 \sigma}{B \log n})$. The extension of the algorithm to a radix base larger than two is straight forward, which can save some I/O in practice. This allows us to compute all required backward extensions for one round of the algorithm in time $O(n \log \sigma)$ and $O(\frac{n \log^2 \sigma}{B \log n})$ I/O.

BACKWARDEXTENDSIGMA(BWT, n, D, σ, I)

```
 1  for i ← 0 to σ − 1 do
 2       (s_i, K_i) ← (D_i, empty)
 3  foreach J = (j_l, j_r) ∈ I do
 4       ▷ Erase symbol counts
 5       for i ← 0 to σ − 1 do
 6            c_i ← 0
 7       ▷ Compute symbol counts in interval
 8       for i ← j_l to j_r − 1 do
 9            c_{BWT[i]} ← c_{BWT[i]} + 1
10       ▷ Output intervals for symbols with non zero count
11       for i ← 0 to σ − 1 do
12            if c_i ≠ 0 then
13                 append (s_i, s_i + c_i) to K_i
14                 s_i ← s_i + c_i
15  I ← concatenation of K_i for i = 0, 1, ..., σ − 1
16  return I
```

Fig. 1. Backward extension of a partition I by all symbols in Σ (bucketing approach)

4.3   Computing LCP Difference Bits

We store the set of ranks set so far using a bit vector S in external memory. Having computed the backward extensions we can determine the vector T of ranks to be newly set in this round by extracting the interval lower bounds and combining them with the S vector. This takes linear time and I/O $O(n/(B \log n))$. To determine which ranks need to be activated we need to compute which ranks r are not set in S but $LF(r)$ will be set in this round, i.e. rank $LF(r)$ is set in the T vector. To this end we reorder T along an inverse LF mapping. Like the forward LF mapping corresponds to a sorting operation on the BWT the inverse

LF mapping corresponds to an inverse sorting operation on the BWT, i.e. taking the sorted BWT and performing reorder operations to obtain the original BWT while avoiding to swap equal elements. Like the forward sorting this can be implemented using a bucket/radix sorting method. Figure 3 shows an inverse sorting procedure for binary keys based on bucket sorting. The key sequence is scanned from left to right and values from the blocks corresponding to the 0 and 1 key areas in the data array are extracted as required. Radix inverse sorting can be implemented by storing the binary key sequences as they are observed during a forward radix sorting based on binary bucket sorting and performing the inverse sorting in reverse order. Binary bucket inverse sorting requires three input streams scanned sequentially (one for the key sequence, two for the 0 and 1 areas of the value sequence) and a single output stream.

```
BACKWARDEXTENDLOGSIGMA(BWT, σ, I)
 1  for d ← 0 to ⌈log₂(σ − 1)⌉ do
 2      (s₀, K₀, E₀, s₁, K₁, E₁) ← (0, empty, empty, 0, empty, empty)
 3      foreach J = (jₗ, jᵣ) ∈ I do
 4          (c₀, c₁) ← (0, 0)
 5          for i ← jₗ to jᵣ − 1 do
 6              a ← BWT[i]
 7              if a&2ᵈ = 0 then
 8                  c₀ ← c₀ + 1
 9                  append a to E₀
10              else c₁ ← c₁ + 1
11                  append a to E₁
12          for i ← 0 to 1 do
13              if cᵢ ≠ 0 then
14                  append (sᵢ, sᵢ + cᵢ) to Kᵢ
15                  sᵢ ← sᵢ + cᵢ
16      (I, BWT) ← (concatenation of K₀ and K₁, concatenation of E₀ and E₁)
17  return I
```

Fig. 2. Backward extension of a partition I by all symbols in Σ (radix sort version)

We inverse sort the T bit vector to T^i based on the BWT. For this note that the inverse sorting key bits can be computed once and do not need to be recomputed for each round of the algorithm. The inverse sorting requires time $O(n \log \sigma)$ and $O(\frac{n \log^2 \sigma}{B \log n})$ I/O. Then we scan T^i and S to find ranks which are set in T^i but not in S. These ranks are marked in a bit vector we use to denote the set of active ranks. This takes time $O(n)$ and I/O $O(n/(B \log n))$. Using this updated vector of active ranks we can update the vector PD storing the information we have accumulated about the succinct LCP array so far. This vector is initialised by n one bits before the first round. If rank r is present (i.e. set, but not necessarily added to the active bit vector in this round) in the active bit vector, then we add one zero bit ahead of the one bit for rank r. The vector PD will grow from n to $2n$ bits during the algorithms progress. After updating the PD vector we deactivate

```
BININVERSEBUCKETSORT(K, A, m)
1   (cnt[0], cnt[1]) ← (0, 0)
2   for i ← 0 to m − 1 do
3       cnt[K[i]] ← cnt[K[i]] + 1
4   (cnt[0], cnt[1]) ← (0, cnt[1])
5   for i ← 0 to m − 1 do
6       B[i] ← A[cnt[K[i]]]
7       cnt[K[i]] ← cnt[K[i]] + 1
8   return B
```

Fig. 3. Inverse binary bucket sorting for key vector K and data vector A of length n

all ranks marked in the T vector which are set in the active bit vector and mark all ranks set in the T vector in the S vector. These operations all together take time $O(n)$ and $O(n)$ elements of I/O $O(n/(B \log n))$. In total one round of the algorithm takes time $O(n \log \sigma)$ and I/O $O(\frac{n \log^2 \sigma}{B \log n})$. The algorithm runs as long as the S vector has unset ranks. This can in the worst case take $n − 1$ rounds, but on average ends after $O(\log_\sigma n)$ rounds (cf. [26]).

4.4   Reordering from Rank to Position Order

When all ranks are marked in S we still have to take care of the task of transforming the vector from rank to position order. To this end assume we have a sampled inverse suffix array of sampling rate $ʃ$, i.e. the sequence ISA[0], ISA[$ʃ$], ... We generate the list P of tuples (ISA[p], p, 0, empty, 1) for $p = 0, ʃ, \ldots$ and sort it by the first (rank) component. The third component stores how many LCP difference values we have attached so far, the fourth stores those values (we encode them in γ code so the total space is bounded by $O(n)$ bits, cf. [28]) and the fifth component marks the tuple's activity. This takes time $O(\frac{n}{ʃ} \log n)$ and I/O $O(\frac{n \log n}{ʃB})$. Then we run $ʃ$ rounds of the following sequence of steps. Perform an LF operation on the tuples turning the first two components from (r, p) to $(\text{LF}(r), p − 1 \mod n)$. This can be done by sorting as stated above in time $O(n \log \sigma)$ and $O(\frac{\log \sigma}{B}(\frac{n}{ʃ} + \frac{n}{\log n}))$ I/O operations. Scan PD and P and add values from PD to P for matching ranks in P (first component). For adding to a tuple the third component of the tuple is incremented by one and the new value is prepended to the list of the fourth component. Tuples which have reached a position s.t. $p \mod ʃ = 0$ are deactivated to avoid copying the same value several times. In total the $ʃ$ rounds require time $O(n \log \sigma ʃ)$ and I/O $O(\frac{\log \sigma}{B}(n + \frac{nʃ}{\log n}))$. Afterwards the list P is sorted by the second (position) component and subsequently the values in the fourth component are extracted in order to obtain the final succinct LCP array. This takes time $O(\frac{n \log n}{ʃ})$ and I/O $O((\frac{n}{ʃ} + \frac{n}{\log n})\frac{1}{B})$.

Assuming an ISA sampling rate $ʃ \in \Omega(\log n / \log \sigma)$ the algorithm has a space requirement of $O(n \log \sigma)$ in external memory and an I/O volume of $O(\frac{n \log \sigma}{B}(1 + \frac{\ell_m \log \sigma}{\log n}))$. If the largest appearing LCP value is ℓ_m. We summarise this section in the following theorem.

Theorem 1. *The succinct 2n bit PLCP representation for a string s of length n can, given its BWT and sampled suffix array of sampling rate $\int \in \Theta(\log n)$, be constructed in worst cast time $O(n^2 \log \sigma)$ and I/O $O(\frac{n^2 \log^2 \sigma}{B \log n})$ and average time $O(n \log n \log \sigma)$ using I/O $O(\frac{n \log^2 \sigma}{B})$. In both cases it uses space $O(n \log \sigma)$ bits of space in EM.*

5   Improvement of Worst Case

While on average our algorithm has a run-time of $O(n \log n \log \sigma)$ as the LCP values are $O(\log n)$ on average, we often see cases in practice where, while most of the LCP values are small (in the order of $\log n$), there are some significantly larger values as well. In this case an easy adaption of our algorithm is to stop the computation of the PD vector after a certain number of rounds (say $3 \log n$) and compute the missing values using the algorithm presented in [16]. This adaption can be performed using the following steps before reordering the LCP difference values from rank to position order.

First we remove all values from PD for ranks which are still in the active set by setting them back to zero as they are incomplete. Then we mark all ranks in S which are reducible. Such ranks can be safely ignored because their value in PD will be zero in the end (cf. [17]). Then compute the bit vector U (unset) by inverting S. Let u denote the number of bits set in U. These steps can be performed in linear time and I/O $O(n/(B \log n))$. Compute the bit vector U_{LF} by setting $LF(r)$ in U_{LF} for each rank r in U, then merge U_{LF} into U. We add those ranks because we need them to compute LCP differences between a position and the previous position. This takes time $O(n \log \sigma)$ and $O(\frac{n \log^2 \sigma}{B \log n})$ I/O. For each rank r in U add rank $r - 1 \bmod n$ in linear time and I/O $O(n/(B \log n))$. These ranks are required for computing LCP values by definition. Similar to transforming rank to position order above use $\int$ rounds of LF operations to annotate the ranks produced with the respective positions in time $O(n \log \sigma \int)$ and I/O $O(\frac{1}{B}(n + \frac{n \int \log \sigma}{\log n}))$. Extract tuples $(r-1 \bmod n, \mathsf{SA}[r-1 \bmod n], r, \mathsf{SA}[r])$ from the pairs generated by checking for neighbouring ranks in the pairs in time $O(u)$ and $O(u/B)$ I/O. Sort those tuples by the last component in time $O(u \log n)$ and $O(\frac{u \log n}{B})$ I/O operations. Compute LCP values based on these tuples using the algorithm in [16], which gives us new tuples $(r, p, \mathsf{LCP}[r])$. It takes time $O(n^2/(M \log_\sigma n) + n \log_{M/B}(n/B))$ if $M \log n$ bits of memory are available and $O(n^2/(MB(\log_\sigma n)^2) + n/B \log_{M/B} n/B)$ (see [16]) I/O. Sort the resulting tuples by the position component in time $O(u \log n)$ and I/O $O(\frac{u \log n}{B})$. Combine tuples for position p with such for position $p - 1 \bmod n$ if present and drop the position component leaving us with tuples (r, δ_l) in time $O(u)$ and I/O $O(u/B)$. Sort these tuples by rank in time $O(u \log n)$ and I/O $O(\frac{u \log n}{B})$. Finally merge the information computed into the PD vector in time $O(n)$ using I/O $O(n/(B \log n) + u/B)$ and continue the algorithm above. Assuming $\int \in O(\log n)$ and $\sigma \in o(n)$ the total I/O for this alternative path inside the algorithm is $O(\frac{n \log \sigma + u \log n}{B} + n^2/(MB(\log_\sigma n)^2) + n/B \log_{M/B} n/B)$. The first term is even in the average case dominated by the I/O required in the earlier stage of the algorithm (Sect. 4).

Using this hybrid algorithm we can obtain a trade off between the faster worst case run-time of the algorithm presented in [16] given sufficient IM and the reduced EM space usage of our algorithm presented above. In this second stage of the hybrid algorithm we are generally only interested in computing values for so called irreducible LCP values (cf. [17]) as only such values produce 0 bits in the succinct PLCP vector. The sum over all irreducible LCP values for any string of length n is bounded by $2n \log n$ (see [17]). This bound is reached for de Bruijn strings (cf. [17]), however in this setting each irreducible LCP value is $\Theta(\log n)$. If we run the algorithm from the previous Sect. 4 for $O(\log^2 n)$ rounds, then all LCP values which remain unset must have a value of $\Omega(\log^2 n)$, which means there are $O(\frac{n}{\log n})$ such values and consequently the hybrid algorithm runs in worst case time $O(n \log^2 n \log \sigma)$ while using $O(n \log \sigma)$ space in EM and $O(\frac{n}{\log n})$ bits in IM.

Theorem 2. *Given the BWT and sampled inverse suffix array of sampling rate $\int \in \Theta(\log n)$ for a string s of length n over an alphabet of size σ the succinct permuted LCP array for s can be computed in time $O(n \log^2 n \log \sigma)$ and I/O $O(\frac{n \log n \log^2 \sigma}{B})$ while using $O(n \log \sigma)$ bits of space in EM and $O(\frac{n}{\log n})$ bits of space in IM.*

As the bound of $2n \log n$ for the sum over the irreducible LCP values of a string is obtained for LCP values which are all of length $O(\log n)$ the interesting question remains whether there is a smaller upper bound for the sum of the irreducible LCP values when only LCP values in $\omega(\log n)$ are considered in the sum.

6   Experimental Results

Presenting an algorithm with quadratic worst case behaviour, even one which can be hybridised with another algorithm to obtain subquadratic run-time, begs the question whether the algorithm is in any way practical, i.e. whether the lower space usage of $O(n \log \sigma)$ bits in EM is bought at the cost of an intolerable run-time. We have implemented a version of the hybrid algorithm. It is available in the lcpbit program of the bwtb3m suite on github[1]. The program contains a parallelised version of the algorithm. We omit the details of the parallelisation from this paper due to lack of space. The first stage of the algorithm was run until less then $n/(2 \log n)$ irreducible ranks were unset in S. Tests were performed on 48 core machines equipped with Intel Xeon E5-2670v3 CPUs running at 2.30 GHz and 1 TB of RAM, the actual amount of memory we used was however set to much lower limits. The machines were running CentOS Linux 7.2.1511 (kernel version 3.10.0). Files were stored on a network file system (Lustre). For testing we used two data sets. The first, called DNA, consists of 20 mammalian genomes

---

[1] https://github.com/gt1/bwtb3m.

downloaded from the NCBI reference database[2]. We extracted the sequences and replaced indeterminate N bases randomly by determinate (A,C,G,T) bases. We added the reverse complement. The second data set, called enwiki, was an XML dump of the English Wikipedia[3]. Note that for enwiki we operate on the full alphabet encoded in Unicode, not the byte sequence used to encode it. The BWT and sampled ISA were computed using bwtb3m (which is an implementation of the concepts described in [27,28]). Details about the data sets can be found in Table 1. For enwiki we reduced the alphabet size by removing unused symbol values, which can be performed by a sorting (which we do anyway for producing the unsort keys) and unsorting loop. This step is contained in the stated run-time and obviously does not change the LCP array obtained.

Table 1. Datasets used, columns: name, size of input alphabet (maximum symbol appearing plus one), size of reduced alphabet $0, 1, \ldots, \sigma_{red} - 1$ with no symbol unused, length of sequence in symbols divided by 2^{30}, size of compressed BWT file in GB, fraction of irreducible ranks, ISA sampling rate used

Name	σ_{in}	σ_{red}	$n/2^{30}$	BWT$/2^{30}$	f_{irred}	f
DNA	4	4	99.678	26.010	0.565671	32
enwiki	1114112	58488	52.410	7.560	0.203	64

Run-time and space usage for the sample data sets are shown in Table 2. The table also shows the run-time of the program presented in [16] for the DNA data set. The algorithm is named KK in the table. A comparison for the enwiki data set was not possible as the implementation presented in [16] does not handle alphabets of the size we consider here. The space usage of KK in external memory is stated as given by the authors in [16]. The fact that this value is computed from the input length and not measured is denotes by an asterisk in the table. The program does not output the amount of space used. Table 2 shows that the algorithm presented in this paper is feasible for large data sets. The I/O volume is higher than what is reported for comparable data sets reported in [16], however the average I/O data rate is easily handled by modern solid state drives (M.2 or SATA Express). Our current implementation is also not optimal in terms of minimising I/O yet. The implementation of the algorithm in Fig. 2 for instance really sorts the BWT instead of only reading the key bits already stored for the BWT anyway. For enwiki the run-time suffers somewhat from the dependence on the alphabet size, but the approach is still functional for larger alphabets. In addition the run-time for enwiki is higher due to the lower sampling frequency of the ISA. However both cases show that a succinct LCP array can be computed in external memory without using an amount of disk space which is an order of magnitude larger than the length of the original text. On these examples we

---

[2] ftp://ftp.ncbi.nlm.nih.gov/genomes/refseq/vertebrate_mammalian.

[3] https://dumps.wikipedia.org.

use about $3n$ bytes per symbol in EM which is significantly less than the $21n$ bytes per symbol reported in [16] for computing an uncompressed LCP array. Our internal memory use on these examples may look high, but it is mainly due to compression and decompression support data structures which are allocated per thread. These can be assumed to scale with the alphabet size and thread number but not with the length of the input sequence. We could reduce them at the expense of sacrificing compression efficiency. An interesting property of the algorithm is that it reads more data than it writes. This has little bearing for conventional hard disks on which both operations usually have the same speed, however it is beneficial for solid state drives which are often faster at reading data than writing it. The reduced space requirement in external memory comes at the cost of increased CPU requirements for encoding and decoding compressed data structures. This however is alleviated by the use of multi threading. And as the steady increase of CPU frequency over a long time has all but ceased and is increasingly replaced by an increase of parallelism (more CPU cores) multi core machines are now very common.

Table 2. Experimental data, columns: name of data set, run-time (wall clock, d:h:m:s) to produce succinct LCP bit vector, run-time per symbol (seconds per 2^{20} symbols), input in bytes per symbol, output in bytes per symbol, I/O in bytes per symbol (in+out), average I/O rate (MB/s), EM used in GB including input (BWT+ISA) and output (succinct LCP bit vector), EM (bytes) per input symbol, peak IM allocation in GB, peak IM allocation per thread in GB

algo	data	run-time	t/s	in	out	I/O	(I/O)/s	EM	EM/n	IM	IM/thr
lcpbit	DNA	0:18:03:25	0.636	319	100	419	658.24	319.220	3.203	8.340	.174
kk	DNA	2:06:04:20	1.907	n/a	n/a	n/a	n/a	2093*	21*	10.855	10.855
lcpbit	enwiki	2:06:31:07	3.657	1700	294	2064	545.23	162.55	3.15	11.323	.236

Our lcpbit program is capable of computing a succinct PLCP bit vector the complete NCBI reference sequence database (forward plus reverse complement, together about 1.44 TB) in about a week using a system similar to the test system described but with half the number of CPU cores while employing about 100 GB of RAM to speed up the semi external part derived from [16].

7 Conclusion and Future Work

We have presented a practical algorithm for computing the succinct PLCP array in external memory while using $O(n \log \sigma)$ bits of space in external memory. The algorithm can be extended to operate on circular strings and to compute the LCP data structure described in [11] as well as supporting data structures for related queries like previous/next smaller value (cf. [12]) and range minimum queries (cf. [10]) on the non permuted LCP array. We will describe these extensions elsewhere.

References

1. Beller, T., Gog, S., Ohlebusch, E., Schnattinger, T.: Computing the longest common prefix array based on the Burrows-Wheeler transform. J. Discrete Algorithms **18**, 22–31 (2013). http://dx.doi.org/10.1016/j.jda.2012.07.007
2. Bingmann, T., Fischer, J., Osipov, V.: Inducing suffix and LCP arrays in external memory. In: Sanders, P., Zeh, N. (eds.) Proceedings of ALENEX 2013, pp. 88–102. SIAM (2013). http://dx.doi.org/10.1137/1.9781611972931.8
3. Burrows, M., Wheeler, D.: A block-sorting lossless data compression algorithm. Technical report 124, Digital Equipment Corporation (1994)
4. Crochemore, M., Hancart, C., Lecroq, T.: Algorithms on Strings, 392 p. Cambridge University Press, Cambridge (2007)
5. Dementiev, R., Kärkkäinen, J., Mehnert, J., Sanders, P.: Better external memory suffix array construction. ACM J. Exp. Algorithmics **12**, 1–24 (2008). http://doi.acm.org/10.1145/1227161.1402296
6. Ferragina, P., Gagie, T., Manzini, G.: Lightweight data indexing and compression in external memory. Algorithmica **63**(3), 707–730 (2012). http://dx.doi.org/10.1007/s00453-011-9535-0
7. Ferragina, P., Manzini, G.: Opportunistic data structures with applications. In: Proceedings FOCS 2000, pp. 390–398. IEEE Computer Society (2000). http://dx.doi.org/10.1109/SFCS.2000.892127
8. Ferragina, P., Manzini, G.: An experimental study of a compressed index. Inf. Sci. **135**(1–2), 13–28 (2001). http://dx.doi.org/10.1016/S0020-0255(01)00098-6
9. Ferragina, P., Manzini, G., Mäkinen, V., Navarro, G.: An alphabet-friendly FM-index. In: Apostolico, A., Melucci, M. (eds.) SPIRE 2004. LNCS, vol. 3246, pp. 150–160. Springer, Heidelberg (2004). http://dx.doi.org/10.1007/978-3-540-30213-1_23
10. Fischer, J.: Optimal succinctness for range minimum queries. In: López-Ortiz, A. (ed.) LATIN 2010. LNCS, vol. 6034, pp. 158–169. Springer, Heidelberg (2010). http://dx.doi.org/10.1007/978-3-642-12200-2_16
11. Fischer, J.: Wee LCP. Inf. Process. Lett. **110**(8–9), 317–320 (2010). http://dx.doi.org/10.1016/j.ipl.2010.02.010
12. Fischer, J., Mäkinen, V., Navarro, G.: Faster entropy-bounded compressed suffix trees. Theor. Comput. Sci. **410**(51), 5354–5364 (2009). http://dx.doi.org/10.1016/j.tcs.2009.09.012
13. Grossi, R., Gupta, A., Vitter, J.S.: High-order entropy-compressed text indexes. In: Proceedings SODA 2003, pp. 841–850. ACM/SIAM (2003). http://dl.acm.org/citation.cfm?id=644108.644250
14. Grossi, R., Vitter, J.S.: Compressed suffix arrays and suffix trees with applications to text indexing and string matching. In: Yao, F.F., Luks, E.M. (eds.) Proceedings of STOC 2000, pp. 397–406. ACM (2000). http://doi.acm.org/10.1145/335305.335351
15. Hon, W., Sadakane, K., Sung, W.: Breaking a time-and-space barrier in constructing full-text indices. SIAM J. Comput. **38**(6), 2162–2178 (2009). http://dx.doi.org/10.1137/070685373
16. Kärkkäinen, J., Kempa, D.: LCP array construction in external memory. In: Gudmundsson, J., Katajainen, J. (eds.) SEA 2014. LNCS, vol. 8504, pp. 412–423. Springer, Heidelberg (2014). http://dx.doi.org/10.1007/978-3-319-07959-2_35
17. Kärkkäinen, J., Manzini, G., Puglisi, S.J.: Permuted longest-common-prefix array. In: Kucherov, G., Ukkonen, E. (eds.) CPM 2009 Lille. LNCS, vol. 5577, pp. 181–192. Springer, Heidelberg (2009). http://dx.doi.org/10.1007/978-3-642-02441-2_17

18. Kärkkäinen, J., Sanders, P., Burkhardt, S.: Linear work suffix array construction. J. ACM **53**(6), 918–936 (2006). http://doi.acm.org/10.1145/1217856.1217858

19. Kasai, T., Lee, G.H., Arimura, H., Arikawa, S., Park, K.: Linear-time longest-common-prefix computation in suffix arrays and its applications. In: Amir, A., Landau, G.M. (eds.) CPM 2001. LNCS, vol. 2089, pp. 181–192. Springer, Heidelberg (2001). http://dx.doi.org/10.1007/3-540-48194-X_17

20. Manber, U., Myers, E.W.: Suffix arrays: a new method for on-line string searches. SIAM J. Comput. **22**(5), 935–948 (1993). http://dx.doi.org/10.1137/0222058

21. Munro, J.I.: Tables. In: Chandru, V., Vinay, V. (eds.) FOCS 1996. LNCS, vol. 1180, pp. 37–42. Springer, Heidelberg (1996). http://dx.doi.org/10.1007/3-540-62034-6_35

22. Okanohara, D., Sadakane, K.: A linear-time Burrows-Wheeler transform using induced sorting. In: Karlgren, J., Tarhio, J., Hyyrö, H. (eds.) SPIRE 2009. LNCS, vol. 5721, pp. 90–101. Springer, Heidelberg (2009). http://dx.doi.org/10.1007/978-3-642-03784-9_9

23. Sadakane, K.: New text indexing functionalities of the compressed suffix arrays. J. Algorithms **48**(2), 294–313 (2003). http://dx.doi.org/10.1016/S0196-6774(03)00087-7

24. Sadakane, K.: Compressed suffix trees with full functionality. Theory Comput. Syst. **41**(4), 589–607 (2007). http://dx.doi.org/10.1007/s00224-006-1198-x

25. Sirén, J.: Sampled longest common prefix array. In: Amir, A., Parida, L. (eds.) CPM 2010. LNCS, vol. 6129, pp. 227–237. Springer, Heidelberg (2010). http://dx.doi.org/10.1007/978-3-642-13509-5_21

26. Szpankowski, W.: On the height of digital trees and related problems. Algorithmica **6**(1–6), 256–277 (1991)

27. Tischler, G.: Faster average case low memory semi-external construction of the Burrows-Wheeler transform. In: Iliopoulos, C.S., Langiu, A. (eds.) Proceedings of ICABD 2014. CEUR Workshop Proceedings, vol. 1146, pp. 61–68 (2014). http://ceur-ws.org/Vol-1146/paper10.pdf

28. Tischler, G.: Faster average case low memory semi-external construction of the Burrows-Wheeler transform. Mathematics in Computer Science (2014, accepted)

29. Vitter, J.S.: Algorithms and data structures for external memory. Found. Trends Theor. Comput. Sci. **2**(4), 305–474 (2008). http://dx.doi.org/10.1561/0400000014

30. Weiner, P.: Linear pattern matching algorithms. In: Proceedings of FOCS 1973, pp. 1–11. IEEE Computer Society (1973). http://dx.doi.org/10.1109/SWAT.1973.13

Efficient Representation of Multidimensional Data over Hierarchical Domains

Nieves R. Brisaboa[1], Ana Cerdeira-Pena[1], Narciso López-López[1],
Gonzalo Navarro[2], Miguel R. Penabad[1(✉)], and Fernando Silva-Coira[1]

[1] Database Laboratory, University of A Coruña, A Coruña, Spain
{brisaboa,acerdeira,narciso.lopez,penabad,fernando.silva}@udc.es
[2] Department of Computer Science, University of Chile, Santiago, Chile
gnavarro@dcc.uchile.cl

Abstract. We consider the problem of representing multidimensional data where the domain of each dimension is organized hierarchically, and the queries require summary information at a different node in the hierarchy of each dimension. This is the typical case of OLAP databases. A basic approach is to represent each hierarchy as a one-dimensional line and recast the queries as multidimensional range queries. This approach can be implemented compactly by generalizing to more dimensions the k^2-treap, a compact representation of two-dimensional points that allows for efficient summarization queries along generic ranges. Instead, we propose a more flexible generalization, which instead of a generic quadtree-like partition of the space, follows the domain hierarchies across each dimension to organize the partitioning. The resulting structure is much more efficient than a generic multidimensional structure, since queries are resolved by aggregating much fewer nodes of the tree.

1 Introduction

In many application domains the data is organized into multidimensional matrices. In some cases, like GIS and 3D modelling, the data are actually points that lie in a two- or three-dimensional discretized space. There are, however, other domains such as OLAP systems [5,7] where the data are sets of tuples that are regarded as entries in a multidimensional cube, with one dimension per attribute. The domains of those attributes are not necessarily numeric, but may have richer semantics. A typical case in OLAP [10], in particular in snowflake schemes [12], is that each tuple contains a numeric summary (e.g., amount of sales), which

Founded in part by Fondecyt 1-140796 (for Gonzalo Navarro); and, for the Spanish group, by MINECO (PGE and FEDER) [TIN2013-46238-C4-3-R]; CDTI, AGI, MINECO [IDI-20141259/ITC-20151305/ITC-20151247]; ICT COST Action IC1302; and by Xunta de Galicia (co-founded with FEDER) [GRC2013/053]. This article was elaborated in the context of BIRDS, a European project that has received funding from the European Union's Horizon 2020 research and innovation programme under the Marie Sklodowska-Curie GA No. 690941.

© Springer International Publishing AG 2016
S. Inenaga et al. (Eds.): SPIRE 2016, LNCS 9954, pp. 191–203, 2016.
DOI: 10.1007/978-3-319-46049-9_19

is regarded as the value of a cell in the data cube. The domain of each dimension is hierarchical, so that each value in the dimension corresponds to a leaf in a hierarchy (e.g., countries, cities, and branches in one dimension, and years, months, and days in another). Queries ask for summaries (sums, maxima, etc.) of all the cells that are below some node of the hierarchy across each dimension (e.g., total sales in New York during the previous month).

A way to handle OLAP data cubes is to linearize the hierarchy of the domain of each dimension, so that each internal node corresponds to a range. Summarization queries are then transformed into multidimensional range queries, which are solved with multidimensional indexes [14]. Such a structure is, however, more powerful than necessary, because it is able to handle *any* multidimensional range, whereas the OLAP application will only be interested in queries corresponding to combinations of nodes of the hierarchies. There are well-known cases, in one dimension, of problems that are more difficult for general ranges than if the possible queries form a hierarchy. For example, categorical range counting queries (i.e., count the number of different values in a range) requires in general $\Omega(\log n / \log \log n)$ time if using $O(n \operatorname{polylog} n)$ space [11], where n is the array size, but if queries form a hierarchy it is easily solved in constant time and $O(n)$ bits [13]. A second example is the range mode problem (i.e., find the most frequent value in a range), which is believed to require time $\Omega(n^{1.188})$ if using $O(n^{1.188})$ space [4], but if queries form a hierarchy it is easily solved in constant time and linear space [8].

In this paper we aim at a compact data structure to represent data cubes where the domains in each dimension are hierarchical. Following the general idea of the tailored solutions to the problems we mentioned [8,13], our structure partitions the space according to the hierarchies, instead of performing a regular partition like generic multidimensional structures. Therefore, the queries of interest for OLAP applications, which combine nodes of the different hierarchies, will require aggregating the information of just a few nodes in our partitions, much fewer than if we used a generic space partitioning method.

Since we aim at compact representations, our baseline will be an extension to multiple dimensions of a two-dimensional compact summarization structure known as k^2-treap [1], a k^2-tree [3] enriched with summary information on the internal nodes. This n-dimensional treap, called k^n-treap, will then be extended so that it can follow an arbitrary hierarchy, not only a regular one. The topology of each hierarchy will be represented using a compact tree representation, precisely LOUDS [9]. This new structure is called CMHD (Compact representation of Multidimensional data on Hierarchical Domains). Although we focus on sum queries in this paper, it is easy to extend our results to other kinds of aggregations.

The rest of this paper is organized as follows. Sections 2 and 3 describe our compact baseline and then how it is extended to obtain our new data structure. An experimental evaluation is given in Sect. 4. Finally, we offer some conclusions and guidelines for future work.

2   Our Baseline: k^n-treaps

The k^n-treap is a straightforward extension of the k^2-treap to manage multiple dimensions. It uses a k^n-tree (in turn a straightforward extension of the k^2-tree) to store its topology, and stores separately the list of aggregate values obtained from the sum of all values in the corresponding submatrix. Figure 1 shows a matrix and the corresponding k^n-treap. The example uses two dimensions, but the same algorithms are used for more dimensions.

Consider a hypercube of n dimensions, where the length of each dimension is $len = k^i$ for some i. If the length of the dimensions are different, we can artificially extend the hypercube with empty cells, with a minimum impact in the k^n-treap size. The k^n-trees, which will be used to represent the k^n-treap topology, are very efficient to represent wide empty areas. The algorithm to build the k^n-treap starts storing on its root level the sum of all values on the matrix[1]. It also splits each dimension into k equal-sized parts, thus giving a total of k^n submatrices. We define an ordering to traverse all the submatrices (in the example, rows left-to-right, columns top-to-bottom). Following this ordering, we add a child node to the root for each submatrix. The algorithm works recursively for each child node that represents a nonempty submatrix, storing the sum of the cells in this submatrix, splitting it and adding child nodes. For empty submatrices, the node stores a sum of 0.

As we can see in Fig. 1, the root node stores 51, the sum of all values in the matrix, and it is decomposed into 4 matrices of size 4 × 4, thus adding 4 children to the root node. Notice that the second submatrix (top-right) is full of zeroes, so this node just stores a sum of 0 and is not further decomposed. The algorithm proceeds recursively for the remaining 3 children of the root node.

The final data structures used to represent the k^n-treap are the following:

- *Values (V):* Contains the aggregated values (sums) for each (sub)matrix, as they would be obtained by a levelwise traversal of the k^n-treap. It is encoded using DACs [2], which compress small values while allowing direct access.
- *Tree structure (T):* It is a k^n-tree that stores a bitmap T for the whole tree except its leaves. In this case, the usual bitmap L for the leaves in a standard k^n-tree is not used, because the information about which cells have or not a value is already represented in V. Therefore L is not needed.

The navigation through the k^n-treap is basically a depth first traversal. Finding the child of a node can be done very efficiently by using *rank* and *select* operations [9] as in the standard k^2-tree. The typical queries in this context are: finding the value of an individual cell and finding the sum of the values in a given range of cells, specified by the initial and final coordinates that define the submatrix of interest.

---

[1] The implemented algorithm is recursive and each sum is actually computed only once, when returning from the recursive calls.

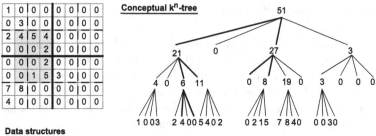

Fig. 1. k^n-treap with a highlighted range query

Finding the value of a specific cell by its coordinates. To find the value of the cell, for example the cell at coordinates $(4, 3)$ in the figure, the search starts at the root node and in each step goes down trough the children of the matrix overlapping the searched cell. In this example, the search goes through the first child node (with value 21 in the figure), then through its third child (with value 6) and finally through the second child, reaching the leaf node with value 4, which is the value returned by the query.

Finding the sum of the cells in a submatrix. The second type of query looks for the aggregated value of a range of cells, like the shaded area in Fig. 1. This is implemented as a depth-first multi-branch traversal of the tree. If the algorithm finds that the range specified in the query fully contains a submatrix of the k^n-treap that has a precomputed sum, it will use this sum and will not descend to its child nodes. The figure highlights the branches of the k^n-treap that are used. Notice that this query completely includes the sumatrices of values $\{5, 4, 0, 2\}$ and $\{0, 2, 1, 5\}$, that have their sums (11 and 8) explicitly stored on the third level of the tree. Therefore, the algorithm does not need to reach the leaf levels of the tree for these matrices. Notice also that there is an empty submatrix that intersects with the region of the query (the first child of the third child of the root), so the algorithm also stops before reaching the leaf levels in this submatrix. Only for cells $(3, 2)$ (with a value of 4) and $(4, 2)$ (with a value of 0) needs the algorithm to reach the leaf levels.

3   Our Proposal: CMHD

As previously stated, CMHD divides the matrix following the natural hierarchy of the elements in each dimension. In this way we allow the efficient answer of queries that consider the semantic of the dimensions.

3.1 Conceptual Description

Consider an n-dimensional matrix where each cell contains a weight (e.g., product sales, credit card movements, ad views, etc.). The CMHD recursively divides the matrix into several submatrices, taking into account the limits imposed by the hierarchy levels of each dimension.

Figure 2 depicts an example of a CMHD representation for two dimensions. The matrix records the number of product sales in different locations. For each dimension, a hierarchy of three levels is considered. In particular, cities are aggregated into countries and continents, while products are grouped into sections and good categories. The tree at the right side of the image shows the resulting conceptual CMHD for that matrix. Observe that each hierarchy level leads to an irregular partition of the grid into submatrices (each of them defined by the limits of its elements), having as associated value the sum of product sales of the individual cells inside it. Thus, the root of the tree stores the total amount of sales in the complete matrix. Then the matrix is subdivided by considering the partition corresponding to the first level of the dimension hierarchies (see the bold lines). Each of the submatrices will become a child node of the root, keeping the sum of values of the cells in the corresponding submatrix. The decomposition procedure is repeated for each child, considering subsequent levels of the hierarchies (see the dotted lines), as explained, until reaching the last one. Also notice that, as happens in the k^n-treap, the decomposition concludes in all branches when empty submatrices are reached (that is, in this scenario, when a submatrix with no sales is found). See, for example, the second child of the root.

Note that CMHD assumes the same height in all the hierarchies that correspond to the different dimensions. Observe that, for each crossing of elements of the same level from different dimensions, an aggregate value is stored. Notice also that artificial levels can be easily added to a hierarchy of one dimension by subdividing all the elements of a level in just one element (itself), thus creating a new level identical to the previous one. This feature allows us to arbitrarily match the levels of the different hierarchies, and thus to flexibly adapt the representation of aggregated data to particular query needs. That is, by introducing artificial intermediate levels where required, more interesting aggregated values will be precomputed and stored. For example, assume we have two dimensions: (d_1) with levels for *department*, *section* and *product*; and (d_2) with levels for *year*, *season*, *month* and *day*. If we were interested in obtaining the number of sales per *section* for *season*s, but also for *month*s, we could devise a new level arrangement for d_1, that will have now the levels *department, section, section', product*; where each particular *section* of the second hierarchy level results into just one *section'* child, which is actually itself. In this way aggregated values will be computed and stored considering sales for *section* in each *season*, but also sales for *section'* in each *month*.

3.2 Data Structures

The conceptual tree that defines the CMHD is represented compactly with different data structures, for the domain hierarchies and for the matrix data itself.

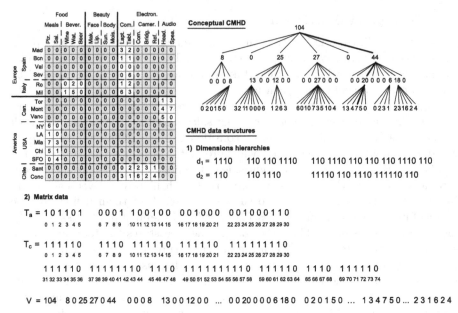

Fig. 2. Example of CMHD construction for a two-dimensional matrix.

Domain hierarchy representation. The hierarchy of a dimension domain is essentially a tree of C nodes. We represent this tree using LOUDS [9], a tree representation that uses $2C$ bits, and can efficiently navigate it. Using LOUDS, a tree representing the hierarchy of the elements of a dimension is encoded by appending the degree r of each node in (left-to-right) level-order, in unary: 1^r0. Figure 2 illustrates the hierarchy encoding of the dimensions used in that example (see d_1 and d_2). For instance, the degree of the first node for the products dimension (d_1) is 3, so its unary encoding is 1110. Note that each node (i.e., element of a dimension placed at any level of its hierarchy) is associated with one 1 in the encoded representation of the degree of its parent. LOUDS is navigated using *rank* and *select* queries: $rank_b(i)$ is the number of bits b up to position i, and $select_b(j)$ is the position of the jth occurrence of bit b. Both queries are computed in constant time using $o(C)$ additional bits [6]. For example, given a node whose unary representation starts at position i, its parent is $p = select_0(t - j) + 1$, where $t = select_1(j)$ and $j = rank_0(i)$; and i is the $(t - p + 1)$th child of p. On the other hand, the kth child of i is $select_0(rank_1(i) + k - 1) + 1$. We also use a hash table to associate the domain nodes (i.e., labels such as "USA" in Fig. 2) with the corresponding LOUDS node position.

Data representation. To represent the n-dimensional matrix, we use the following data structures:

– *Tree structures (T_a and T_c):* to navigate the CMHD, we need to use two different data structures in conjunction. First, T_a, a bit array that, similarly to the k^n-treap, provides a compact representation of the conceptual tree independently of the node values, for all the tree levels, except the last one[2]. That is, internal nodes whose associated value is greater than 0, will be represented with a 1. In other case, they will be labeled with a 0. Observe that, for the k^n-treap, the use of this data structure is enough to navigate the tree, taking advantage of the regular partition of the matrix into equal-sized submatrices. Instead, CMHD follows different hierarchy partitions, which results into irregular submatrices. Therefore, a second data structure, T_c, is also required to traverse the CMHD. This is a bit array aligned to T_a, which marks the limits of each tree node in T_a (this time, it also considers the last tree level). If the next tree node in T_a has z children, we append $1^{z-1}0$ to T_c. Notice that each node of T_a is associated with a 0 in T_c, which allows navigating the trees using $rank$ and $select$ on T_a and T_c: say we are at a node in T_a that starts at position i; then it has a kth child iff $T_a[i + k - 1] = 1$, and if so this child starts at position $select_0(T_c, rank_1(T_a, i + k - 1)) + 1$.

– *Values (V):* the CMHD is traversed levelwise storing the values associated with each node (either corresponding to original matrix cells, or to data aggregations) in a single sequence, which is then represented with DACs [2].

3.3   Queries

Queries in this context give the names of elements of the different dimensions and ask for the sum of the cells defined for those values. Depending on the query, we can answer it by just reporting a single aggregated value already kept in V, or by retrieving several stored values, and then adding them up. The first scenario arises when the elements (labels) of the different dimensions specified in the query are all at the same level in their respective hierarchies. The second situation arises from queries using labels of different levels. In both contexts, top-down traversals of the conceptual CMHD are required to fetch the values. The algorithm always starts searching the hash tables for the labels provided by the query for the different dimensions, to locate the corresponding LOUDS nodes. From the LOUDS nodes, we traverse each hierarchy upwards to find out its depth and the child that must be followed at each level to reach it.

This information is then used to find the desired nodes in T_a. For example, with two dimensions, we start at the root of T_a and descend to the child number $k_1 + a_1 \cdot k_2$, where k_i is the child that must be followed in the ith dimension to reach the queried node, and a_i is the number of children of the root in the ith dimension (a_i is easily computed with the LOUDS tree of its dimension). We continue similarly to the node at level 2, and so on, until we reach one of the query nodes in a dimension, say in the first. Now, to reach the other (deeper) node in the second dimension, we must descend by every child in the first dimension, at every level, until reaching the second queried node. Finally,

---

[2] We do not actually need to represent the nodes of the last level in T_a. This data structure will be used to first identify a node whose children will be later located in another bit array (T_c). But these already constitute matrix cells, with no children.

when we have reached all the nodes, we collect and sum up the corresponding values from V. Note that, if all the queried nodes are in the same level, we perform a single traversal in T_a. Note also that, if we find any zero in a node of T_a along this traversal, we immediately prune that branch, as the submatrix contains no data.

Example. Assume we want to retrieve the total amount of *speaker* sales in *Montreal*, in Fig. 2. Since both labels belong to the same level in both dimension hierarchies (the last one), we will have to retrieve a single stored value in that level. The path to reach it has been highlighted in the conceptual tree of the image. To perform the navigation we must start at the root of the tree (position 0 in T_a). In the first level, we need to fetch the sixth child (offset 5), as it corresponds to the submatrix including the element to search, in that level. Hence we access position 5 in T_a. Since $T_a[5] = 1$, we must continue descending to the next level. Recall that we have a 1 in T_a for each node with children, and that each node is associated with just one 0 in T_c. So the child starts at position $select_0(T_c, rank_1(T_a, 5)) + 1 = select_0(T_c, 4) + 1 = 22$ in T_a. In this level we must access the third child (offset 2), so we check $T_a[24] = 1$. Again, as we are in an internal node, we know that its children are located at position $select_0(T_c, rank_1(T_a, 24)) + 1 = select_0(T_c, 9) + 1 = 59$. Finally, we reach the third and last level of the tree, where we know that the corresponding child is the fourth one (at $T_a[59 + 3] = T_a[62]$). Recall, however, that this last level is not represented in T_a. To perform this final step, we directly look into the array V: $V[62 + 1] = V[63] = 7$ is the answer.

In case of queries combining labels of different levels, the same procedure would apply, but having to get the values corresponding to all the possible combinations with the element of the lowest hierarchy level (e.g., if we want to obtain the number of *meal* sales in *America*, we must first recover the values associated with *meal-Canada*, *meal-USA*, and *meal-Chile*, and then sum them up).

4   Experimental Evaluation

This section presents the empirical evaluation of the two previously described data structures. Both representations have been implemented in C/C++, and the compiler used was GCC 4.6.1. (option -O9). We ran our experiments in a dedicated Intel (R) Core (TM) i7-3820 CPU @ 3.60 GHz (4 cores) with 10 MB of cache, and 64 GB of RAM. The machine runs Ubuntu 12.04.5 LTS with kernel 3.2.0-99 (64 bits).

We generate different datasets, all of them synthetic, to evaluate the performance of the two data structures, varying the number of dimensions and the number of items on each dimension. These datasets have been labeled as <dim#>D_<item#>, thus referring to their size specifications in the own name. For example, dataset 5D_16 has 5 dimensions, and the number of items on each dimension is 16. The total size of this dataset is $16^5 = 1048576$ elements.

In order to show the CMHD advantage of considering the domain semantics, and computing the aggregate values according to the natural limits imposed by the hierarchy of elements in each dimension, the dimensions hierarchies have been generated in two different ways for each dataset. First, the *binary* organization, that corresponds to a regular partition. That is, the hierarchies of each dimension are exactly the same as those produced by a k^n-treap matrix partition into equal-sized submatrices. In this way both data structures store exactly the same aggregated values. We named it *binary* because we use a value of $k = 2$. Second, the *irregular* organization, which arbitrary groups data, on each dimension, into different and irregular hierarchies (different number of divisions, and also different size at each level). The last scenario simulates what would be a matrix partition following the semantic needs of a given domain. In this case the aggregated values stored by the CMHD will be different from those stored by the k^n-treap, and therefore more appropriated to answer queries using the same "semantic". That means, in our context, queries considering regions that exactly match the natural divisions of each dimension at some level of the hierarchies.

To test the structures behavior, we have also considered three different datasets, with a different number of empty cells, for each size specification: with no empty cells, and with 25 % and 50 % of empty cells, respectively.

First we analyze the space requirements of both data structures for all the datasets (see Table 1). Of course, the size decreases as the number of empty cells increases, in both cases. Moreover, we can also observe that the k^n-treap size is slightly lower than the CMHD. This is expected, because CMHD has to store the LOUDS representation of each dimension hierarchy, while dimensions are implicit for the k^n-treap. Additionally, CMHD uses a second bitmap (T_c) to navigate the conceptual tree, which is not necessary when using the k^n-treap.

We must also clarify a small issue about the sizes of the k^n-treaps: the size of a standard k^n-treap for a specific dataset is always the same, regardless of the organization of its dimensions (binary or irregular). However, Table 1 shows some difference in the sizes. For example, for 4D_16, the size for the binary organization is 44.84, but it is 44.42 for the irregular one. The reason for this variation is that all queries are performed by taking dimension labels as input, so we need a vocabulary to translate each label into a range of cells. We have included that vocabulary (dimension labels and cell ranges) into the size of the k^n-treaps, and the vocabulary for the irregular organization is usually smaller, as it has less levels and less dimension labels (because each node in the conceptual tree can have more than 2 children in the irregular organization, while the binary organization always has 2).

Regarding query times, we have run several sets of queries for all the datasets. As previously mentioned, queries are posed in this context by giving one element name (label) for each different dimension, as it is the natural way to query a multidimensional matrix defined by hierarchical dimensions. Since the k^n-treap does not directly work with labels, each query has been translated into the equivalent ranged query, establishing the initial and final coordinates for each dimension. The following types of queries have been considered:

- *Finding one precomputed value.* This value can be a specific cell of the matrix (so forcing the algorithms to reach the last level of the tree), or a precomputed value that corresponds to an internal node of the conceptual tree.

Table 1. Space requirements of k^n-treap and CMHD data structures (in KB)

Name	0 % Zeroes				25 % Zeroes				50 % Zeroes			
	Binary		Irregular		Binary		Irregular		Binary		Irregular	
	kn-treap	CMHD	kn-treap	CMHD	kn-treap	CMHD	kn-treap	CMHD	kn-treap	CMHD	kn-treap	CMHD
4D_16	44.84	55.56	44.22	47.82	38.16	47.54	37.54	43.04	29.51	37.09	28.89	34.18
4D_32	680.45	864.17	679.08	750.17	552.15	710.05	550.78	640.80	400.13	527.57	398.76	501.01
5D_16	631.10	793.34	630.41	729.48	527.23	667.69	526.54	653.25	408.10	523.27	407.41	509.80
5D_32	20098.99	25328.26	20097.43	23344.18	16272.61	20691.87	16271.04	20167.62	11776.94	15237.57	11775.37	15471.61
6D_16	9663.37	12073.82	9662.48	11456.36	8180.31	10278.04	8179.43	10532.61	6419.99	8135.86	6419.11	8279.60

Following the example of Fig. 2, a query asking for the amount of *speakers* sales in *Montreal* or the total number of *beberages* sales in *Italy* would be queries of this type, the former accessing an individual cell and the later obtaining a precomputed value in the penultimate level of the tree.

- *Finding the sum of several precomputed values.* This kind of query must obtain a sum that is not precomputed and stored in the data structure itself. In turn, it must access several of these aggregated values and then add them up. Given that we are specifying the queries by dimension labels, this type of query is defined by using labels that belong to different levels of the hierarchies across the dimensions. The lowest level, which corresponds to individual cells, is not used for this scenario.

An example of this query type would be to find the total number of sales of *electronic* products in Chile. Note that *electronic* is located at the first level of its dimension hierarchy, but *Chile* is at the second level of the second dimension (see Fig. 2). Hence, the values corresponding to *computers-Chile*, *cameras-Chile*, and *audio-Chile* must be first retrieved to finally sum them up.

Each created set contains 10, 000 queries, randomly generated, of the two previous types, for each dataset. The following tables show the average query times (in microseconds per query) for both data structures, taking into account the two different matrix partitions of the datasets (*binary* or *irregular*) and also the percentage of empty cells.

We first show the results obtained for queries that just need to retrieve one precomputed value, at different levels. On the one hand, Table 2 displays query times for specific matrix cells, that is, located at the last level of the conceptual tree. In this case, the k^n-treap performs better than the CMHD in almost all cases. This is an expected outcome as both data structures must reach the leaf level to get the answer, and the depth first navigation of the tree is simpler in the k^n-treap (just products and *rank* operations). In any case, CMHD also performs quite well, using just a few microseconds to answer any of the queries.

On the other hand, Table 3 shows the average query times for queries of the same type, but now considering precomputed values stored in nodes of an intermediate level of the tree (in particular, the penultimate level). Note that this fact holds for both data structures when working with a regular partition of the matrix (that is, the *binary* scenario). Thus, in this case, the k^n-treap gets better results than CMHD, but with slight time differences. Yet, observe

Table 2. Average query times (in μs) for queries finding one precomputed value (original matrix cells)

| Dataset | 0 % Zeroes | | | | 25 % Zeroes | | | | 50 % Zeroes | | | |
| | Binary | | Irregular | | Binary | | Irregular | | Binary | | Irregular | |
	kn-treap	CMHD	kn-treap	CMHD	kn-treap	CMHD	kn-treap	CMHD	kn-treap	CMHD	kn-treap	CMHD
4D_16	2	4	2	3	2	4	2	2	2	4	2	3
4D_32	2	5	3	4	2	4	2	1	2	4	1	4
5D_16	2	4	3	4	2	5	3	2	2	5	2	2
5D_32	3	6	3	5	2	4	3	3	2	6	3	2
6D_16	3	4	3	4	3	6	4	2	4	5	4	4

that this is not the actual scenario when dealing with meaningful application domains, where rich semantics arise. This situation is that corresponding to what we called *irregular* datasets. In this case, CMHD excels, as expected, given that this data structure has been particularly designed to manage hierarchical domains. Results show that CMHD is able to perform up to 12 times faster than k^n-treap (for the best case).

Table 3. Average query times (in μs) for queries finding one precomputed value (penultimate tree level)

| Dataset | 0 % Zeroes | | | | 25 % Zeroes | | | | 50 % Zeroes | | | |
| | Binary | | Irregular | | Binary | | Irregular | | Binary | | Irregular | |
	kn-treap	CMHD	kn-treap	CMHD	kn-treap	CMHD	kn-treap	CMHD	kn-treap	CMHD	kn-treap	CMHD
4D_16	1	4	7	3	1	3	6	2	2	4	5	2
4D_32	1	3	9	3	1	3	7	2	1	3	6	1
5D_16	2	4	11	1	2	3	9	1	2	4	7	3
5D_32	3	4	23	2	2	4	18	3	2	3	12	2
6D_16	2	4	35	2	2	2	28	3	3	4	21	1

To check whether the observed differences are significative (in the cases where times were closer) we performed a statistical significance test. We checked the 4D_16 and 5D_16 datasets, for the irregular organization, with all the different configurations of empty cells.

We show here, as a proof, the details for 4D_16 with 50 % of empty cells, which took 5 μs to the k^n-treap, and 2 μs to CMHD. We ran 20 sets of 10,000 queries, and measured both the average time and the standard deviation for the k^n-treap (5.100 and 0.447, respectively) and for the CMHD (1.750 and 0.550, respectively). With these results, we obtain a critical value of 4.725, which is greater than 2.580, so the difference is significative with a 99 % of confidence level. The remaining tests also proved the same significance results.

Finally, Table 4 presents the average query times for the second type of queries (that is, those having to recover several precomputed values and then adding them up to provide the final answer). As results show, the k^n-treap displays a better performance than CMHD for the *binary* scenario. However, again this is not the most interesting situation in real domains. If we observe the results

Table 4. Average query times (in μs) for queries finding a sum of precomputed values

Dataset	0 % Zeroes				25 % Zeroes				50 % Zeroes			
	Binary		Irregular		Binary		Irregular		Binary		Irregular	
	kn-treap	CMHD	kn-treap	CMHD	kn-treap	CMHD	kn-treap	CMHD	kn-treap	CMHD	kn-treap	CMHD
4D_16	3	10	20	4	3	6	16	3	2	6	12	6
4D_32	6	21	21	3	4	20	17	4	4	20	12	5
5D_16	4	8	30	8	4	8	25	3	3	7	19	1
5D_32	6	26	49	3	8	23	39	5	6	21	27	2
6D_16	5	15	106	7	5	10	82	6	4	9	63	8

obtained for the *irregular* datasets, we will appreciate that CMHD clearly outperforms the k^n-treap in this scenario, thus demonstrating the good capabilities of our proposal to cope with the aim of this work.

5   Conclusions and Future Work

We have presented a multidimensional compact data structure that is tailored to perform aggregate queries on data cubes over hierarchical domains, rather than general range queries. The structure represents each hierarchy with a succinct tree representation, and then partitions the data cube according to the product of the hierarchies. This partition is represented with an extension of the k^2-treap to higher dimensions and to non-regular partitions. The resulting structure, dubbed CMHD, is much faster than a regular multidimensional k^2-treap when the queries follow the hierarchical domains. This makes it particularly attractive to represent OLAP data cubes compactly and efficiently answer meaningful aggregate queries.

As future work, we plan to experiment on much larger collections. This would make the vocabulary of hierarchy nodes much less significant compared to the data itself (especially for the CMHD). We also plan to test real datasets (for example, coming from data warehouses) and real query workloads. We also expect to compare our results with established OLAP database management systems, and to enrich our prototype with other kinds of queries and data.

References

1. Brisaboa, N.R., de Bernardo, G., Konow, R., Navarro, G., Seco, D.: Aggregated 2d range queries on clustered points. Inf. Syst. **60**, 34–49 (2016)
2. Brisaboa, N.R., Ladra, S., Navarro, G.: DACs: bringing direct access to variable-length codes. Inf. Process. Manag. **49**, 392–404 (2013)
3. Brisaboa, N.R., Ladra, S., Navarro, G.: Compact representation of web graphs with extended functionality. Inf. Syst. **39**, 152–174 (2014)
4. Chan, T., Durocher, S., Larsen, K., Morrison, J., Wilkinson, B.: Linear-space data structures for range mode query in arrays. In: Proceedings of 29th International Symposium on Theoretical Aspects of Computer Science (STACS), pp. 290–301 (2012)

5. Chaudhuri, S., Dayal, U.: An overview of data warehousing and OLAP technology. SIGMOD Rec. **26**(1), 65–74 (1997)

6. Clark, D.: Compact PAT Trees. Ph.D. thesis, University of Waterloo, Canada (1996)

7. Codd, E.F., Codd, S.B., Salley, C.T.: Providing OLAP. On-Line Analytical Processing to User-Analysts: An IT Mandate. E. F. Codd and Associates (1993)

8. Hon, W., Shah, R., Thankachan, S.V., Vitter, J.S.: Space-efficient frameworks for top-k string retrieval. J. ACM **61**(2), 9:1–9:36 (2014)

9. Jacobson, G.: Space-efficient static trees and graphs. In: Proceedings of the 30th Annual Symposium on Foundations of Computer Science, SFCS 1989, pp. 549–554. IEEE Computer Society, Washington, DC (1989)

10. Kimball, R., Ross, M.: The Data Warehouse Toolkit: The Complete Guide to Dimensional Modeling, 2nd edn. Wiley, New York (2002)

11. Larsen, K., van Walderveen, F.: Near-optimal range reporting structures for categorical data. In: Proceedings of 24th Symposium on Discrete Algorithms (SODA), pp. 265–276 (2013)

12. Levene, M., Loizou, G.: Why is the snowflake schema a good data warehouse design? Inf. Syst. **28**(3), 225–240 (2003)

13. Sadakane, K.: Succinct data structures for flexible text retrieval systems. J. Discrete Algorithms **5**, 12–22 (2007)

14. Samet, H.: Foundations of Multidimensional and Metric Data Structures. Morgan Kaufmann, San Francisco (2006)

LCP Array Construction Using O(sort(n)) (or Less) I/Os

Juha Kärkkäinen$^{(\boxtimes)}$ and Dominik Kempa

Department of Computer Science and Helsinki Institute for Information
Technology HIIT, University of Helsinki, Helsinki, Finland
{juha.karkkainen,dominik.kempa}@cs.helsinki.fi

Abstract. The suffix array, one of the most important data structures
in modern string processing, needs to be augmented with the longest-
common-prefix (LCP) array in many applications. Their construction is
often a major bottleneck especially when the data is too big for internal
memory. While there are external memory algorithms that construct
the suffix array and the LCP array simultaneously in the optimal I/O
complexity of $\mathcal{O}(\text{sort}(n))$, for several reasons it would be desirable to
construct the suffix array first and then the LCP array from the suffix
array in a separate stage. In this paper we describe the first algorithm
that achieves $\mathcal{O}(\text{sort}(n))$ I/O complexity for the LCP array construction
stage and is not an extension of a suffix sorting algorithm. As a variant,
we obtain a Monte Carlo algorithm that, given a sparse suffix array
containing $m < n$ suffixes in sorted order, computes the corresponding
LCP array in $\mathcal{O}(\text{scan}(n) + \text{sort}(m)\log(n/m))$ I/Os if the suffix positions
are evenly spaced, and in $\mathcal{O}(\text{scan}(n) + \text{sort}(m)\log(n))$ I/Os in general.

1 Introduction

The suffix array [13,28], a lexicographically sorted list of the suffixes of a text,
is one of the most important data structures in modern string processing. It is
frequently augmented with the longest-common-prefix (LCP) array, which stores
the lengths of the longest common prefixes between lexicographically adjacent
suffixes. Together they are the basis of powerful text indexes such as enhanced
suffix arrays [1] and many compressed full-text indexes [30]. Modern textbooks
spend dozens of pages in describing their applications, see e.g. [27,33].

The construction of the two arrays is a bottleneck in many applications. They
can be constructed either simultaneously using a single algorithm, a SLACA (suffix and LCP array construction algorithm), or separately constructing the suffix
array first using a SACA (suffix array construction algorithm) and then the LCP
array from the suffix array and the text using a LACA (LCP array construction algorithm). The latter option is preferred because the separate algorithms
are simpler, enable separate development and optimization, and allow many different combinations. The best SACA+LACA combinations are also both faster
and more space efficient than the best SLACAs in practice. This is true even in
external memory computation as shown in [17,18].

© Springer International Publishing AG 2016
S. Inenaga et al. (Eds.): SPIRE 2016, LNCS 9954, pp. 204–217, 2016.
DOI: 10.1007/978-3-319-46049-9_20

However, asymptotically the external memory LACAs are inferior to the best SLACAs. In the standard external memory (EM) model, with RAM size M and disk block size B, common I/O complexities are scan$(n) = n/B$, which is the complexity of scanning a sequence of n elements, and sort$(n) = (n/B) \log_{M/B}(n/B)$, which is the complexity of sorting n elements. The I/O complexity of the best SACAs and SLACAs is $\mathcal{O}(\text{sort}(n))$, where n is the length of the text, which is clearly optimal since the construction involves sorting. The I/O complexity of the external memory LACAs is $\mathcal{O}(\text{sort}(n) + (n^2/(MB \log_\sigma^2 n)))$ (or worse), where σ is the size of the alphabet. This leaves open the existence of a LACA with I/O complexity $\mathcal{O}(\text{sort}(n))$.

Our Contribution. We describe the first LACA with I/O complexity $\mathcal{O}(\text{sort}(n))$. It is based on two sampling schemes, difference covers and sparse PLCP arrays, both of which have been previously used in LCP array construction, but never together and never in the way we use them.

Difference cover sampling has been used in SACAs [15], SLACAs [21] and LACAs [34] as well as in a data structure for answering longest common extension (LCE) queries [6], which ask for lcp(i, j), the length of the longest common prefix of the suffixes starting at positions i and j. A difference cover sample defines a subset of text positions with specific properties. All of the above applications compute a sparse suffix array containing in lexicographical order the suffixes starting at the difference cover positions. The corresponding sparse LCP array is used in the LACA and the LCE data structure. The SACA DC3 [15] also involves $\mathcal{O}(n)$ substrings defined by recursive difference covers in the early stages of its computation, and it is these DC-substrings that form the central data structure of our new algorithm. Each DC-substring is assigned a name so that we can compare the equality of two DC-substrings by comparing their names. The names can be computed in $\mathcal{O}(\text{sort}(n))$ I/Os similarly to DC3. Alternatively, in $\mathcal{O}(\text{scan}(n))$ I/Os we can compute Karp-Rabin fingerprints for the DC-substrings, which results in a Monte Carlo algorithm that works correctly with high probability.

We show that given the substring names, we can answer an LCE query lcp(i, j) in $\mathcal{O}(\log \text{lcp}(i, j))$ time. We can answer informed, approximate LCE queries even faster, where informed means that we are given lower and upper bounds on lcp(i, j) as input, and approximate means that the output consists of (tighter) lower and upper bounds instead of the exact value. In external memory, we can answer batches of such LCE queries efficiently. Specifically, we can answer a batch of d exact queries in $\mathcal{O}(\text{scan}(n) + \text{sort}(d) \log \ell)$ I/Os, where ℓ is the average value of the results, and informed approximate queries even faster.

The second sampling technique, sparse PLCP array, has been used for LCP array construction in [20]. The PLCP array is a permutation of the LCP array into the text order instead of the lexicographical order, and a sparse PLCP array is a subsequence of the full PLCP array. A sparse PLCP array allows computing lower and upper bounds for the other PLCP array entries. In [20], a simple sparse PLCP was used in a space efficient LCP array construction algorithm. Our new algorithm involves a recursive hierarchy of sparse PLCP arrays, which are used for obtaining input bounds for informed approximate LCE queries.

A careful combination of the two sampling techniques produces the full PLCP array, and thus the LCP array, using $\mathcal{O}(\text{sort}(n))$ I/Os. Furthermore, given an arbitrary sparse suffix array of size $m < n$, the associated LCP array can be computed in $\mathcal{O}(\text{scan}(n) + \text{sort}(m)\log(n))$ I/Os, excluding the computation of the DC-substring names. Using Karp-Rabin fingerprints as DC-substring names results in a Monte Carlo algorithm with the same I/O complexity. If the suffixes are evenly spaced in the text, the I/O complexity can be improved to $\mathcal{O}(\text{scan}(n) + \text{sort}(m)\log(n/m))$.

Related Work. The first SLACA appeared already in the seminal paper by Manber and Myers [28], but the LCP array did not really become popular until Kasai et al. [23] introduced the first LACA. Since then several new LACAs have been developed with a particular emphasis on reducing the space requirements [4,12,20,26,29,34,35]. Some of the algorithms are even semi-external, i.e., they keep most of the data structures on disk but need to have at least the full text in RAM [20,34].

The first I/O-optimal external memory SACA, DC3 [15], came right away with a modification into a SLACA [21, Sect. 4]. Other external memory SLACAs are eSAIS [7], which is I/O optimal, and eGSA [25], which does not have useful worst case bounds on the I/O complexity. Several recent external memory SACAs are based on induced sorting [24,31,32] and could probably be converted into SLACAs using the same technique (introduced in [9]) as eSAIS. For practical purposes, the best SACAs are probably SAscan [16] and pSAscan [19], even though their I/O complexity is a non-optimal $\mathcal{O}\big(\text{sort}(n) + (n^2/(MB\log_\sigma n))\big)$.

The external memory LACAs in [17,18] have an I/O-complexity similar to SAscan and pSAscan, $\mathcal{O}\big(\text{sort}(n) + (n^2/(MB\log_\sigma^2 n))\big)$. Despite the apparently quadratic I/O complexity, the SACA+LACA combination of these algorithms is probably the best way to construct the suffix and LCP arrays for large texts in most practical situations. Based on the analysis and experiments in [18], the text would have to be more than 100 times the size of the available RAM before the quadratic part becomes dominant, and we do not believe our new algorithm would be competitive for any smaller texts. Beyond that limit though, a well engineered implementation of the new algorithm could become the algorithm of choice.

We are not aware of previous results directly comparable to our results on LCE queries and sparse LCP array construction, but there exists tangentially related recent work on external memory range minimum queries [2,3] (since LCE queries can be answered as range minimum queries on the LCP array), as well as on LCE queries [5,11,36] and sparse suffix and LCP array construction [10,14] in internal memory.

2   Preliminaries

Throughout we consider a string $X = X[0..n) = X[0]X[1]...X[n-1]$ of $|X| = n$ symbols drawn from the alphabet $[0..\sigma)$. Here and elsewhere we use $[i..j)$ as a shorthand for $[i..j-1]$. For $i \in [0..n]$, we write $X[i..n)$ to denote the *suffix* of X of length $n - i$, that is $X[i..n) = X[i]X[i+1]...X[n-1]$. We will often refer to suffix $X[i..n)$ simply as "suffix i". Similarly, we write $X[0..i)$ to denote the *prefix* of X of length i. $X[i..j)$ is the *substring* $X[i]X[i+1]...X[j-1]$ of X that starts at position i and ends at position $j - 1$.

The *suffix array* SA of X is an array $SA[0..n]$ which contains a permutation of the integers $[0..n]$ such that $X[SA[0]..n) < X[SA[1]..n) < \cdots < X[SA[n]..n)$. In other words, $SA[j] = i$ iff $X[i..n)$ is the $(j+1)^{\text{th}}$ suffix of X in ascending lexicographical order. The *inverse suffix array* SA^{-1} is the inverse permutation of SA, that is $SA^{-1}[i] = j$ iff $SA[j] = i$. Conceptually, $SA^{-1}[i]$ tells the position of suffix i in SA. Another representation of the permutation is the Φ *array* [20] $\Phi[0..n)$ defined by $\Phi[SA[j]] = SA[j-1]$ for $j \in [1..n]$. In other words, the suffix $\Phi[i]$ is the immediate lexicographical predecessor of the suffix i.

Let $lcp(i,j)$ denote the length of the longest-common-prefix (LCP) of suffix i and suffix j. For example, in the string $X = cccccatcat$, $lcp(0,3) = 2 = |cc|$, and $lcp(4,7) = 3 = |cat|$. The *longest-common-prefix array*, $LCP[1..n]$, is defined such that $LCP[i] = lcp(SA[i], SA[i-1])$ for $i \in [1..n]$. The *permuted LCP array* [20] $PLCP[0..n)$ is the LCP array permuted from the lexicographical order into the text order, i.e., $PLCP[SA[j]] = LCP[j]$ for $j \in [1..n]$. Then $PLCP[i] = lcp(i, \Phi[i])$ for all $i \in [0..n)$. The following property of the PLCP array is the basis of all efficient LACAs.

Lemma 1 ([18]). *Let $i, j \in [0..n)$. If $i \leq j$, then $i + PLCP[i] \leq j + PLCP[j]$. Symmetrically, if $\Phi[i] \leq \Phi[j]$, then $\Phi[i] + PLCP[i] \leq \Phi[j] + PLCP[j]$.*

Let p be a prime and choose $s \in [0..p-1]$ uniformly at random. The *Karp-Rabin fingerprint* [22] for a substring $X[i..j]$ of X is defined as $FP[i..j] = \sum_{k=i}^{j} X[k] \cdot s^{j-k} \bmod p$. Clearly, if $X[i..i+\ell] = X[j..j+\ell]$ then $FP[i..i+\ell] = FP[j..j+\ell]$. On the other hand, if $X[i..i+\ell] \neq X[j..j+\ell]$ then $FP[i..i+\ell] \neq FP[j..j+\ell]$ with probability at least $1 - n^{-c}$ for any constant $c > 0$ [8] (assuming $p > n^{c+4}$). The fingerprint of a concatenation can be computed efficiently using $FP[i..k] = FP[i..j] \cdot s^{k-j} + FP[j+1..k] \bmod p$ for any $i \leq j < k$.

3   LCE Queries Using DC-substrings

In this section we develop the basic machinery that is used to compute (or approximate) the LCE queries. Assume for simplicity that n is a power of 3. Let $b_{k-1}...b_0$ be the binary representation of integer b. For $k \geq 0$ let

$$S_k = \{a3^k + \sum_{i=0}^{k-1}(b_i+1)3^i \mid a \in [0..n/3^k), b \in [0..2^k)\}.$$

Note that for any $k \geq 0$, $S_k \subset [0 \mathinner{.\,.} n)$. The set of *DC-substrings* of X is defined as

$$\bigcup_{k=0}^{\log_3 n} \{ \mathsf{X}[i \mathinner{.\,.} i + 3^k) \mid i \in S_k \}.$$

In the above definition we implicitly assume that X is followed by a sequence of infinitely repeated special symbol that is smaller than any symbol in the alphabet. From the definition of S_k we have $|S_k| = 2^k (n/3^k)$. Thus, the total number of DC-substrings is $n \sum_{k=0}^{\log_3 n} (2/3)^k = \mathcal{O}(n)$.

We want to assign a *name* to each DC-substring such that any two substrings of the same length are equal (or equal with high probability) if and only if their names are equal. We now describe a procedure for computing *deterministic names* (that when compared guarantee the equality of corresponding substrings) and *Monte-Carlo names* (that guarantee the equality with high probability) in external memory. For any $i \in S_k$ we denote the name (of any kind) of DC-substring $\mathsf{X}[i \mathinner{.\,.} i + 3^k)$ by $\alpha_k(i)$. We will assume that names for DC-substrings of different lengths are stored in different files on disk, so that accessing the names of all DC-substrings of length 3^k takes $\mathcal{O}\big(\operatorname{scan}(n(2/3)^k)\big)$ I/Os.

Lemma 2. *The deterministic names of all DC-substrings can be computed using $\mathcal{O}(\operatorname{sort}(n))$ I/Os in the standard EM model.*

Proof. For $k = 0$ we sort all letters of X and assign a rank of each letter (in sorted order) as the name of the substring. For larger k we observe that $S_{k+1} \subset S_k$. Furthermore, if $i \in S_k$ and $i + 3^k < n$ then $i + 3^k \in S_k$. Thus, given the names of DC-substrings of length 3^k we can compute the names for DC-substrings of length 3^{k+1} by sorting the set of triples $\{(\alpha_k(i), \alpha_k(i + 3^k), \alpha_k(i + 2 \cdot 3^k)) \mid i \in S_{k+1}\}$ lexicographically and again assigning a rank of each triple as the name of the corresponding substring (if either of the positions $i + 3^k$ and $i + 2 \cdot 3^k$ are outside the range $[0..n)$ we use -1 as the name of the corresponding substring). A single sorting step takes $\mathcal{O}(\operatorname{scan}(|S_k|) + \operatorname{sort}(|S_{k+1}|)) = \mathcal{O}\big(\operatorname{sort}(n(2/3)^k)\big)$ I/Os which over all lengths of DC-substrings sums up to $\mathcal{O}(\operatorname{sort}(n))$ I/Os.   □

Lemma 3. *The Monte-Carlo names of all DC-substrings can be computed using $\mathcal{O}(\operatorname{scan}(n))$ I/Os in the standard EM model.*

Proof. The goal is to compute Karp-Rabin fingerprint for every DC-substring. The general scheme of the naming procedure follows the one from Lemma 2. However, unlike in Lemma 2 the Monte-Carlo name of substring $\mathsf{X}[i \mathinner{.\,.} i + 3^{k+1})$ can be directly computed from the names of substrings $\mathsf{X}[i \mathinner{.\,.} i + 3^k)$, $\mathsf{X}[i + 3^k \mathinner{.\,.} i + 2 \cdot 3^k)$, and $\mathsf{X}[i + 2 \cdot 3^k \mathinner{.\,.} i + 3 \cdot 3^k)$ in $\mathcal{O}(1)$ time. Thus, we only need to scan the file containing the names of DC-substrings of length 3^k which takes $\mathcal{O}\big(\operatorname{scan}(n(2/3)^k)\big)$ I/Os. Over all lengths of DC-substrings we spend $\mathcal{O}(\operatorname{scan}(n))$ I/Os.   □

Note that to efficiently collect the names during scans in the above lemmas, within a single file we need to additionally group the names of DC-substrings according to the value $i \bmod 3^k$, where i is the starting position of the substring.

We will next show how to use names of DC-substrings to efficiently compute or approximate LCE queries. For simplicity we now describe the internal-memory versions of basic procedures and later explain how to modify them for external memory. Figure 1 gives a pseudo-code of an algorithm to answer an LCE query for an arbitrary pair of suffixes. The number of iterations of the while loop in lines 3–7 is bounded using the following lemma.

Lemma 4. *Let* $i, j \in S_k$ *and assume that* $\max\{i, j\} + 2 \cdot 3^k < n$. *Then either* $\{i, j\} \subset S_{k+1}$ *or* $\{i + 3^k, j + 3^k\} \subset S_{k+1}$ *or* $\{i + 2 \cdot 3^k, j + 2 \cdot 3^k\} \subset S_{k+1}$.

Proof. Let a, b be such that $i = a3^k + \sum_{i=0}^{k-1}(b_i + 1)3^i$ where $a \in [0 .. n/3^k)$ and $b \in [0 .. 2^k)$ (which exist from the definition of S_k). It is easy to check that $i \in S_{k+1}$ iff $a \bmod 3 \neq 0$. Thus, exactly two out of $\{i, i + 3^k, i + 2 \cdot 3^k\}$ are in S_{k+1}. Since the analogous property holds for j, the claim follows. □

Function lcp(i, j)
1: $i_{\text{init}} \leftarrow i$
2: $k \leftarrow 0$
3: **while** $\alpha_k(i) = \alpha_k(j)$ **do**
4:     $i \leftarrow i + 3^k$
5:     $j \leftarrow j + 3^k$
6:     **if** $\{i, j\} \subset S_{k+1}$ **then**
7:         $k \leftarrow k + 1$
8: **while** $k > 0$ **do**
— Invariant: $\alpha_k(i) \neq \alpha_k(j)$
9:     $k \leftarrow k - 1$
10:     **while** $\alpha_k(i) = \alpha_k(j)$ **do**
11:         $i \leftarrow i + 3^k$
12:         $j \leftarrow j + 3^k$
13: **return** $i - i_{\text{init}}$

Function lcp$_h(i, j, \check{\ell}, \hat{\ell})$
1: **if** $\hat{\ell} - \check{\ell} \leq 3^h$ **then**
2:     **return** $(\check{\ell}, \hat{\ell})$
3: $i_{\text{init}} \leftarrow i$
4: $k \leftarrow \min(\lfloor \log_3(\check{\ell} + 1) \rfloor, \lfloor \log_3(\hat{\ell} - \check{\ell}) \rfloor)$
5: $\delta \leftarrow \delta_k^-(i + \check{\ell}, j + \check{\ell})$
6: $i \leftarrow i + \check{\ell} - \delta$, $j \leftarrow j + \check{\ell} - \delta$
7: **while** $\alpha_k(i) = \alpha_k(j)$ **do**
8:     $i \leftarrow i + 3^k$
9:     $j \leftarrow j + 3^k$
10:     **if** $\{i, j\} \subset S_{k+1}$ **then**
11:         $k \leftarrow k + 1$
12: **while** $k > h$ **do**
— Invariant: $\alpha_k(i) \neq \alpha_k(j)$
13:     $k \leftarrow k - 1$
14:     **while** $\alpha_k(i) = \alpha_k(j)$ **do**
15:         $i \leftarrow i + 3^k$
16:         $j \leftarrow j + 3^k$
17: $\check{\ell} \leftarrow \max(\check{\ell}, i - i_{\text{init}})$
18: $\hat{\ell} \leftarrow \min(\hat{\ell}, i - i_{\text{init}} + 3^k)$
19: **return** $(\check{\ell}, \hat{\ell})$

Fig. 1. Left: Computation of lcp(i, j) using DC-substrings. Right: Approximating the value of lcp(i, j). Given $\check{\ell}$ and $\hat{\ell}$ such that $\check{\ell} \leq$ lcp$(i, j) < \hat{\ell}$ the function lcp$_h$ returns a pair $(\check{\ell}, \hat{\ell})$ that in addition satisfies $\hat{\ell} - \check{\ell} \leq 3^h$. In both functions we assume $i \neq j$

To perform the check $\{i, j\} \subset S_{k+1}$ efficiently, we identify the DC-substrings of length 3^k starting at positions i and j using triples (k, a, b) where a, b are as in the definition of S_k. This representation supports the check in constant time. Every update of i and j (represented in this way) in Fig. 1 also takes constant time. Thus, since by Lemma 4 the lcp algorithm uses $\mathcal{O}(1)$ DC-substrings of each length it runs in $\mathcal{O}(\log n)$ time. A more careful analysis shows that the algorithm only inspects DC-substrings up to length $\Theta(\mathrm{lcp}(i,j))$, and thus its running time is in fact $\mathcal{O}(\log \mathrm{lcp}(i,j))$.

Given $h \geq 0$, $\check{\ell}$, and $\hat{\ell}$ such that $\check{\ell} \leq \mathrm{lcp}(i,j) < \hat{\ell}$ we define the *informed approximate LCE query with accuracy* 3^h as the task of refining the *slack* defined as $\hat{\ell} - \check{\ell}$, so that in addition to initial assumption, it satisfies $\hat{\ell} - \check{\ell} \leq 3^h$. We now describe a method of answering approximate LCE queries using DC-substrings. We start by introducing useful auxiliary functions.

Lemma 5. *For any $k \geq 0$ and any $i,j \in [0..n)$, $\max\{i,j\} + 3^k \leq n$, there exists $\delta \in [0..3^k)$ such that $\{i+\delta, j+\delta\} \subset S_k$. We denote such δ by $\delta_k^+(i,j)$. Symmetrically, if $i,j \geq 3^k - 1$, there exists $\delta \in [0..3^k)$ such that $\{i-\delta, j-\delta\} \subset S_k$. We denote such δ by $\delta_k^-(i,j)$.*

Proof. Clearly $\{i,j\} \subset S_0$. By Lemma 4 we can find $\delta_0 \leq 2$ such that $\{i+\delta_0, j+\delta_0\} \subset S_1$. Iteratively applying Lemma 4 gives $\delta = \delta_0 + \ldots + \delta_{k-1}$ such that $\{i+\delta, j+\delta\} \subset S_k$. Since $\delta_t \leq 2 \cdot 3^t$, we have $\delta \leq 2\sum_{t=0}^{k-1} 3^t < 3^k$. The proof for $\delta_k^-(i,j)$ is analogous. □

The pseudo-code of the function approximating lcp is given in Fig. 1. It works essentially the same as the exact version except we start (lines 4–5) by computing k and δ such that $0 \leq \delta < 3^k \leq \check{\ell} + 1$ and $3^k \leq \hat{\ell} - \check{\ell}$. The first condition ensures that i and j are increased by a value in the interval $[0..\check{\ell}]$ in line 6 (which is correct from the definition of $\check{\ell}$). The second condition guarantees that the algorithm does not use DC-substrings longer than $\Theta(\hat{\ell} - \check{\ell})$. This is necessary in the case $\hat{\ell} - \check{\ell} \ll \check{\ell}$ because otherwise the algorithm would perform $\Theta(\log(\check{\ell}/(\hat{\ell} - \check{\ell})))$ comparisons of DC-substrings in the while loop in lines 12–16 which are guaranteed (from the definition of $\hat{\ell}$) to not be equal. The shortest DC-substrings used in the algorithm are of length $\Omega(\min(\check{\ell}, 3^h))$. Thus, the number of compared DC-substrings is $\mathcal{O}(\log((\hat{\ell} - \check{\ell})/\min(\check{\ell}, 3^h)))$.

In the remainder of the paper we focus on a special type of informed approximate LCE queries for which the bounds provided as input additionally satisfy $3^h \leq \check{\ell}, \hat{\ell} - \check{\ell}$, where 3^h is the required accuracy of the query. We call them 3^h-*LCE queries*. Note that from the discussion above a 3^h-LCE query can be answered using $\mathcal{O}(\log((\hat{\ell} - \check{\ell})/3^h))$ comparisons.

4   Answering Batches of LCE Queries

Assume we are given a sequence of d LCE queries $R = [(i_1, j_1), \ldots, (i_d, j_d)]$. We can answer a single LCE query (i,j) using $\mathcal{O}(\log \mathrm{lcp}(i,j))$ comparisons of DC-substrings. Thus, to answer a batch of d queries we need $\mathcal{O}(\sum_{t=1}^{d} \log \mathrm{lcp}(i_t, j_t))$ comparisons. By Jensen's inequality this is bounded by $\mathcal{O}(d \log \ell)$ where $\ell = (\sum_{t=1}^{d} \mathrm{lcp}(i_t, j_t))/d$ is the average lcp value. Thus, we obtain the following lemma.

Lemma 6. *It suffices to compare $\mathcal{O}(d \log \ell)$ DC-substrings to answer a batch of d LCE queries with an average value ℓ.*

Consider now the task of answering a batch of d 3^h-LCE queries $R = [(i_1, j_1, \check{\ell}_1, \hat{\ell}_1), \ldots, (i_d, j_d, \check{\ell}_d, \hat{\ell}_d)]$. As shown in the previous section, answering a single 3^h-LCE query takes $\mathcal{O}(\log((\hat{\ell} - \check{\ell})/3^h))$ comparisons, thus a batch of d queries needs $\mathcal{O}(\sum_{t=1}^{d} \log((\hat{\ell}_t - \check{\ell}_t)/3^h))$ comparisons. Again, by Jensen's inequality this is bounded by $\mathcal{O}(d \log(\ell/3^h))$, where $\ell = (\sum_{t=1}^{d} (\hat{\ell}_t - \check{\ell}_t))/d$ is the average slack in R.

Lemma 7. *It suffices to compare $\mathcal{O}(d \log(\ell/3^h))$ DC-substrings to answer a batch of d 3^h-LCE queries with an average slack of ℓ.*

Suppose now we want to answer a batch of d LCE queries in external memory. Assume that both the set of queries R and names of all DC-substrings are stored on disk. We divide the lcp algorithm in Fig. 1 into two phases corresponding to loops in lines 3–7 and 8–12.

Consider the first phase. During the algorithm we maintain $\log_3 n$ files on disk and at any given moment each LCE query is stored in exactly one of the files. The k-th file stores all triples (i_{init}, i, j) such that i_{init} corresponds to the value that we store in line 1, and $\{i, j\} \subset S_k$ stores the current state of the query. We process files in increasing order of k. To process k-th file we scan all triples and for every (i_{init}, i, j) we generate requests to retrieve the names $\alpha_k(i)$, $\alpha_k(i+3^k)$, $\alpha_k(i+2 \cdot 3^k)$, $\alpha_k(j)$, $\alpha_k(j+3^k)$, $\alpha_k(j+2 \cdot 3^k)$. By Lemma 4 these are the only names of DC-substrings of length 3^k used by the lcp algorithm. All name requests are first sorted by the starting position and then the corresponding names are retrieved with a single scan of the file containing k-level names. The name requests are then sorted back to the original order and each of the LCE queries is now updated. Depending on the result of the name comparison, the query either stays in the current file (mismatch) or is moved to the $(k + 1)$-th file (match) and the values i, j are updated.

If by d_k we denote the number of triples in the k-th file then executing the k-th step takes $\mathcal{O}(\text{scan}(n(2/3)^k) + \text{sort}(d_k))$ I/Os. Over all steps the I/O is $\mathcal{O}(\text{scan}(n) + \sum_{k=0}^{\log_3 n} \text{sort}(d_k))$. By Lemma 6 we have $\sum_{k=0}^{\log_3 n} d_k = \mathcal{O}(d \log \ell)$ where ℓ is the average lcp value. Thus by Jensen's inequality the total I/O volume is bounded by $\mathcal{O}(\text{scan}(n) + \text{sort}(d) \log \ell)$.

To execute the second stage of the algorithm (lines 8–12) the algorithm proceeds analogously, except now we process the remaining items in all files in the decreasing order of k. The I/O complexity does not change.

Lemma 8. *A batch of d LCE queries with an average value ℓ can be answered using $\mathcal{O}(\text{scan}(n) + \text{sort}(d) \log \ell)$ I/Os in the standard EM model.*

Answering a batch of 3^h-LCE queries in external memory works analogously and the result follows from Lemma 7. We don't access DC-substrings shorter than 3^h and thus the scanning time is reduced.

Function Reduce(k, R)
  — *Step 1: For all i determine if* $\text{lcp}(i, \Phi(i)) \geq 3^{k+1} - 1$
1: **foreach** $(i, \Phi(i), \check{\ell}_i, \hat{\ell}_i) \in R$ **do**
2:    $\delta \leftarrow \delta_k^+(i, \Phi(i))$
3:    **if** $\alpha_k(i + \delta) \neq \alpha_k(\Phi(i) + \delta)$ **then** $\hat{\ell}_i \leftarrow \delta + 3^k$
4:    **elsif** $\alpha_k(i + \delta + 3^k) \neq \alpha_k(\Phi(i) + \delta + 3^k)$ **then** $(\check{\ell}_i, \hat{\ell}_i) \leftarrow (\delta + 3^k, \delta + 2 \cdot 3^k)$
5:    **elsif** $\alpha_k(i + \delta + 2 \cdot 3^k) \neq \alpha_k(\Phi(i) + \delta + 2 \cdot 3^k)$ **then** $(\check{\ell}_i, \hat{\ell}_i) \leftarrow (\delta + 2 \cdot 3^k, \delta + 3 \cdot 3^k)$
6:    **else** $\check{\ell}_i \leftarrow \delta + 3 \cdot 3^k$
  — *Step 2: Create a sample of* $\mathcal{O}(|R|/3)$ *tuples*
7: $R' \leftarrow [(i, \Phi(i), \check{\ell}_i, \hat{\ell}_i) \in R \mid \check{\ell}_i \geq 3^{k+1} - 1]$ (sorted by i)
8: $S \leftarrow [R'[3i] \mid i \in [0 .. \lceil |R'|/3 \rceil)]$
  — *Step 3: Recursively reduce the sample slacks to at most* 3^{k+1}
9: **if** $S \neq \emptyset$ **then** Reduce($k + 1, S$)
  — *Step 4: Using sample slacks reduce the total slack to* $\mathcal{O}(n)$
10: **foreach** $(i, \Phi(i), \check{\ell}_i, \hat{\ell}_i) \in R' \setminus S$ **do**
11:   Let $(p, \Phi(p), \check{\ell}_p, \hat{\ell}_p)$ be the predecessor (w.r.t. i) of $(i, \Phi(i), \check{\ell}_i, \hat{\ell}_i)$ in S
12:   Let $(s, \Phi(s), \check{\ell}_s, \hat{\ell}_s)$ be the successor (w.r.t. i) of $(i, \Phi(i), \check{\ell}_i, \hat{\ell}_i)$ in S
13:   $\check{\ell}_i \leftarrow \max(\check{\ell}_i, \check{\ell}_p - (i - p))$
14:   $\hat{\ell}_i \leftarrow \min(\hat{\ell}_i, \hat{\ell}_s + (s - i))$
  — *Step 5: Reduce all slacks to at most* 3^k
15: **foreach** $(i, \Phi(i), \check{\ell}_i, \hat{\ell}_i) \in R$ **do**
16:   **if** $\hat{\ell}_i - \check{\ell}_i > 3^k$ **then**
17:     $(\check{\ell}_i, \hat{\ell}_i) \leftarrow \text{lcp}_k(i, \Phi(i), \check{\ell}_i, \hat{\ell}_i)$

Fig. 2. Given $R = [(i, \Phi(i), \check{\ell}_i, \hat{\ell}_i)]$, $|R| \leq n/3^k$ such that $3^k - 1 \leq \check{\ell}_i \leq \text{lcp}(i, \Phi(i)) < \hat{\ell}_i$, refine all $\check{\ell}_i, \hat{\ell}_i$ so that in addition to initial assumptions they satisfy $\hat{\ell}_i - \check{\ell}_i \leq 3^k$

Lemma 9. *A batch of d 3^h-LCE queries with an average slack ℓ can be answered using $\mathcal{O}\big(\text{scan}\big(n(2/3)^h\big) + \text{sort}(d) \log(\ell/3^h)\big)$ I/Os in the standard EM model.*

Note that all the complexities stated in this section exclude the time needed to compute the names of DC-substrings. Thus, to solve an arbitrary set of LCE queries we need to additionally spend $\mathcal{O}(\text{sort}(n))$ I/Os to compute deterministic names (Lemma 2) or $\mathcal{O}(\text{scan}(n))$ I/Os to compute Monte-Carlo names (Lemma 3).

5   LCP Array Construction

Let $k \geq 0$ and consider an arbitrary subset P of at most $n/3^k$ text positions from $[0 .. n)$ such that $3^k - 1 \leq \text{lcp}(i, \Phi(i))$ for all $i \in P$. Let $R = \{(i, \Phi(i), \check{\ell}_i, \hat{\ell}_i) \mid i \in P\}$ be such that $3^k - 1 \leq \check{\ell}_i \leq \text{lcp}(i, \Phi(i)) < \hat{\ell}_i$ for all i. The main ingredient of the new LCP array construction algorithm is the procedure to reduce all slacks in R to at most 3^k. It uses batched 3^k-LCE queries in the final step after first improving the bounds with a different technique.

The pseudo-code of the procedure is given in Fig. 2. For simplicity we use the standard notation for internal-memory algorithms. Below we outline all steps and explain how to implement them in external memory using scanning and sorting.

We start by checking, for every i, whether $\mathrm{lcp}(i, \Phi(i)) \geq 3^{k+1} - 1$. This requires fetching $\mathcal{O}(1)$ names of DC-substrings of length 3^k for each tuple, and thus takes $\mathcal{O}\big(\mathrm{scan}(n(2/3)^k) + \mathrm{sort}(|R|)\big)$ I/Os. Next, we create a sample consisting of every third tuple (we assume they are sorted by i) for which $\mathrm{lcp}(i, \Phi(i)) \geq 3^{k+1} - 1$ and recursively reduce the slacks of the sample to at most 3^{k+1}. Excluding the cost of recursive call, it takes $\mathcal{O}(\mathrm{sort}(|R|))$ I/Os. In the next step we use the slacks of the sample set to reduce all remaining slacks. The correctness of the reduction follows from Lemma 1. The reduced slacks satisfy the following property.

Lemma 10. *The total slack in R after step 4 is $\mathcal{O}(n)$.*

Proof. Denote the elements of S after returning from recursion (line 9) by $(i_j^S, \Phi(i_j^S), \check{\ell}_j^S, \hat{\ell}_j^S)$, where $j \in [1\mathinner{.\,.}|S|]$ and assume $i_j^S < i_{j+1}^S$ for $j \in [1\mathinner{.\,.}|S|)$. For $j = |S|+1$ we set $i_j^S = n$, $\hat{\ell}_j^S = 0$. Let $(i, \Phi(i), \check{\ell}_i, \hat{\ell}_i) \in R' \backslash S$ be one of the elements processed in line 10. Let $(i_j^S, \Phi(i_j^S), \check{\ell}_j^S, \hat{\ell}_j^S)$ be the predecessor of $(i, \Phi(i), \check{\ell}_i, \hat{\ell}_i)$ in S (which always exists since S contains the smallest item of R'). Then the successor of $(i, \Phi(i), \check{\ell}_i, \hat{\ell}_i)$ in S is $(i_{j+1}^S, \Phi(i_{j+1}^S), \check{\ell}_{j+1}^S, \hat{\ell}_{j+1}^S)$. The slack of the processed tuple after the update satisfies $\hat{\ell}_i - \check{\ell}_i \leq (\hat{\ell}_{j+1}^S + (i_{j+1}^S - i)) - (\check{\ell}_j^S - (i - i_j^S)) = (\hat{\ell}_{j+1}^S - \check{\ell}_j^S) + (i_{j+1}^S - i_j^S)$. Since each tuple in S can be the predecessor of at most two elements in $R' \backslash S$ (from definition of S), the total slack in $R' \backslash S$ is bounded by

$$2 \sum_{j=1}^{|S|} ((\hat{\ell}_{j+1}^S - \check{\ell}_j^S) + (i_{j+1}^S - i_j^S)) \leq 2n + 2 \sum_{j=1}^{|S|} (\hat{\ell}_j^S - \check{\ell}_j^S).$$

Since the total slack in S does not exceed $|S| \cdot 3^{k+1} = \mathcal{O}(n)$, the total slack in $R' \backslash S$ is also $\mathcal{O}(n)$. Finally, by Step 1 and definition of R', the slack in $R \backslash R'$ is not greater than $|R| \cdot 3^{k+1} = \mathcal{O}(n)$.   □

As a last step we apply the algorithm from the previous section to answer a batch of at most $|R|$ approximate lcp queries. The average slack in R at this point is $\mathcal{O}(n/|R|)$, and thus by Lemma 9 answering all approximate lcp queries takes $\mathcal{O}\big(\mathrm{scan}(n(2/3)^k) + \mathrm{sort}(|R|) \log(n/3^k|R|)\big)$ I/Os. Excluding the cost of the recursive call, this step dominates the I/O complexity.

Consider the call of Reduce processing the sample S (i.e., the first level of recursion). The number of performed I/Os (excluding deeper recursive calls) is $\mathcal{O}\big(\mathrm{scan}(n(2/3)^{k+1}) + \mathrm{sort}(|S|) \log(n/3^{k+1}|S|)\big)$. We have $|S| \leq |R|/3$, but since $\mathcal{O}(\mathrm{sort}(d) \log(n/3^{k+1}d))$ as a function of d is non-decreasing (assuming $d = \mathcal{O}(n/3^{k+1})$), the I/O complexity is maximized for $|S| = |R|/3$ and hence $\mathrm{sort}(|S|) \log(n/3^{k+1}|S|) = \mathcal{O}\big(\mathrm{sort}(|R|/3) \log(n/3^k|R|)\big)$. Thus, since I/O decreases exponentially with every level of recursion, the total I/O complexity of Reduce(k,R) is not greater than the complexity at zero-level recursion.

Lemma 11. *Reduce uses* $\mathcal{O}\big(\text{scan}\big(n(2/3)^k\big) + \text{sort}(|R|)\log(n/3^k|R|)\big)$ *I/Os in the standard EM model.*

Using Reduce we can compute the LCP array as follows. We scan the suffix array and for every $i > 0$ we create a tuple $(i, \text{SA}[i], \text{SA}[i-1])$. We then sort these tuples by the second component to obtain a sequence $(\text{SA}^{-1}[i], i, \Phi[i])$. Using that we create a sequence of tuples $(i, \Phi[i], 0, n)$ which is then used as an input to Reduce with $k = 0$. As a result we obtain a sequence $(i, \cdot, \text{PLCP}[i], \cdot)$, which we can now permute into LCP using SA^{-1} values. The total I/O complexity (including the computation of deterministic names for DC-substrings) is $\mathcal{O}(\text{sort}(n))$.

Theorem 1. *Using DC-substrings the LCP array can be computed correctly in the standard EM model from the text and its suffix array using* $\mathcal{O}(\text{sort}(n))$ *I/Os.*

6 $o(\text{sort}(n))$ I/Os

Finally, we look at some variants of the LCP array construction problem, where we can achieve an I/O complexity of $o(\text{sort}(n))$. In all of these, we use the Monte Carlo names of DC-substrings, which can be computed in $\mathcal{O}(\text{scan}(n))$ I/Os, and thus the results are correct with high probability.

A sparse suffix array contains some subset of suffixes in the lexicographical order and the associated LCP array contains lcps of suffixes that are lexicographically adjacent in that subset. The LCP array can be computed as a batch of LCE queries and thus by Lemma 8 we obtain the following result.

Theorem 2. *Given a sparse suffix array containing* $m < n$ *suffixes of a text of length* n *in sorted order it takes*

$$\mathcal{O}(\text{scan}(n) + \text{sort}(m)\log(\ell)) = \mathcal{O}(\text{scan}(n) + \text{sort}(m)\log(n))$$

I/Os in the standard EM model to compute the corresponding LCP array correctly with high probability, where ℓ *is the average value in the LCP array.*

We can do better in the special case of evenly spaced sparse suffix array that contains exactly every q^{th} suffix of the text for some $q \geq 1$. In this case, we can use the Reduce algorithm and give as input the set of pairs $(i, \Phi(i))$, where i and $\Phi(i)$ are divisible by q and the suffix $\Phi(i)$ is the immediate lexicographical predecessor of the suffix i among the sparse set. Notice that this approach does not work correctly for an arbitrary sparse set (because of Step 4 in Reduce).

Theorem 3. *Given an evenly spaced sparse suffix array containing every* q^{th} *suffix of a text of length* n *in sorted order it takes* $\mathcal{O}(\text{scan}(n) + \text{sort}(n/q)\log(q))$ *I/Os in the standard EM model to compute the corresponding LCP array correctly with high probability.*

If $q = \omega(1)$, the number of performed I/Os is $o(\text{sort}(n))$. For example, if $q = \Omega((\log_{M/B}(n/B))^2)$, the I/O complexity is $\mathcal{O}(\text{scan}(n))$.

Finally, we can sometimes compute the full PLCP array using $o(\text{sort}(n))$ I/Os by computing first a subset of so called *irreducible* lcp values, from which the other lcp values are easy to derive (see [18,20]). For highly repetitive texts, the number r of the irreducible lcp values can be much smaller than n. To identify the irreducible positions quickly we need the Burrows–Wheeler transform as an additional input. We can compute the irreducible entries of the PLCP array using Reduce in $\mathcal{O}(\text{scan}(n) + \text{sort}(r)\log(n/r))$ I/Os and then the other entries by a simple scan in $\mathcal{O}(\text{scan}(n))$ I/Os.

Theorem 4. *Given the suffix array and the Burrows-Wheeler transform of a text of length n, it takes $\mathcal{O}(\text{scan}(n) + \text{sort}(r)\log(n/r))$ I/Os in the standard EM model to compute the PLCP array correctly with high probability, where r is the number of irreducible lcp values.*

If $r = o(n)$, the complexity is $o(\text{sort}(n))$. Transforming the PLCP array into an LCP array still needs $\Theta(\text{sort}(n))$ I/Os.

References

1. Abouelhoda, M.I., Kurtz, S., Ohlebusch, E.: Replacing suffix trees with enhanced suffix arrays. J. Discrete Algorithms **2**(1), 53–86 (2004)
2. Afshani, P., Sitchinava, N.: I/O-efficient range minima queries. In: Ravi, R., Gørtz, I.L. (eds.) SWAT 2014. LNCS, vol. 8503, pp. 1–12. Springer, Heidelberg (2014)
3. Arge, L., Fischer, J., Sanders, P., Sitchinava, N.: On (dynamic) range minimum queries in external memory. In: Dehne, F., Solis-Oba, R., Sack, J.-R. (eds.) WADS 2013. LNCS, vol. 8037, pp. 37–48. Springer, Heidelberg (2013)
4. Beller, T., Gog, S., Ohlebusch, E., Schnattinger, T.: Computing the longest common prefix array based on the Burrows-Wheeler transform. J. Discrete Algorithms **18**, 22–31 (2013)
5. Bille, P., Gørtz, I.L., Knudsen, M.B.T., Lewenstein, M., Vildhøj, H.W.: Longest common extensions in sublinear space. In: Cicalese, F., Porat, E., Vaccaro, U. (eds.) CPM 2015. LNCS, vol. 9133, pp. 65–76. Springer, Heidelberg (2015)
6. Bille, P., Gørtz, I.L., Sach, B., Vildhøj, H.W.: Time-space trade-offs for longest common extensions. J. Discrete Algorithms **25**, 42–50 (2014)
7. Bingmann, T., Fischer, J., Osipov, V.: Inducing suffix and LCP arrays in external memory. In: Sanders, P., Zeh, N. (eds.) ALENEX 2013. pp. 88–102. SIAM (2013)
8. Dietzfelbinger, M., Gil, J., Matias, Y., Pippenger, N.: Polynomial hash functions are reliable. In: Kuich, W. (ed.) ICALP 1992. LNCS, vol. 623, pp. 235–246. Springer, Heidelberg (1992)
9. Fischer, J.: Inducing the LCP-array. In: Dehne, F., Iacono, J., Sack, J.-R. (eds.) WADS 2011. LNCS, vol. 6844, pp. 374–385. Springer, Heidelberg (2011)
10. Fischer, J., I, T., Köppl, D.: Deterministic sparse suffix sorting on rewritable texts. In: Kranakis, E., Navarro, G., Chávez, E. (eds.) LATIN 2016. LNCS, vol. 9644, pp. 483–496. Springer, Heidelberg (2016)

11. Gawrychowski, P., Kociumaka, T., Rytter, W., Walen, T.: Faster longest common extension queries in strings over general alphabets. In: Grossi, R., Lewenstein, M. (eds.) CPM 2016. LIPIcs, vol. 54. Schloss Dagstuhl - Leibniz-Zentrum fuer Informatik (2016)

12. Gog, S., Ohlebusch, E.: Fast and lightweight LCP-array construction algorithms. In: Müller-Hannemann, M., Werneck, R.F.F. (eds.) ALENEX 2011. pp. 25–34. SIAM (2011)

13. Gonnet, G.H., Baeza-Yates, R.A., Snider, T.: New indices for text: PAT trees and PAT arrays. In: Frakes, W.B., Baeza-Yates, R. (eds.) Information Retrieval: Data Structures & Algorithms, pp. 66–82. Prentice-Hall, Englewood Cliffs (1992)

14. I, T., Kärkkäinen, J., Kempa, D.: Faster sparse suffix sorting. In: Mayr, E.W., Portier, N. (eds.) STACS 2014. LIPIcs, vol. 25, pp. 386–396. Schloss Dagstuhl - Leibniz-Zentrum fuer Informatik (2014)

15. Kärkkäinen, J., Sanders, P., Burkhardt, S.: Linear work suffix array construction. J. ACM 53(6), 918–936 (2006)

16. Kärkkäinen, J., Kempa, D.: Engineering a lightweight external memory suffix array construction algorithm. In: Iliopoulos, C.S., Langiu, A. (eds.) ICABD 2014. pp. 53–60 (2014)

17. Kärkkäinen, J., Kempa, D.: Faster external memory LCP array construction. In: Sankowski, P., Zaroliagis, C. (eds.) ESA 2016. LIPIcs, Schloss Dagstuhl - Leibniz-Zentrum fuer Informatik (2016)

18. Kärkkäinen, J., Kempa, D.: LCP array construction in external memory. J. Exp. Algorithmics 21(1), 1.7:1–1.7:22 (2016)

19. Kärkkäinen, J., Kempa, D., Puglisi, S.J.: Parallel external memory suffix sorting. In: Cicalese, F., Porat, E., Vaccaro, U. (eds.) CPM 2015. LNCS, vol. 9133, pp. 329–342. Springer, Heidelberg (2015)

20. Kärkkäinen, J., Manzini, G., Puglisi, S.J.: Permuted longest-common-prefix array. In: Kucherov, G., Ukkonen, E. (eds.) CPM 2009. LNCS, vol. 5577, pp. 181–192. Springer, Heidelberg (2009)

21. Kärkkäinen, J., Sanders, P.: Simple linear work suffix array construction. In: Baeten, J.C.M., Lenstra, J.K., Parrow, J., Woeginger, G.J. (eds.) ICALP 2003. LNCS, vol. 2719, pp. 943–955. Springer, Heidelberg (2003)

22. Karp, R.M., Rabin, M.O.: Efficient randomized pattern-matching algorithms. IBM J. Res. Dev. 31(2), 249–260 (1987)

23. Kasai, T., Lee, G.H., Arimura, H., Arikawa, S., Park, K.: Linear-time longest-common-prefix computation in suffix arrays and its applications. In: Amir, A., Landau, G.M. (eds.) CPM 2001. LNCS, vol. 2089, pp. 181–192. Springer, Heidelberg (2001)

24. Liu, W.J., Nong, G., Chan, W.H., Wu, Y.: Induced sorting suffixes in external memory with better design and less space. In: Iliopoulos, C., Puglisi, S., Yilmaz, E. (eds.) SPIRE 2015. LNCS, vol. 9309, pp. 83–94. Springer, Heidelberg (2015)

25. Louza, F.A., Telles, G.P., De Aguiar Ciferri, C.D.: External memory generalized suffix and LCP arrays construction. In: Fischer, J., Sanders, P. (eds.) CPM 2013. LNCS, vol. 7922, pp. 201–210. Springer, Heidelberg (2013)

26. Mäkinen, V.: Compact suffix array – a space efficient full-text index. Fund. Inform. 56(1–2), 191–210 (2003)

27. Mäkinen, V., Belazzougui, D., Cunial, F., Tomescu, A.I.: Genome-Scale Algorithm Design: Biological Sequence Analysis in the Era of High-Throughput Sequencing. Cambridge University Press, Cambridge (2015)

28. Manber, U., Myers, G.W.: Suffix arrays: a new method for on-line string searches. SIAM J. Comput. 22(5), 935–948 (1993)

29. Manzini, G.: Two space saving tricks for linear time LCP array computation. In: Hagerup, T., Katajainen, J. (eds.) SWAT 2004. LNCS, vol. 3111, pp. 372–383. Springer, Heidelberg (2004)
30. Navarro, G., Mäkinen, V.: Compressed full-text indexes. ACM Comput. Surv. **39**(1), 2 (2007)
31. Nong, G., Chan, W.H., Hu, S.Q., Wu, Y.: Induced sorting suffixes in external memory. ACM Trans. Inf. Syst. **33**(3), 12:1–12:15 (2015)
32. Nong, G., Chan, W.H., Zhang, S., Guan, X.F.: Suffix array construction in external memory using d-critical substrings. ACM Trans. Inf. Syst. **32**(1), 1:1–1:15 (2014)
33. Ohlebusch, E.: Bioinformatics Algorithms: Sequence Analysis, Genome Rearrangements, and Phylogenetic Reconstruction. Oldenbusch Verlag, Bremen (2013)
34. Puglisi, S.J., Turpin, A.: Space-time tradeoffs for longest-common-prefix array computation. In: Hong, S.-H., Nagamochi, H., Fukunaga, T. (eds.) ISAAC 2008. LNCS, vol. 5369, pp. 124–135. Springer, Heidelberg (2008)
35. Sirén, J.: Sampled longest common prefix array. In: Amir, A., Parida, L. (eds.) CPM 2010. LNCS, vol. 6129, pp. 227–237. Springer, Heidelberg (2010)
36. Tanimura, Y., I, T., Bannai, H., Inenaga, S., Puglisi, S.J., Takeda, M.: Deterministic sub-linear space LCE data structures with efficient construction. In: Grossi, R., Lewenstein, M. (eds.) CPM 2016. LIPIcs, vol. 54, pp. 1:1–1:10. Schloss Dagstuhl - Leibniz-Zentrum fuer Informatik (2016)

GraCT: A Grammar Based Compressed Representation of Trajectories

Nieves R. Brisaboa[1], Adrián Gómez-Brandón[1], Gonzalo Navarro[2],
and José R. Paramá[1(✉)]

[1] Depto. de Computación, Universidade da Coruña, Coruña, Spain
{brisaboa,adrian.gbrandon,jose.parama}@udc.es
[2] Department of Computer Science, University of Chile, Santiago, Chile
gnavarro@dcc.uchile.cl

Abstract. We present a compressed data structure to store free trajectories of moving objects (ships over the sea, for example) allowing spatio-temporal queries. Our method, GraCT, uses a k^2-tree to store the absolute positions of all objects at regular time intervals (snapshots), whereas the positions between snapshots are represented as logs of relative movements compressed with Re-Pair. Our experimental evaluation shows important savings in space and time with respect to a fair baseline.

1 Introduction

After more than two decades of research on moving objects, this field still presents interesting problems that represent a topic of active research. The renewed interest to represent and exploit data about moving objects is mainly due to the new context in which large amounts of data (from, for example, cellular phones informing about the GPS coordinates of their position in real time) need to be stored and analyzed. Therefore, new big data sets and new application domains demand more efficient technology to manage moving objects.

Traditional spatio-temporal indexes can be classified into two families, *space-based* indexes and *trajectory-based* indexes. Each type of index is adapted to answer different types of queries. Indexes in the first family usually are modifications of the classical spatial R-tree, like for example the RT-tree [17], the HR-tree [11], the 3DR-tree [14], the MV3R-Tree [13], or the SEST-Index [5]. Those indexes efficiently answer queries which return the ids or the number of objects into a given spatial region at a specific time instant (time-slice queries) or at a specific time interval (time-interval queries), but they cannot efficiently return the position of an object at a time instant or which was its trajectory[1] during a time interval.

This work was funded in part by European Unions Horizon 2020 Marie Skłodowska-Curie grant agreement No. 690941; Ministerio de Economía y Competitividad under grants [TIN2013-46238-C4-3-R], [CDTI IDI-20141259], [CDTI ITC-20151247], and [CDTI ITC-20151305]; Xunta de Galicia (co-founded with FEDER) under grant [GRC2013/053]; and Fondecyt Grant 1-140796, Chile.

[1] We informally define trajectory as a list of positions in consecutive time instants.

© Springer International Publishing AG 2016
S. Inenaga et al. (Eds.): SPIRE 2016, LNCS 9954, pp. 218–230, 2016.
DOI: 10.1007/978-3-319-46049-9_21

The second family of indexes were designed to improve the management of trajectories, like SETI [4], the CSE-tree [15], and trajectory splitting strategies [6,12]. Those indexes can describe trajectories of individual objects but cannot answer efficiently time-slice or time-interval queries over objects in a specific region of the space.

Those indexes maintain the bulk of the data on disk, while the index structures reside in main memory. They rarely use compression to reduce disk or memory usage, or to reduce the disk transfer time.

In this paper we introduce an in-memory representation called *Grammar based Compressed representation of Trajectories (GraCT)*. GraCT is a trajectory-oriented technique, that is, it belongs to the second family. However, it structures the index into snapshots of the objects taken at regular time instants, and logs of their movements between snapshots. This allows GraCT to efficiently answer *time-slice* and *time-interval* queries as well, by processing the logs between two snapshots. Besides, GraCT represents data and index together, and uses grammar compression on the logs. This not only reduce the size of the representation, but also the nonterminals are enriched to allow processing long parts of the log files without decompressing them, and *faster* than with a plain representation. Its space savings allow GraCT fitting much larger datasets in main memory, where they can be queried much faster than on disk.

2   Background

2.1 K^2-tree

The k^2-tree is a compact data structure originally designed for representing Web graphs in little space, allowing its manipulation directly in compressed form [3]. The k^2-tree is used to represent the adjacency matrix of the graphs, and it can also be used to represent any type of binary matrices.

The k^2-tree is conceptually a non-balanced k^2-ary tree built from a binary matrix by recursively subdividing the matrix into k^2 submatrices of the same size. It starts by subdividing the original matrix into k^2 submatrices of size n^2/k^2, being $n \times n$ the size of the matrix. The submatrices are ordered from left to right and from top to bottom. Each of those submatrices generates a child of the root node whose value is 1 if there is at least one 1 in the cells of that submatrix, and 0 otherwise. The subdivision proceeds recursively for each child with value 1 until it reaches a submatrix full of 0s, or it reaches the cells of the original matrix (i.e., submatrices of size 1×1). Figure 1 shows an example of this subdivision (left) and the resulting conceptual k^2-ary tree (right up) for $k = 2$.

Instead of using a pointer-based representation, the k^2-tree is compactly stored using two bitmaps T and L (see Fig. 1). T stores all the bits of the k^2-tree except those in the last level. The bits are placed following a levelwise traversal: first the k^2 binary values of the root node, then the values of the second level, and so on. L stores the last level of the tree. Thus, it represents the value of original cells of the binary matrix.

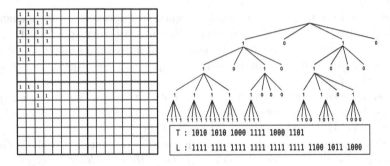

Fig. 1. Example of binary matrix (left) and resulting k^2-tree conceptual representation (right up), and the compact representation (right down), with $k = 2$.

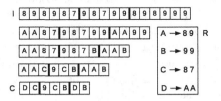

Fig. 2. An example of Re-pair compression.

It is possible to obtain any cell, row, column, or region of the matrix very efficiently, by just running *rank* and *select* operations [7] over the bitmap T: $rank_b(T, p)$ is the number of occurrences of bit $b \in \{0, 1\}$ in T up to position p, and $select_b(T, j)$ is the position in T of the jth occurrence of the bit b. For example, given a value 1 at position p in T, its k^2 children will start at position $p_{children} = rank_1(T, p) \times k^2$ of T, except when the position of the children of a node returns a position $p_{children} > |T|$; in that case we access instead $L[p_{children} - |T|]$ to retrieve the actual value of the cells. Similarly, the parent of a position p in $T : L$ is $q - (q \bmod k^2)$, where $q = select_1(T, \lfloor p/k^2 \rfloor)$, and $q \bmod k^2$ indicates which is the submatrix of p within its parent's.

2.2 Re-Pair

Re-pair [9] is a grammar-based compression method. Given a sequence of integers I (called *terminals*), it proceeds as follows: (1) it obtains the most frequent pair of integers ab in I, (2) it adds the rule $s \rightarrow ab$ to a dictionary R, where s is a new symbol not appearing in I (called a *nonterminal*), (3) every occurrence of ab in I is replaced by s, and (4) it repeats steps 1–3 until every pair in I appears only once (see Fig. 2). The resulting sequence after compressing I is called C. Every symbol in C represents a phrase (a sequence of 1 or more of the integers in I). If the length of the represented phrase is 1, then the phrases consists of an original (terminal) symbol, otherwise it is a new (nonterminal) symbol. Re-Pair can be implemented in linear time, and a phrase can be recursively expanded in optimal time (that is, proportional to its length).

3 Our Approach

GraCT represents moving objects that follow free trajectories on the space. We consider the time as discrete, therefore each time instant actually corresponds to a short period of time. We assume that in each time instant, each object informs its position (e.g., international regulations require that ships inform their GPS position at regular intervals). We use a raster model to represent the space, therefore the space is divided into cells of a fixed size, and objects are assumed to fit in one cell. The size of the cells and the period used to sample the time are parameters that can be adapted to the specific domain.

Every s time instants, GraCT uses a data structure based on k^2-trees to represent the absolute positions of all objects. We call those time instants snapshots. The distance s between snapshots is another parameter of the system. Between two consecutive snapshots the trajectory of each moving object is represented as a log, which is an array of movements, that is, relative positions with respect to the previous time instant.

Snapshots. Each snapshot uses a k^2-tree where a cell set to 1 indicates that one or more objects are placed in that cell, whereas a 0 means that no object is in that cell. However, we still need to know which objects are in a cell set to 1. Observe that each 1 in the binary matrix corresponds to a bit set to 1 in the bitmap L of the k^2-tree. We store the list of object identifiers corresponding to each of those bits set to 1 in an array, where the objects identifiers are sorted following the order of appearance in L. We call that array *perm*, since that array is a permutation [8]. In addition, we need a bitmap, called Q, aligned with *perm*, that informs with a 0 that the object identifier aligned in *perm* is the last object of a leaf, whereas a 1 signals that more objects exist. Observe in Fig. 3, the object identifiers corresponding to the first 1 in L (which is at position 3 of L) are stored starting at position 1 of *perm*. In order to know how many objects are in the corresponding cell, we access Q starting at position 1 searching for the first 0, which is at position 2, therefore there are two objects in the inspected cell. By accessing positions 1 and 2 of *perm*, we obtain the object identifiers 4 and 2. Now, in position 3 of *perm* starts the object identifiers corresponding to the second 1 in L, and so on.

With these structures used to represent the absolute positions of all the moving objects at snapshots we can answer two types of queries:

- *Find the objects in a given cell*: First, using the procedure shown in Sect. 2.1 to navigate downwards the k^2-tree, we traverse the tree from the root until reaching the position n in L corresponding to that cell. Next, we count the number of 1 s in the array of leaves L until the position n; this gives us the number of leaves with objects up to the nth leaf, $x = rank_1(L, n)$. Then we

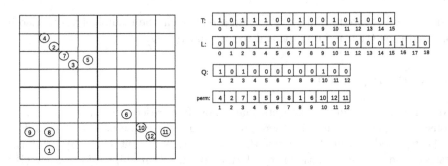

Fig. 3. The position of objects in the space (left), and the representing snapshot (right).

calculate the position of the $(x-1)$th 0 in Q, which indicates the last bit of the previous leaf (with objects), and we add 1 to get the first position of our leaf, $p = select_0(Q, x-1) + 1$. Then p is the position in *perm* of the first object identifier corresponding to the searched position. From p, we read all the object identifiers aligned with 1s in Q, until we reach a 0, which signals the last object identifier of that leaf.

- *Find the position in the space of a given object.* First, we need to obtain the position k in *perm* of the searched object. In order to avoid a sequential search over *perm* to obtain that position, we add additional structures to compute cells of the inverse permutation of *perm* [10]. Then, we have to find the leaf in L corresponding to the kth position of *perm*. For this sake, we calculate the number of leaves before the object in position k of *perm*, that is, we need to count the number of 0s until the position before k, $y = rank_0(Q, k-1)$. Then we find in L the position of the $(y+1)^{th}$ 1, that is, $select_1(L, y+1)$. With that position of L, we can traverse the k^2-tree upwards in order to obtain the position in the space of that cell, and thus the position of the object.

Log of Relative Movements. The changes that occur between snapshots are tracked using a log file per object. The use of snapshots and logs is not new [16], but in previous works log values are stored according the appearance of "events" (such as objects that appear in or disappear from an area).

The log stores relative movements with respect to the last known position of an object, that is, to its position in the preceding time instant. Objects can change their positions along the two Cartesian axes, so every movement in the log can be described with two integers. Instead, in order to save space, we encode the two values with a unique positive integer. For this sake, we enumerate the cells around the actual position of an object, following a spiral where the origin is the initial object position, as it is shown in Fig. 4 (left). Let us suppose that an object moves with respect to the previous known position one cell to the East in the x-axis, and one cell to the North in the y-axis. Instead of encoding the movement as the pair $(1, 1)$, we encode it as an 8. In Fig. 4 (right) we show the trajectory of an object starting at cell $(0, 2)$. Each number indicates a movement

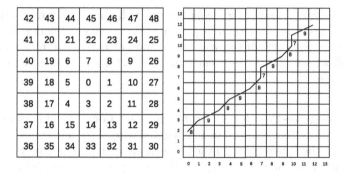

Fig. 4. Encoding object's movements.

between two consecutive time instants. Since most relative movements involve short distances, this technique produces a sequence of usually small numbers.

Sometimes real objects stop emitting signals during periods of time. This forces us to add two new possible movements inside a log: *relative reappearance* and *absolute reappearance*. We reserve two codewords to signal these events. We use a relative reappearance when an object disappears and reappears between the same snapshots, and an absolute reappearance otherwise. Relative reappearances are followed by the time elapsed from the disappearance and a relative movement from that time instant, whereas absolute reappearances are followed by the number of time instants that elapsed since the disappearance and the absolute values of the (x, y) coordinates of the new position of the object.

4 Compressing the Log

The log not only saves much space compared to using k^2-trees for every instant, but it also offers important opportunities for further compression. A first choice is statistical compression, since as said, most movements are short-distanced and thus our spiral encoding uses mostly small numbers. We exploit this fact using (s, c)-Dense Codes (SCDC) [2], a very fast-to-decode statistical compressor that has a low redundancy over the zero-order empirical entropy of the sequence. We will use this approach as fair baseline.

The second approach, which gives the title to this paper, uses grammar compression on the set of all the log files. Our aim is to exploit the fact that there are typical trajectories followed by many objects, which translate into long sequences of identical movements that grammar compression can convert into single non-terminals. This includes, in particular, long straight trajectories in any direction.

ScdcCT: Using SCDC for Compressing the Logs. The size of the cells and the time elapsed between consecutive time instants must be carefully chosen to represent properly the typical speed of moving objects, so that short movements to contiguous cells are more frequent than movements to distant cells. Instead of sorting the spiral codes by frequency, we will simply assume that smaller numbers are more frequent than larger ones. Since the (s, c)-codes depend only on the relative frequency of the symbols, we do not need to store any statistical model. Still, we will use the frequencies to optimize s and c in order to minimize the space usage.

GraCT: Using Re-Pair for Compressing the Logs. Moving objects spend most of the time either stopped or moving following a specific course and speed. In both cases, the logs will present longs sections with numbers representing the same or contiguous values of the spiral. For example, the moving object in Fig. 4 follows a NE trajectory moving one or two cells per time instant. Therefore its log represents the series of relative movements $8, 9, 8, 9, 8, 7, 9, 8, 7, 9$; see the array I of Fig. 2. Those series of similar movements are very efficiently compressed using a grammar compressor such as Re-Pair. To avoid having to decompress the log before processing it, we enrich the rules of the grammar R with further data apart from the two symbols to be replaced. Specifically, each rule in R will have the following information: $s \rightarrow a, b, \#t, x, y, MBR$, where s, a and b are the components of a normal rule of Re-Pair, $\#t$ is the number of instants covered by the rule, (x, y) are the relative coordinates of the final position of the object after the application of the rule, and MBR is the minimum bounding rectangle enclosing the movements of the rule.

For example, the rules of Fig. 2 are enriched as follows. The first rule of R is $A \rightarrow 8, 9, 2, (3, 2), (0, 0, 3, 2)$: 8 and 9 are the substituted symbols, 2 indicates that the rule represents a sequence of 2 movements, $(3, 2)$ indicates the position of the object after the application of the rule if we start at $(0, 0)$, and the last four values are two points defining a rectangle that encloses all the movements encoded by the rule. The other rules are $B \rightarrow 9, 9, 2, (4, 2), (0, 0, 4, 2)$, $C \rightarrow 8, 7, 2, (1, 2), (0, 0, 1, 2)$, and $D \rightarrow A, A, 4, (6, 4), (0, 0, 6, 4)$.

Thanks to this additional information, to obtain the position of an object at any time instant between two snapshots, the nonterminal symbols of array C do not need to be decompressed in most cases. Assume we want to know the position of the object at the 5th time instant, which is when the object in Fig. 4 (right) is at position $(7, 7)$. The preceding snapshot informs that the absolute position of the object at the beginning of the log is $(0, 2)$. Next, we inspect the log (the C array of Fig. 2) from the beginning. The first value is a D. The enriched rule indicates that such symbol represents 4 time instants, and after it, the object is displaced 6 columns to the East and 4 rows to the North, that is, starting at $(0, 2)$, after the application of this rule, the object will be at $(6, 6)$. Since our target time instant is later than the final time instant of this rule, we do not have to decompress it, and this is the usual case. The next symbol is C, which lasts 2 time instants. This would take us to time instant 6, but this surpasses our target time instant(5). Therefore, in this case, that is, only in the last step of the search, we have to decompress the rule, and process its components: $C \rightarrow 8\ 7$. The 8 is a terminal symbol that lasts 1 time instant, and thus it is enough to reach our target time instant. An 8 moves 1 column to the

East and 1 to the North, which applied to the previous position $(6, 6)$ takes us to the position $(7, 7)$.

The MBR component aids during the computation of time-slice and time-interval queries, as we will see soon.

The additional elements enriching the rules are compressed with an encoder designed for small integers (Directly Addressable Codes, DAC) that support efficient access to any individual value in the sequence [1]. To obtain better compression, the times of all the rules are compressed with one DAC, separately from the 3 pairs of coordinates of all the rules, which are compressed with another.

5   Querying

Obtain the Position of an Object in a Given Time Instant. This query is solved by accessing the snapshot preceding the queried time instant t_q, where we retrieve the position of the object at the snapshot time instant. We then apply the movements of the log over this position until we reach t_q. In the case of SCDC compression, we follow the log decoding each codeword and applying the relative movement to the previous position. In the case of Re-Pair, we follow the process described in the previous section.

Obtain the Trajectory of an Object Between Two Time Instants. First, we obtain the position of the object in the start time instant t_s, using the same algorithm of the previous query; then we apply the movements of the log until reaching the end time instant t_e. In this case, when using GraCT, we have to decompress C to recover I, since only with I we are capable of describing the trajectory in detail, and thus we cannot take advantage of the enriched nonterminal data. Therefore this query is more time-consuming than the previous one for GraCT, and scdcCT takes over.

Time Slice Query. Given a time instant t_q and a window rectangle r of the space, this query returns the objects that lie within r at time t_q, and their positions. We can distinguish two cases. First, if t_q corresponds to a snapshot, we only need to traverse the k^2-tree until the leaves, inspecting those nodes that intersect r. When we reach the leaves, we know their position and can retrieve from the permutation the objects that are in this area.

The second case occurs when t_q is between two snapshots s_i and s_{i+1}. In this case, we inspect in s_i a region r', which is an enlargement of the query region r. Region r' is defined using using the fastest object of the dataset as an upper bound. Thus, r' is the rectangle containing all the points from where we can reach the region r at t_q if moving at maximum speed along some direction. Then, from s_i, we only track the objects that are within r' in the snapshot, therefore limiting the objects to follow and not wasting time with objects that do not have chances to be in the answer. We follow the movements of those objects from s_i using the log, until reaching t_q. We further prune the tracking as we process the log: a candidate object may follow a direction that takes it away from region r, so we recheck the condition after every movement and discard an object as soon as it loses the chance of reaching r at time t_q.

The tracking of objects is performed with the same algorithm explained for obtaining the position of an object in a given time instant, but in the case of GraCT, when a non terminal in the log corresponds to a rule that brings the object we are following from an instant before to an instant after t_q, instead of decompressing the nonterminal, we intersect the MBR of the rule with r, and disregard the object if the intersection is empty. Otherwise we decompress the nonterminal into two and try again until reaching t_q or discarding the object.

Figure 5 shows an example where we want to find the objects that are located in r at t_q. Assume that the fastest object can move only to an adjacent cell in the period between two consecutive time instants. Let s_i be the last snapshot preceding t_q and let there be 2 time instants between s_i and t_q. The left part of the figure shows the state of the objects at the time instant corresponding to s_i, in the middle to $s_i + 1 = t_q - 1$, and in the right to t_q, where we show the region r. In r' (shown on the left grid) we have four elements $(1, 4, 5, 8)$, which are candidates to be in r at t_q, thus we follow their log movements. In the middle grid, we show the region r'' where the objects still have chances to be within r at t_q. Observe that, from the candidate objects in s_i, object 4 has no further chances to reach r, and thus it is not followed anymore. However, object 1 still have chances, and therefore we keep tracking it.

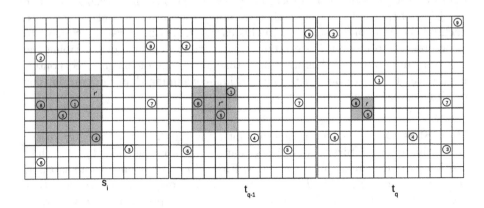

Fig. 5. Example of enlarged region r' and query region r in a time-slice query.

This query is affected by the time elapsed between s_i and t_q. The farther away s_i and t_q are, the larger r' will be, and thus, we will have more candidate objects that have to be followed through the log movements. In addition, with a large period between s_i and t_q, we have to traverse a longer portion of the log. To alleviate this problem, if t_q is closer to s_{i+1} than to s_i, we can start the search at s_{i+1} and follow backwards the movements of the log. For this backward traversal, we need to add before each snapshot the last known position and its corresponding time instant of the objects that are disappeared at the time instant of the snapshot. This applies to both approaches scdcCT and GraCT. Therefore, the maximum distance will be half of the distance between two snapshots.

Time Interval Query. In the time-slice query we have to know which objects are in r at the query time instant and their positions, but in the time-interval query, the target is to know which objects were within r at any time instant of the time interval $[t_s, t_e]$, specified in the query. In this case, we use the expanded region r' again, which is built as in the time-slice query, but using the time t_e.

Using SCDC, the objects within r at t_s are reported as part of the solution, the other objects with chances at t_s are followed until they reach the region r, in which case they are added to the answer; or when they move such that they lose the chance to reach r at t_e, in which case they are not followed anymore.

Table 1. Compression ratio.

Period	GraCT				scdcCT			
	120	240	360	720	120	240	360	720
Size (MB)	196.79	193.31	192.24	179.60	312.27	263.46	273.20	282.95
Ratio	39.27 %	38.58 %	38.36 %	35.84 %	62.32 %	56.47 %	54.52 %	52.58 %
Snapshot (MB)	7.55 (3.83 %)	3.77 (1.95 %)	2.51 (1.31 %)	1.25 (0.70 %)	7.55 (2.42 %)	3.77 (1.33 %)	2.51 (0.92 %)	1.25 (0.48 %)
Log (MB)	189.25 (96.17 %)	189.54 (98.05 %)	189.73 (98.69 %)	178.34 (99.30 %)	304.73 (97.58 %)	279.18 (98.67 %)	270.69 (99.08 %)	262.20 (99.52 %)

In GraCT, we process the log without decompressing nonterminal symbols until the final time of a symbol in the log is equal or larger than t_s. After this moment, for each symbol we read in the log, until the object is selected or the next symbol in the log to read starts after t_e, we follow the next procedure: For each log symbol we check if the final point is inside r. If it is, the object is selected. If not, and the MBR does not intersect r, we go on to the next log symbol. If the final point is not in r but the MBR intersects r, we must apply the same procedure recursively to the pair of symbols represented by the nonterminal, until the object is selected or we process the whole nonterminal.

6   Experimental Evaluation

GraCT and scdcCT were coded in C++ and the experiments were run on a 1.70 GHzx4 Intel Core i5 computer with 8 GBytes of RAM and an operating system Linux of 64 bits.

Datasets Description and Compression Data. We use a real dataset obtained from the site http://marinecadastre.gov/ais/. The dataset provides the location signals of 3,654 ships during a month. Every position emitted by a ship is discretized into a matrix where the cell size is 50 × 50 m. With this data normalization, we obtain a matrix with 100,138,325 cells, 36,775 in the x-axis and 2,723 in the y-axis. Observe that our structure deals with object positions at regular intervals, but in the dataset ship signals are emitted with distinct frequencies, or they can send erroneous information. Therefore, we preprocessed the signals to obtain regular time instants every minute, thus discretizing the time into 44,642 min in one month. With these settings, the original dataset occupies 501 MBs.

We built GraCT and scdcCT data structures over that dataset using different snapshot distances, namely every 120, 240, 360, and 720 time instants. The construction time of the complete structure takes around 1 min. Table 1 shows the results of compression ratio[2], where we can see that GraCT obtains much better compression ratios than scdcCT. The rows *snapshot* and *log* show the size of the snapshot and the log as a percentage of the compressed data structure. We can see that the log is the most space demanding structure. As reference we compress the plain data with *p7zip* and we obtain a compression ratio of 10,97 %, which is better than GraCT, however with this compressed data is impossible to answer any type of query.

Query Types and Answer Times. Table 2 shows the average answer times of 50 random queries of different types: *object* t_q shows searches for the position of an object in a given time instant, *trajectory* searches for the trajectory followed by an object between two time instants, *slice S* are time-slice queries that check for small regions (367 × 272 cells) and *slice L* for large regions (3677 × 2723 cells), *interval S* are time-interval queries with a small region and a small time interval ($\frac{1}{10}$ of the snapshot period) and *interval L* with large regions and a large time interval ($\frac{1}{4}$ of the snapshot period).

Table 2. Time of different queries (ms)

	GraCT				scdcCT			
Period	120	240	360	720	120	240	360	720
Object t_q	0.0157	**0.0169**	**0.0210**	**0.0229**	**0.0125**	0.0169	0.0220	0.0246
Trajectory	0.1582	0.1210	0.1130	0.1153	**0.0881**	**0.0904**	**0.0900**	**0.0960**
Slice S	1.5386	**2.4241**	**2.9580**	**5.8788**	**1.3080**	2.6712	4.0430	7.6636
Slice L	1.7835	**2.8074**	**3.6000**	**6.6430**	**1.5883**	3.8615	5.1130	9.1384
Interval S	2.4435	**3.6005**	**5.0090**	**9.5635**	**1.4882**	3.8612	5,8610	11.1765
Interval L	2.7505	**4.0847**	**6.1330**	**12.1832**	**1.7161**	4.9578	9.8680	16.4471

---

[2] The size of the compressed data structure as a percentage of the original dataset.

GraCT is the overall winner in all queries, except in the *trajectory* query. This is expected, since to recover the trajectory, GraCT has to decode all the symbols in C, given that the enriched information in rules does not have the details of the movements inside each rule. In the rest of the queries the enriched information avoids in many cases to decode the rules in C, and thus, since the logs in GraCT have far fewer values than in scdcCT, the searches are faster. The exception is when the size of the log between two snapshots is small, as the effect is not noticeable. Notice that in this case the nonterminal symbols in the grammar cannot represent arrays of more than 120 terminals.

7 Conclusions

We have presented a grammar based data structure for representing moving objects. It uses snapshots where the objects are represented in the space using a k^2-tree and movement logs that are grammar-compressed. The results of this first experimental evaluation are very promising, as compression yields significant reductions in both space and time performance with respect to the baseline.

One reason why our space results are not even better is that the enriched data pose a significant space overhead per nonterminal. We plan to improve our representation by encoding these data in smarter ways. We also plan to compare GraCT with state of the art indexes aimed at both time-slice and time-interval queries, and trajectories.

References

1. Brisaboa, N., Ladra, S., Navarro, G.: DACs: bringing direct access to variable-length codes. Inf. Process. Manag. **49**(1), 392–404 (2013)
2. Brisaboa, N.R., Fariña, A., Navarro, G., Param, J.R.: Lightweight natural language text compression. Inf. Retrieval **10**(1), 1–33 (2007)
3. Brisaboa, N.R., Ladra, S., Navarro, G.: Compact representation of web graphs with extended functionality. Inf. Syst. **39**(1), 152–174 (2014)
4. Chakka, V.P., Everspaugh, A., Patel, J.M.: Indexing large trajectory data sets with SETI. In: Proceedings of the conference on innovative data systems research, CIDR 2003 (2003). http://www-db.cs.wisc.edu/cidr/cidr2003/program/p.15.pdf
5. Gutiérrez, G.A., Navarro, G., Rodríguez, M.A., González, A.F., Orellana, J.: A spatio-temporal access method based on snapshots and events. In: GIS, pp. 115–124. ACM (2005)
6. Hadjieleftheriou, M., Kollios, G., Tsotras, V.J., Gunopulos, D.: Efficient indexing of spatiotemporal objects. In: Jensen, C.S., Jeffery, K., Pokorný, J., Šaltenis, S., Bertino, E., Böhm, K., Jarke, M. (eds.) EDBT 2002. LNCS, vol. 2287, pp. 251–268. Springer, Heidelberg (2002)
7. Jacobson, G.: Space-efficient static trees and graphs. In: IEEE Symposium on Foundations of Computer Science (FOCS), pp. 549–554 (1989)
8. Knuth, E.: Efficient representation of perm groups. Combinatorica **11**, 33–43 (1991)
9. Larsson, N.J., Moffat, A.: Off-line dictionary-based compression. Proc. IEEE **88**(11), 1722–1732 (2000)

10. Munro, J.I., Raman, R., Raman, V., Rao, S.: Succinct representations of permutations and functions. Theor. Comput. Sci. **438**, 74–88 (2012)
11. Nascimento, M.A., Silva, J.R.O.: Towards historical R-trees. In: George, K.M., Lamont, G.B. (eds.) Proceedings of the 1998 ACM Symposium on Applied Computing, SAC 1998, pp. 235–240. ACM (1998). http://doi.acm.org/10.1145/330560
12. Rasetic, S., Sander, J., Elding, J., Nascimento, M.A.: A trajectory splitting model for efficient spatio-temporal indexing. In: Proceedings of the 31st International Conference on Very Large Data Bases, pp. 934–945. VLDB Endowment (2005)
13. Tao, Y., Papadias, D.: MV3R-tree: a spatio-temporal access method for timestamp and interval queries. In: Apers, P.M.G., Atzeni, P., Ceri, S., Paraboschi, S., Ramamohanarao, K., Snodgrass, R.T. (eds.) Proceedings of 27th International Conference on Very Large Data Bases, VLDB 2001, pp. 431–440. Morgan Kaufmann (2001)
14. Vazirgiannis, M., Theodoridis, Y., Sellis, T.K.: Spatio-temporal composition and indexing for large multimedia applications. ACM Multimedia Syst. J. **6**(4), 284–298 (1998)
15. Wang, L., Zheng, Y., Xie, X., Ma, W.Y.: A flexible spatio-temporal indexing scheme for large-scale GPS track retrieval. In: Mobile Data Management, pp. 1–8. IEEE (2008)
16. Worboys, M.F.: Event-oriented approaches to geographic phenomena. Int. J. Geogr. Inf. Sci. **19**(1), 1–28 (2005)
17. Xu, X., Han, J., Lu, W.: RT-tree: an improved R-tree index structure for spatiotemporal databases. In: Proceedings of the 4th International Symposium on Spatial Data Handling, vol. 2, pp. 1040–1049 (1990)

Lexical Matching of Queries and Ads Bid Terms in Sponsored Search

Ricardo Baeza-Yates[1,2]($\boxtimes$) and Guoqiang Wang[3]

[1] Web Research Group, DTIC, Universitat Pompeu Fabra, Barcelona, Spain
rbaeza@acm.org
[2] DCC, Universidad de Chile, Santiago, Chile
[3] Yahoo Inc., Sunnyvale, USA

Abstract. We present an algorithm that matches queries to a set of advertising bid terms, including lexical variants. We use an enhanced digital tree as index that allows to match misspelled words, word segmentation and splitting, and some stems. The results, both in quality and performance, are competitive and improved revenue in several percentage points. To the best of our knowledge, this is the first pure lexical matching algorithm for sponsored search.

1 Introduction

When a person searches for a query in a web search engine, the answer page has several parts. The main part is composed by the ranked search results, also called organic results. The second important part are the sponsored results, which are small text ads to the right of the organic results and sometimes also above and below them, depending on the search engine. Companies that wish to advertise their products or services in this way, bid for terms that will appear in the queries of their interest, hence called *bid terms*. Advertisers assign bid values to each individual bid term and the search engine will run a special auction with them. This bid value is the maximum value that the advertiser is willing to pay for a click in the ad. This payment mechanism is known as cost per click (CPC), and the most used ranking scheme for the ads is to estimate the probability of a click in an ad and use the product of this probability and the bid value for each ad as ranking score. The most used auction is the well known second price, because the best strategy is to be truthful. For more details and the impact of sponsored search we refer the reader to [9].

In the context of sponsored search, we propose a new lexical matching algorithm that handles misspells, split/merge words, plurals, etc. It is based in running a NDFA of the query with all lexical variations over a digital tree of the ads bid terms, a particular case of [2]. Nodes of the digital tree that matches words and lexical variants of bid terms have inverted lists with all bid terms and

---

This work was carried out when the first author was at Yahoo Labs.

S. Inenaga et al. (Eds.): SPIRE 2016, LNCS 9954, pp. 231–239, 2016.
DOI: 10.1007/978-3-319-46049-9_22

its associated ads in a special order, as done in standard inverted indexes [1]. Results in a real production system showed a significant revenue improvement with good time performance.

This paper is organized as follows. The next section covers the main related work. In Sect. 3 we present the new lexical matching algorithm. In Sect. 4 we present our quality evaluation of the results as well as some performance figures. We end with some concluding remarks.

2   Related Work

In the last decade there was a lot of work on the matching and the economics behind sponsored search [9–11]. To simplify the matching of ads, the initial idea of Goto.com [6] was to ask advertisers for bid terms that should appear in the queries. Hence, in some sense, part of the matching problem was left to people. However, if an ad does not have good bid terms, if the query is not common, or the search engine does not have a large portfolio of ads, the matching phase can produce few or no results at all. For this reason the concept of *broad match* appeared, allowing the search engine to find queries similar to the bid terms of an ad (*e.g.*, see [5] for the case of long tail queries). In this context, lexical matching of ads is crucial, to allow for lexical variants of the bid terms. On the other hand, lexical match is not enough and must be combined by other techniques to deal with semantic similarity (*e.g.*, see [14]). In the work presented in this paper, the semantic similarity of the production system is based in word2vec over the query log stream enhanced with clicked and non-clicked ads, among other innovations [7,8].

Regarding the searching techniques used, our algorithm is inspired in searching regular expressions on a text in average sub-linear time by simulating an NDFA over a digital trie or trie built from the text [4]. This idea has later been adapted and enhanced for searching biological sequences [3] and also inspired applied algorithms to search similar proper names [13]. For the main results on approximate string matching we refer the reader to [12].

Regarding our specific problem, most algorithms used in sponsored search are proprietary and never published. For this reason, to the best of our knowledge, our algorithm is the first pure lexical matching algorithm ever published.

3   A New Lexical Matching Algorithm

3.1   Formal Problem

Given a set of ads bid terms BT, the lexical match for a query q, implies to find and rank a set of ads that have bid terms that are lexically similar to q, allowing the following variations:

- exact match;
- partial match (*e.g.*, a proper subset of words in the bid term);

- misspellings (that is, a few substitutions, deletions, or insertions);
- singular/plural (the number cardinality should not matter);
- word split/merge (to handle missing and extra spaces); and
- other lexical variations (*e.g.*, stemming).

As we can have millions of ads, the solution must keep an index of the bid terms to achieve fast search for the ads to be shown for each query.

For BT, we use only the active bid terms. An active bid term is defined as one that has at least one active entity for every entity in the advertising hierarchy: advertiser, campaign, ad group, and ads (bid terms are associated to the lowest level of the hierarchy, ads). In addition, active bid terms are also restricted to advertisers and campaigns that had budget in the last week.

3.2 Index Design

We start by detailing the index used, which is based in a trie or digital tree for all active bid terms BT. Every bid term has an *id* and we also keep the number of words in the term. Each node in the trie can be either an internal node or a final node that implies that a potential lexical match has been found, depending on several rules that we detail later. Each final node links to an inverted list of the ordered set of bid terms that are associated to that final node. For the order we use first the number of competing advertisers, then the number of words, and finally the *id*.

To improve the search time and considering that many ads have a geographical target (that is, the user should be in a specific region or the query should mention a location in that region), the inverted lists are partitioned by target regions (*e.g.*, USA could be a region). This not only makes the search faster, but also disallows queries that do not have a geographical intent (that is, they do not mention a location) to match a bid term that has a specific geographic target, and vice versa. Hence, inverted lists are implemented as a hash map using the associated geographic location or *woeid*[1] as key.

To further improve the search performance, we use two bigrams tables: standard bigrams as well as skip-bigrams from BT. Bigrams are used in word split, where two consecutive words do appear in the bid terms, while skip-bigrams are used for pruning the search paths in the trie. For each bigram we store the two words and the frequencies associated to when they are a bigram or a skip-bigram. Notice that stopwords are removed while generating the bigrams.

Although bigrams increase the use of space, fast online performance is the primary goal of the system. In addition, the way that we partition the index (see later), drastically reduces the number of bigrams.

---

[1] See https://en.wikipedia.org/wiki/GeoPlanet.

3.3   Index Building

The complete preprocessing has the following phases:

- Filter and clean active bid terms. This includes removing bid terms that contain Unicode characters, convert all letters to lowercase, convert periods to spaces, and convert multiple spaces to a single space.
- Build the bigrams tables with a geographically partitioned hash map.
- Build the trie in parallel. For this we use a producer-consumer approach where:
  - Producers: one reader per producer, and emit tokens of the bid terms into a shared queue; and
  - Consumers: one consumer per first character of each token, such that synchronization of map insertions on leaf nodes is not needed, since they will happen sequentially in the same thread.

  Balancing the work of producers and consumers determines the number of each, as producers do less work than consumers, which are the ones that insert all bid terms in parallel into the trie.
- Build the inverted lists in each final node of the trie.

3.4   Searching Algorithm

The high level process for the search is the following:

- Split a query into tokens;
- For every token, search the trie to discover its lexical variations;
- Convert single-token variations found in final states into all the possible combinations of token sequences;
- For every token sequence, find matched bid terms;
- Intersect the inverted lists of bid terms for all token sequences; and
- Join all the partial results, delete duplicates if any, and return the top candidates.

The search can be modeled as traversing the trie with a non-deterministic finite automata, where

- we have a set of active states that are tuples (token position, trie node, errors), where the initial state is $(0, \text{ trie root, } 0)$;
- we have transitions that can advance the token position, can follow a child link in the trie, or do both; and
- we reach a final state if the tuple satisfies the following condition: (token length, a trie final node, errors < maximum errors), where in practice the maximum number of errors is 1 or 2.

We now detail all possible transitions in each trie node. First, the simple ones that take care of misspellings:

- MATCH: $[position + 1, trie \rightarrow_=, errors]$, for the child labeled with the letter that matches the current token letter, if exists;

- SUBS: [*position* + 1, *trie*→$_{\neq}$, *errors* + 1], for all children labeled with a letter that does not match the current letter;
- INS: [*position*, *trie*→$_{all}$, *errors* + 1], for all children of the current node; and
- DEL: [*position* + 1, *trie*, *errors* + 1].

For matching singular and plurals we have the following standard cases:

1. Query token is singular and trie node is plural. In this case the token ends and if the trie node is a final node, the trie node includes a child with *s* that is also a final node.
2. Query token is plural and trie node is singular. In this case if a token has only an *s* left and the trie node is a final node, we consider that a final state.

Other forms of regular singular/plural matches are also supported, such as: *y* to *ies*, and vice versa; *f* to *ves*, and vice versa; *z* to *zes*, and vice versa; *ch* to *ches*, and vice versa; and *sh* to *shes*.

To handle token splits we add a special transition to the state (*position* + 1, *trie root*, 0) only when the current trie node is a final node (that is, if we had en extra space, we would have a token match). In addition, we allow further splits if the associated bigram frequency must be over a certain threshold. For example, if we have the query `alaskaairlines`, the first valid split would be `alaska airlines` and the second valid split might be `air lines` while the third split `line s` will be certainly invalid. Adding the plural rule, the final splits for the example would be `alaska airline`, `alaska airlines`, `alaska air line`, and `alaska air lines`.

To avoid an exponential number of active states due to the NFA matching many misspelling variations, we store intermediate results in a priority queue using the number of errors as a key, always following first the transitions in the state that has the smallest number of errors (in case of ties we use a FIFO policy). In this way we first find exact matches, followed by the lexical variations. In addition, we do not enqueue intermediate states with the maximum allowed number of error when enough results have been already found (usually a few hundreds).

After all results are found, we also remove lexical variations with frequency less than a certain ratio of maximal term frequency in all the results (in practice we used 0.01). We also remove token sequences that are infrequent or that never appear, based on the skip-bigrams table. For instance, for the query `now yark`, we may have the following variations:

$$now = <new|now> \quad \& \quad york = <york|yark>,$$

that can match the following token sequences:

$$<new york>, <new yark>, <now york> \& <now yark>.$$

Based on skip-bigrams, the final result will be limited to just the first token sequence, as all the other cases are infrequent.

To end, for every token sequence candidate we:

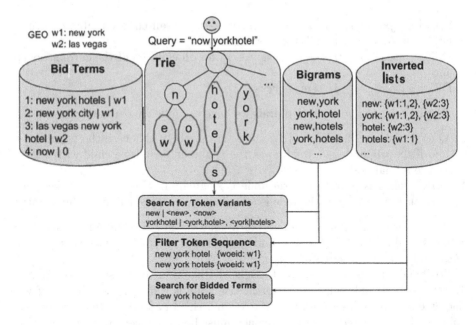

Fig. 1. Overall system example.

- Calculate the *woeid* for the token sequence;
- For every trie final node, we obtain the set of bid terms from the inverted list map using the *woeid* as the map look-up key;
- Compute the intersection of bid terms for all the final nodes, starting with the shortest inverted list as is customary.

To search faster, multi-threading was implemented, with one thread for every token sequence. Figure 1 shows the full system based on the previous examples.

4 Evaluation

4.1 Quality

We evaluated the results for a random sample of 5,000 queries with no coverage or few ads. We first did an editorial evaluation, where expert editors labeled our results with four levels: excellent (E), good (G), fair (F), or bad (B). We obtained that 65.2 % of the results had the levels EGF while 33.4 % were bad. The rest (1.4 %) had no judgments, a situation that happens when the expert cannot assess the right level.

Next, we used the ad relevance scoring of the current system in production, which is a number in the range from 0 to 1, where below 0.2 is considered bad. The production system is trained by a much larger sample of editorial judgments. The results are shown in Table 1. We also compared the agreement between the system and the editorial evaluation and the results are shown in Table 2. Finally, we compared the overall score of the production system (that uses ad relevance as a filter) with our results in Table 3. These results are quite reasonable, specially considering the agreement levels.

Table 1. Quality evaluation.

Judgment	Average # of ads	Good ads in answer	Good ads overall	Average score
EGF	15.9	100 %	63.1 %	0.64
Mixed	22.2	46 %	10.3 %	0.31
Bad	4.7	0 %	14.7 %	0.11

Table 2. Editorial and system agreement.

Judgment level	Agreement	Disagreement
EGF	72.0 %	28.0 %
Bad	49.2 %	50.8 %

Table 3. Overall comparison with the production system score.

Score range	Our results	Editorial relevance	Ad relevance
[0.54, 0.70]	8.30 %	[EGF, B] = [27 %, 73 %]	[EGF, B] = [43 %, 31 %]
(0.70, 0.75]	19.88 %	[EGF, B] = [41 %, 59 %]	[EGF, B] = [49 %, 23 %]
(0.75, 0.80]	33.71 %	[EGF, B] = [65 %, 35 %]	[EGF, B] = [63 %, 14 %]
(0.80, 0.85]	24.89 %	[EGF, B] = [85 %, 15 %]	[EGF, B] = [75 %, 7 %]
(0.85, 1.00]	13.18 %	[EGF, B] = [96 %, 4 %]	[EGF, B] = [75 %, 5 %]

4.2   Performance

We used a sample of about 70 million active ads to test the performance of our system in a 192 GB memory 24-core computing server. To build the trie of this sample we used 5 producers and 39 consumers that finished in 3 min. This is 10 times faster than a sequential approach and implies that the index can be refreshed often as new active ads appear due to new advertisers or budget changes. The final index had about 13.9M nodes, of which 29 % where final nodes.

To test the searching performance, we used a sample of almost 55M queries from Yahoo's search query log, such that in the last 45 days were searched at least 5 times. We searched all the queries with 20 parallel threads in 4.75 h achieving a QPS (queries per second) rate of 3,200. The 95 % of the queries that had at least one result, completes in less than 66 ms. For those queries, 93.3 % of them took less than 50 ms with an average latency of just 17 ms. The rest of the queries had an average latency of 125 ms. These results are quite encouraging considered that were obtained with a prototype, and the production version will easily have a QPS of more than 5,000 which implies 432M queries per day.

5   Conclusions

We have presented a new algorithm for lexical matching of ads and queries that has competitive quality and fast performance by using a combination of algorithmic and engineering techniques. Recent results of the A/B testing of the algorithm showed revenue increases of one digit percentage when integrated with the current sponsored search system that uses a word2vec approach [7,8]. Due to these results, this algorithm is currently being re-implemented to be included as one more of the several heuristics being used in the production system.

This algorithm can be used off-line as well, to generate bid terms that can be merged with the results from other matching algorithms. Possible extensions to our algorithm is to add verbal stemming, to allow different conjugations. We can even match token synonyms at final nodes to have also semantic similarity.

Although lexical matching is an important component of current sponsored search systems, would be much better to remove the need for advertisers to specify bid terms and let the system find semantic matches between ads and queries. Indeed, using new technologies such as deep learning, we can find similar queries and ads by using all the ads metadata as well as the content of the target ad page as input.

Finally, there are still many possible improvements in sponsored search, including semantic similarity, click prediction, budget pacing, and pricing for partial matches, to mention just a few.

References

1. Baeza-Yates, R., Ribeiro-Neto, B.: Modern Information Retrieval: The Concepts and Technology Behind Search, 2nd edn. Addison-Wesley, New York (2011)
2. Baeza-Yates, R.A., Gonnet, G.H.: Fast text searching for regular expressions or automaton searching on tries. J. ACM **43**(6), 915–936 (1996)
3. Baeza-Yates, R.A., Gonnet, G.H.: A fast algorithm on average for all-against-all sequence matching. In: Sixth International Symposium on String Processing and Information Retrieval and Fifth International Workshop on Groupware, SPIRE/CRIWG 1999, Cancun, Mexico, 21–24 September 1999, pp. 16–23 (1999)
4. Baeza-Yates, R.A., Navarro, G.: A faster algorithm for approximate string matching. In: 7th Annual Symposium on Combinatorial Pattern Matching CPM 1996, Laguna Beach, California, USA, 10–12 June 1996, pp. 1–23 (1996)

5. Broder, A., Ciccolo, P., Gabrilovich, E., Josifovski, V., Metzler, D., Riedel, L., Yuan, J.: Online expansion of rare queries for sponsored search. In: Proceedings of the 18th International Conference on World Wide Web, pp. 511–520. ACM (2009)
6. Davis, D.J., Derer, M., Garcia, J., Greco, L., Kurt, T.E., Kwong, T., Lee, J.C., Lee, K.L., Pfarner, P., Skovran, S., et al.: System and method for influencing a position on a search result list generated by a computer network search engine, US Patent 6,269,361, 31 July 2001
7. Grbovic, M., Djuric, N., Radosavljevic, V., Silvestri, F., Baeza-Yates, R., Feng, A., Ordentlich, E., Yang, L., Owens, G.: Scalable semantic matching of search queries to ads in sponsored search advertising. In: Proceedings of the 39th International ACM SIGIR Conference on Research and Development in Information Retrieval. ACM (2016)
8. Grbovic, M., Djuric, N., Radosavljevic, V., Silvestri, F., Bhamidipati, N.: Context- and content-aware embeddings for query rewriting in sponsored search. In: Proceedings of the 38th International ACM SIGIR Conference on Research and Development in Information Retrieval, pp. 383–392. ACM (2015)
9. Jansen, B.J., Flaherty, T., Baeza-Yates, R.A., Hunter, L., Kitts, B., Murphy, J.: The components and impact of sponsored search. IEEE Comput. **42**(5), 98–101 (2009)
10. Jansen, B.J., Mullen, T.: Sponsored search: an overview of the concept, history, and technology. Int. J. Electron. Bus. **6**(2), 114–131 (2008)
11. Levin, J., Milgrom, P.: Online advertising: heterogeneity and conflation in market design. Am. Econ. Rev. **100**(2), 603–607 (2010)
12. Navarro, G.: Approximate string matching. In: Kao, M-Y. (ed.) Encyclopedia of Algorithms, pp. 1–5. Springer, New York (2014)
13. Navarro, G., Baeza-Yates, R.A., Arcoverde, J.M.A.: Matchsimile: a flexible approximate matching tool for searching proper names. JASIST **54**(1), 3–15 (2003)
14. Radlinski, F., Broder, A., Ciccolo, P., Gabrilovich, E., Josifovski, V., Riedel, L.: Optimizing relevance and revenue in ad search: a query substitution approach. In: Proceedings of the 31st Annual International ACM SIGIR Conference on Research and Development in Information Retrieval, pp. 403–410. ACM (2008)

Compact Trip Representation over Networks

Nieves R. Brisaboa[1], Antonio Fariña[1], Daniil Galaktionov[1(✉)],
and M. Andrea Rodríguez[2]

[1] Database Laboratory, University of A Coruña, A Coruña, Spain
d.galaktionov@udc.es
[2] Department of Computer Science, University of Concepción, Concepción, Chile

Abstract. We present a new *Compact Trip Representation* (CTR) that
allows us to manage users' trips (moving objects) over networks. These
could be public transportation networks (buses, subway, trains, and so
on) where nodes are stations or stops, or road networks where nodes are
intersections. CTR represents the sequences of nodes and time instants
in users' trips. The spatial component is handled with a data structure
based on the well-known Compressed Suffix Array (CSA), which provides
both a compact representation and interesting indexing capabilities. We
also represent the temporal component of the trips, that is, the time
instants when users visit nodes in their trips. We create a sequence with
these time instants, which are then self-indexed with a balanced Wavelet
Matrix (WM). This gives us the ability to solve range-interval queries efficiently. We show how CTR can solve relevant spatial and spatio-temporal
queries over large sets of trajectories. Finally, we also provide experimental results to show the space requirements and query efficiency of CTR.

1 Introduction

Current technology allows us to capture data about the usage of transportation networks whose analysis could have an important impact on improving the
quality of services. Data about the origin and destination of passengers of train
services can be directly captured when selling tickets. Using more sophisticated
technology, the movement of people or vehicles over networks of streets or roads
can be collected from the mobile phone signals. Even more, nowadays many
cities (from London to Santiago of Chile) provide smartcards to the users of
their public transportation network. These smartcards (that can be recharged

Funded in part by European Union's Horizon 2020 research and innovation pro-
gramme under the Marie Sklodowska-Curie grant agreement No 690941 (project
BIRDS). N. Brisaboa, A. Fariña, and D. Galaktionov are funded by MINECO (PGE,
CDTI, and FEDER) [TIN2013-46238-C4-3-R, TIN2013-47090-C3-3-P, TIN2015-
69951-R, IDI-20141259, ITC-20151305, ITC-20151247]; by ICT COST Action
IC1302; and by Xunta de Galicia (co-funded with FEDER) [GRC2013/053].
A. Rodríguez is funded by Fondecyt 1140428 and the Complex Engineering Sys-
tems Institute (CONICYT: FBO 16).

© Springer International Publishing AG 2016
S. Inenaga et al. (Eds.): SPIRE 2016, LNCS 9954, pp. 240–253, 2016.
DOI: 10.1007/978-3-319-46049-9_23

with money) allow users to pay the entrance to subways and buses. Even though there typically exists only a card reader in the entrance to the network (i.e., there is no control at exits or in middle stops), it is possible to know how people actually use the public transportation by collecting the entrance and estimating the destination (e.g., as the entry point for the return trip) and the traversed stops (the shorter path among stops used as the entrance and exit) [11]. In all scenarios, the massive data about trips makes the problem of storing and efficiently accessing data about trips a challenging computational problem.

This paper presents a compact and self-indexed data structure to represent trips over networks, which could be public transportation networks where nodes are stations or stops, or road networks where nodes are intersection points.

Although there exist proposals of data structures for moving objects, they have addressed typical spatio-temporal queries such as time slice or time interval queries that retrieve trajectories or objects that were in a spatial region at a time instant or during a time interval. They were not designed to answer queries that are based on counting occurrences such as the number of trips starting or ending at some time instant in specific stops (nodes) or the top-k most used stops of a network during a given time interval, which are more meaningful queries for public-transportation or traffic administrators. Our proposal (CTR) is oriented to efficiently answering these types of queries, and it differs from previous approaches in the use of compact self-indexed data structures to represent the big amount of trips in compact space. It is important to emphasize that our goal is to provide an indexed representation for a static collection of trips in order to allow an efficient batch processing of such data.

CTR combines two well-known data structures. The first one, initially designed for the representation of strings, is Sadakane's Compressed Suffix Array (CSA) [18]. The second one is the Wavelet Matrix (WM) [1]. To make the use of the CSA possible in this domain, we define a trip or trajectory of a moving object over a network as the temporally-ordered sequence of the nodes the trip traverses. An integer id is assigned to each node such that a trip is a string of nodes' ids. Then a CSA, over the concatenation of these strings (trips) is built with some adaptations for this context. In addition, we discretize the time in periods of fixed duration (i.e. timeline split into 5-min instants) and each time segment is identified by an integer id. In this way, it is possible to store the times when trips reach each node by associating the corresponding time id with each node in each trip. The sequence of times for all the nodes within a trip is self-indexed with a WM to efficiently answer spatio-temporal queries.

We experimentally tested our proposal using two sets of synthetic data representing trips over two different real public transportation systems. Our results are promising because the representation uses only around 30 % of its original size and answers spatial and spatio-temporal queries in microseconds. No experimental comparisons with classical spatial or spatio-temporal index structures are possible, because none of them were designed to answer the types of queries in this work. Our approach can be considered as a proof of concept that opens new application domains for the use of CSA and WM, creating a new strategy for exploiting trajectories represented in a self-indexed way.

The organization of this paper is as follows. Section 2 reviews previous works on trip representations. It also makes reference to the CSA and WM, upon which we develop our proposal. Section 3.1 shows how CTR represents the spatial component and Sect. 3.2 the temporal component of trips. Section 4 presents the relevant queries that are solved by CTR and Sect. 5 gives our experimental results. Finally, conclusions and future work are discussed in Sect. 6.

2   Previous Work

Trajectory Indexing. Many data structures have been proposed to support efficient query capabilities on collections of trajectories. We refer to [13, Chapter 4] for a comprehensive and up-to-date survey on data management techniques for trajectories of moving objects. We can broadly classify these data structures into two groups: those that index trajectories in free space and those that index trajectories constrained to a network. The 3D R-tree (an extension of the classical R-tree spatial index [7]), the TB-tree [14], and MV3R-tree [19] are examples of the former, whereas the FNR-tree [4], the MON-tree [2], and PARINET [15] are examples of the latter. While the former type of structures could also apply over networks, the second type exploits the constraints imposed by the topology of the network to optimize the data structure. From them, PARINET is the most efficient alternative [15]. It partitions trajectories into segments from an underlaying road network, and then adds one temporal B^+-tree to index the trajectory segments from each road. Those indexes permit us to filter out candidate trajectory segments matching time constraints at query time.

All previous data structures were designed to answer spatio-temporal queries, where the space and time are the main filtering criteria. Examples of such queries are: *retrieve trajectories that crossed a region within a time interval*, *retrieve trajectories that intersect*, or *retrieve the k-best connected trajectories* (i.e., the most similar trajectories in terms of a distance function). Yet, they could not easily support queries such as *number of trips starting in X and ending at Y*.

The application of data compression techniques has been explored in the context of massive data about trajectories. The work by Meratnia and de By [10] adapts a classical simplification algorithm by Douglas and Peucker to reduce the number of points in a curve and, in consequence, the space use to represent trajectories. Ptomaias *et al.* [16] use concepts, such as speed and orientation, to improve compression. Both techniques work for trajectories in free space.

In [5,8,17], they focus mainly on how to represent trajectories constrained to networks, and in how to gather the location of one or more given moving objects from those trajectories. Yet, these works are also out of our scope as they would poorly support queries oriented to exploit the data about the network usage such as those oriented to aggregate the number of trips with a specific spatio-temporal pattern (e.g. Count the trips starting at stop X and ending at stop Y in working days between 7:00 and 9:00).

In [9], authors use a representation of trajectories where for each edge in a trajectory both the starting and ending times are kept, and present an index called *NETTRA*. They used a relational database where those data are stored in a table and indexes are created in order to support a particular type of queries called *Strict Path Queries*. Although our CTR could also deal with those types of queries, this database-oriented representation is out of our scope as they do not consider space constraints (they do not compress data nor do they consider the size of the indexes used).

Underlying Compact Structures of CTR. Our proposal is based on two well-known compact structures: a Compressed Suffix Array (CSA) [18] and a Wavelet Matrix (WM) [1]. We used the variant of CSA from [3], where authors adapted CSA to deal with large (integer-based) alphabets and created the *integer-based CSA* (iCSA). They also showed that the best compression of Ψ was obtained when combining differential encoding of runs with Huffman and run-length encoding.

WM is a data structure originated from the Wavelet Tree [6], but requires less space and permits to make an efficient occurrence count of a continuous range of values [1] (see Sect. 3.2 for details). WM provides, as the Wavelet Tree a self-indexed representation of symbols based on the rearrangement of their bits in different bit maps at different levels. WM allows us to perform efficient operations over the sequence, among other operations: $access(i)$ returns the symbol at the position i, $rank_\alpha(i)$ counts the number of occurrences of a symbol α up to position i; and $select_\alpha(j)$ gives the position of the j-th α. Those operations are implemented using the classical bit operations $rank$ and $select$ on the underlying bitmaps and they need $O(\log \sigma)$ time, being σ the number of encoded symbols.

3   Compact Trip Representation (CTR)

Trips on networks are temporally-ordered sequences of nodes (referred to as the spatial component) tagged with timestamps (referred to as the temporal component). We show how the proposed *Compact Trip Representation* (CTR) combines a Compact Suffix Array (CSA) to represent the spatial component and a Wavelet Matrix (WM) to represent the temporal one.

3.1   Representing the Spatial Component of CTR with a CSA

In CTR, integer IDs identify stops of the network. The first step to build the CSA is to sort the trips. They are sorted by the first stop, then by the last stop, then by the start time of the trip, and finally by the second, third, and successive stops. For example, we have a dataset $\mathcal{T}$ with the following set of trips: $\{\langle 2, 3, 10, 6 \rangle,$ $\langle 2, 3, 10, 4, 7 \rangle, \langle 1, 2, 3 \rangle \langle 3, 10, 5 \rangle, \langle 1, 2, 3 \rangle \langle 9, 8, 7 \rangle\}$. Let us assume that these trips start at time instants 10, 2, 0, 9, 5, 12, respectively. Following lexicographic order, the trip $\langle 2, 3, 10, 4, 7 \rangle$ should be before the trip $\langle 2, 3, 10, 6 \rangle$. However, because after the first stop, we consider the last stop, the trip $\langle 2, 3, 10, 6 \rangle$ goes before the trip $\langle 2, 3, 10, 4, 7 \rangle$. In addition, the two trips $\langle 1, 2, 3 \rangle$ are sorted by their starting time instants (0 and 5 respectively). This sorting of the trips will allow us to answer a useful query very efficiently (i.e., trips starting at X and ending at Y).

We concatenate the sorted trips and construct an array S where trips are separated with a symbol $. We also add an additional ending $. Figure 1 shows the array S for the running example. Despite the standard suffix array construction in the CSA that compares two suffixes by their lexicographical order until the end of S, we introduced a modification so that two suffixes are now compared considering their trips as a cycle.

Figure 1 depicts the structures Ψ and D used by the CTR over the trips in the dataset $\mathcal{T}$. There is also the vocabulary V containing all the stops in their lexicographic order, as well as the $ symbol. We include the sequence S, the suffix array A, and Ψ' only for clarity (they are not needed in the CTR). Ψ' contains the first entries of Ψ from a regular CSA, just to explain the difference of how we build Ψ. For example, $A[8] = 1$ points to the first stop of the first trip $S[1]$. $\Psi[8] = 10$ and $A[10] = 2$ points to the second stop. $\Psi[10] = 14$ and $A[14] = 3$ points to the third stop. $\Psi[14] = 2$ and $A[2] = 4$ points to the ending $ of the first trip. Therefore, in the standard CSA, $\Psi'[2] = 9$ and $A[9] = 5$ points to the first stop of the second trip. However, in CTR, $\Psi[2] = 8$ and $A[8] = 1$ points to the first stop of the first trip. Thus, subsequent applications of Ψ will allow us to cyclically traverse the stops of the trip. Finally, note that aligned with sequence S, we could keep the times associated with the stops in each trip with the structures I and $Icode$, which are explained in the following subsection.

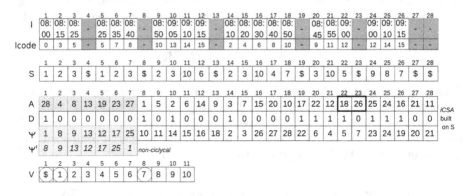

Fig. 1. Structures involved in the creation of a CTR.

The definition of a *suffix* proposed above explains why $A[22] = 18$ is placed before $A[23] = 26$. Note that the suffix starting at $S[18]$ is "$7 \cdot $ \cdot 2 \cdot 3 \ldots$" and that suffix at $S[26]$ is "$7 \cdot $ \cdot 9 \cdot \ldots$". Therefore, it holds that $A[22] \prec A[23]$. However, considering the traditional definition of a *suffix*, these suffixes would be "$7 \cdot $ \cdot 3 \cdots$" and "$7 \cdot $ \cdot $ \cdots$" respectively, and $A[22] \prec A[23]$ would not hold.

Note also that, in the shaded range $\Psi[1, 7]$, the first entry is related to terminator $, whereas the next six entries correspond to the $ symbols that mark the end of each trip in S (sorted by the starting stop, then by the ending stop, then by their initial time, and finally by the second, third and following up stops). This property makes it very simple to find starting stops. For example, the ending \$ of the 4th trip is at the 5th position (because the first \$ corresponds to the final \$ at $S[28]$). Therefore, its starting stop can be obtained by $\Psi[5] = 12$ and $rank_1(D,12) = 3$; that is, the starting stop is the 3th entry in the vocabulary. The next stop of that trip would be obtained by $\Psi[12] = 16$ and $rank_1(D,16) = 4$, and so on.

We expect to obtain good compressibility in CTR due to the structure of the network, and the fact that trips that start in a given stop or simply those going through that stop will probably share the same sequence of "next" stops. This will lead us to obtain many $runs$ in Ψ [12], and consequently, good compression.

3.2   Representing the Temporal Component of CTR with a WM

To exploit usage patterns of a network, we need to represent and query the time component of trips, which indicates when a moving object reaches each node along its trip. To represent this time component, we discretize time and assign an integer code to each resulting time interval. The size of the time interval is a parameter that can be adjusted to fit the required precision in each domain. For example, in a public transportation network, if we had data about five years of trips, a possibility would be to divide that five-years period into 10-min intervals, or in cyclical annual periods resulting in a vocabulary of roughly $365 \times 24 \times 60/10 = 52,560$ different codes. However, in public transportation networks queries such as *"Number of trips using the stop X on May 10 between 9:15 and 10:00"* may be not as useful as queries such as *"Number of trips using stop X on Sundays between 9:15 and 10:00"*. For this reason, CTR can adapt how the time component is encoded depending on the queries that the system must answer.

In Fig. 1, sequence I contains the time associated with each stop in a trip, and $Icode$ a possible encoding of times. In CTR we use a similar encoding to that in $Icode$, yet aligned to Ψ rather than to S.

	1	2	3	4	5	6	7	8	9	10	11	12	13	14	15	16	17	18	19	20	21	22	23	24	25	26	27	28
Ψ	1	8	9	13	12	17	25	10	11	14	15	16	18	2	3	26	27	28	22	6	4	5	7	23	24	19	20	21
Times	0	0	5	10	2	9	12	0	5	3	7	2	10	5	8	4	9	13	8	12	15	10	15	14	12	6	11	14
Bit1	0	0	0	1	0	1	1	0	0	0	0	0	0	1	0	1	0	1	1	1	1	1	1	1	1	0	1	1
Times	0	0	5	2	0	5	3	7	2	5	4	6	10	9	12	10	8	9	13	8	12	15	10	15	14	12	11	14
Bit2	0	0	1	0	1	0	1	0	1	0	1	1	1	0	0	1	0	0	0	1	0	1	1	0	1	1	1	1
Times	0	0	2	0	3	2	10	9	10	8	9	8	10	11	5	5	7	5	4	6	12	13	12	15	15	14	12	14
Bit3	0	0	1	0	1	1	1	1	0	1	0	0	0	1	1	0	0	1	0	0	1	0	0	0	1	1	1	0
Times	0	0	0	9	8	9	8	5	5	5	4	12	13	12	12	2	3	2	10	10	10	11	7	6	15	15	14	14
Bit4	0	0	0	1	0	1	0	1	1	1	1	0	0	1	0	0	0	1	0	0	0	0	1	1	0	1	1	0

Fig. 2. WM representation for the times associated with the trips in Fig. 1.

Those entries in $Icodes$ are given a fixed-length binary code and are represented with a balanced Wavelet Matrix (WM) [1]. That is, for any stop in a trip at the position i in Ψ, its timestamp t_i can be recovered by accessing the WM at position i. Recall [1] that a WM is a grid of $n \times m$ bits. In our case n is the number of entries in the CSA and $m = \log \sigma_t$ are the bits needed to represent the different σ_t codes for the time instants of interest.

Besides the typical $access(i)$, $rank_\alpha(i)$ and $select_\alpha(i)$, the WM provides a $count$ operation that CTR heavily relies on. $count(x1, x2, y1, y2)$ returns the number of occurrences of symbols between y1 and y2 in the range of positions $[x1, x2]$ from the encoded sequence in $O(m)$ time. While its implementation details can be found in [1], we include an example of how to solve $count(20, 28, 10, 15)$ over the sequence shown in Fig. 2. The algorithm starts from the upper level ($Bit1$) of the WM and iterates downwards, refining the searching range. In $Bit1$ we are only interested in positions from $[20, 28]$ that have a 1, because none of the symbols between 10 and 15 starts with a 0. Also, since $rank_0(Bit1, 28) = 12$, in $Bit2$ we will have to search in the positions between $12 + rank_1(Bit1, 20) = 21$ and $12 + rank_1(Bit1, 28) = 28$. Now, while the second bit for 10 and 11 is 0, it is 1 for the symbols between 12 and 15. Because of this, we need to perform both $rank_0$ and $rank_1$ on the limits of $[21, 28]$ in $Bit2$[1], and split the search in two subranges for $Bit3$: $[10, 11]$ using $rank_0$ and $[23, 28]$ using $rank_1$. As the second subrange may only contain symbols from 12 to 15 (11xx), further refinement is not needed. In the case of the range $[10, 11]$, it could contain symbols from 8 to 11, depending on their third bits, so we need to perform $rank_1$ over its limits in $Bit3$, which leads to $[21, 22]$ in $Bit4$. The number of 10 and 11 symbols is the size of this last range.

If we wanted to return the positions of the results in the original sequence, we could do that with a simple algorithm, using $select$ of bits over bitmaps, that iterates upwards from the level where each result is found until the first level where its position in the original sequence can be retrieved.

Summarizing, CTR takes the advantage of the WM to count and report the occurrences of a continuous range of values. The starting positions in the CSA belonging to the $ symbols have no time by themselves, but it is useful to answer some queries to store the starting time instants of the corresponding trip in these positions too.

The time intervals could be mapped to a variable-length code, instead of a fixed length codes, where the most frequent intervals would be represented by less bits and, therefore, requiring less levels in a Wavelet Tree. In the future we will explore this possibility.

4   Query Processing

We distinguish two types of queries to be answered by the CTR: spatial and spatio-temporal queries. We briefly sketch the algorithms to process these queries.

---

[1] $rank_1(Bit2, i) = i - rank_0(Bit2, i)$, and vice versa.

Spatial Queries. The following queries can be solved by only using the CSA that represents the spatial component of trips.

- *Number of trips starting at stop X.* Because Ψ was cyclically built in such a way that every $ symbol is followed by the first stop of its trip, this query is solved by performing the binary search of the pattern $X over the section of Ψ corresponding to $. The size of the resulting range gives the number of trips starting at X.
- *Number of trips ending at stop X.* In a similar way to the previous query, this one can be answered with a binary search for pattern $X$$ over the section of Ψ corresponding to stop X.
- *Number of trips starting at X and ending at Y.* Combining both ideas from above, this query is solved directly by searching for the YX pattern.
- *Number of trips using stop X.* Instead of performing a binary search over Ψ, we operate on bitmap D. Assuming that X is at position p in the vocabulary V of CTR, its total frequency is obtained by $occs_X \leftarrow select_1(D, p+1) - select_1(D, p)$. If p is the last entry in V, we set $occs_X \leftarrow n + 1 - select_1(D, p)$.
- *Top-k most used stops.* We provide two possible solutions for these queries: sequential and binary-partition approaches.
  - To return the k most used stops using a sequential approach, we can apply $select_1$ operation sequentially for every stop from 1 to δ, returning the k stops with highest frequency. We use a min-heap that is initialized with the first k stops, and for every stop s from $k+1$ to δ, we compare its frequency with the frequency of the minimum stop in the heap. In case the new one is higher, the root of the heap is replaced and moved down to comply with the heap ordering. At the end of the process, the heap will contain the top-k most used stops, which can be sorted with the heapsort algorithm if needed. Note that this approach always performs δ $select_1$ operations on D.
  - A binary-partition approach to solve queries about the top-k most used stops takes advantage of the skewed distribution of the stops that trips visit. Working over D and V, D is recursively split into segments of D after each iteration. Each partition must, if possible, leave the same number of different stops in each side of the partition. The segments created after the partitioning step are pushed into a priority queue Q, storing the initial and the final positions of the segment in D, and also the initial and final corresponding entries in V. The priority of each segment in Q is directly its size. The priority queue Q is initialized with a segment covering the whole D (without its initial range of δ $ symbols). When a segment extracted from the queue Q represents the instance of only one stop, that stop is returned as a result of the top-k algorithm. The algorithm stops when the first k stops are found.

    For example, when searching for the top-1 most used stops in the running example, Q is initialized with the segment $[8, 28]$, corresponding to stops from 1 to 10 (positions from 2 to 11 in V). Note that the entries of D from 1 to 7 and $V[1]$ represent the $ symbol. These are not stops and must be skipped. Then $[8, 28]$ is split producing the segments $[8, 20]$ for stops 1 to 5 and $[21, 28]$ for stops 6 to 10. After three more iterations, we extract the segment $[14, 18]$ for the single stop 3, concluding that the top-1 most used stop is 3 with a frequency equal to 5.

Spatio-Temporal Queries. These queries combine both the CSA and WM. The idea is to restrict spatial queries to a time interval $[t_1, t_2]$. An example of this type of query is to return the *number of trips starting at stop X between t_1 and t_2*, which we solve by relying on the *count* operation of the WM. The following are the spatio-temporal queries solved by the CTR:

- *Number of trips starting at stop X during the time interval $[t_1, t_2]$.* Remember that in the WM we also have timestamps associated with the area of $-symbols in Ψ; each $ has associated the time of the first stop of its trip and, therefore, we can use the WM in that area of Ψ. Using the range in Ψ obtained by searching the X pattern, as done in a regular spatial query, a *count* operation is performed over these positions in the WM searching for the limits of the interval. That is, we count the number of entries in the obtained range that have a timestamp in the WM inside $[t_1, t_2]$.
- *Number of trips ending at stop X during the time interval $[t_1, t_2]$.* As before, we use a *count* operation in the WM, restricted to the range in Ψ that corresponds to the pattern X found in the spatial query.
- *Number of trips using stop X during the time interval $[t_1, t_2]$.* As in the spatial query, the range in Ψ is obtained with two $select_1$ on D. Then, a *count* operation is done over the WM to find the occurrences inside the time interval $[t_1, t_2]$.
- *Number of trips starting at X and ending at Y occurring during time interval $[t_1, t_2]$.* We consider two different semantics. A query with *strong semantics* will obtain trips that start and end inside $[t_1, t_2]$. Whereas, a query with *weak semantics* will obtain trips whose time intervals overlap $[t_1, t_2]$ and, therefore, they could actually start before t_1 or end after t_2.

  We can binary search Ψ for the pattern YX, hence obtaining the corresponding continuous range of positions in the section of Ψ devoted to Y. We know that the range for YX in Ψ has pointers to the section $ in Ψ. But, note that taking into account the considerations in the sorting of trips when building the CSA, this section XY is a continuous range of the same size than the range YX, and it also preserves the same order of the trips.

  Note that, the range YX of Ψ has associated the final time of each trip in the WM, whereas the range XY has associated the timestamps of the starting time of each trip in increasing order (due to how we sorted the trips). Therefore, we can use these ranges, respectively, to check time constraints related to the ending stop (Y) and to the starting stop (X) of each trip.

  - *Strong semantics.* Since time instants within the range XY are in increasing order, we can use the WM to obtain a continuous subrange (inside XY) of trips starting during the interval $[t_1, t_2]$. That subrange has a matching subrange inside of the range $Y\$X$ corresponding to the final stop of those trips (in the same order). We can again use the WM to count the number of those trips with valid ending times. That is, we can perform a *count* operation in the WM over the subrange of $Y\$X$ corresponding to the subrange of $\$XY$ with valid starting times.

- *Weak semantics.* In this case we need to consider all the trips in the range $\$XY$ starting within $[t_1, t_2]$, as well as the ones starting before t_1 but ending after t_1.

5 Experimental Evaluation

In this section we provide experimental results to show how CTR handles a large collection of trips. We discuss both the space requirements of our representation and also show its performance at query time. Although due to legal issues we could not provide experiments over real trips gathered from transport companies, we managed to use real data of the Madrid's public transportation network[2] (in the GTFS[3] format) to generate two datasets of synthetic trips:

- **Subway trips.** This combines the subway network with the Spanish commuter rail system called "Cercanías". In total, there are 313 different stations organized in 23 lines. They are open to the public from 6:00 AM to 2:00 AM, thus trips were always generated within 20 h a day.
- **Bus trips.** It uses 4648 bus stops, organized in 206 lines. Some of these lines are from special night services, so we generated trips using 24 h a day.

Trip generation process choses a starting stop and an ending stop, and uses the network description to generate every stop that the trip must traverse. We generated 50 million trips in both datasets, whose lengths vary from 2 to 31 stops following a binomial distribution with a mean length of 11.81 stops. Based on the GTFS data, we also generated realistic timestamps along each stop, and built the WM-based time index in CTR discretizing these timestamps into 5-min intervals. We distinguished four kinds of days in a week: regular working days, Fridays/holiday eves, Saturdays, and Sundays/holidays; and two kinds of weeks for high and low season representations. In total, a time interval may belong to eight types of day.

Below, we show the space/time tradeoff for both datasets obtained by three settings of CTR. We tune its Ψ sample rate parameter to values 16, 64, and 256, respectively. All tests were run on a machine with an Intel(R) Core(TM) i5-4440@3.1 GHz CPU, and 8 GB DDR3 RAM. The operating system was Ubuntu 15.04 and the compiler gcc 4.9.2 (options -O3).

We compare the space usage of the stops representation in the CTR with the space required by two baseline compressors: *gzip* and *bzip2*. To measure the

---

[2] Data from the EMT corporation https://www.emtmadrid.es/movilidad20/googlet. html.

[3] GTFS is a well-known specification for representing an urban transportation network. See https://developers.google.com/transit/gtfs/reference?hl=en.

compression, we assume, as a reference, a plain representation that uses the least amount of bits needed to represent every stop with a fixed width[4]. The sizes of these plain representations are 687.28 MiB for the subway trips dataset, and 992.59 MiB for the bus trips. Note that we ignore the space needed for the representation of time intervals, as WM does not offer any compression by itself, and needs 866.27 and 944.88 MiB for subway and bus trips, respectively.

Results regarding space usage are given in Table 1. Note that an iCSA built on English text [3] typically reached the compression of *gzip* (around 35 % in compression ratio). As expected, the high compressibility of our sorted dataset of trips permits CTR to improve those numbers with compression ratios under 30 % in the most sparse setup, much better than *gzip*, and even than *bzip2*. Yet, CTR offers also indexing features that allow us to perform efficient searches.

To provide a rough comparison with a database solution similar to *NET-TRA* [9] we included in a table a row containing each trip ID (represented with 4 bytes), stop ID (represented with 2 bytes), and time interval (represented with 2 bytes instead of a full `datetime`). The size of the whole table was around 4505 MiB, without taking any indexes into account. Therefore, such representation would use at least more than twice the space of CTR while it could not efficiently support the queries discussed in this paper.

Table 1. Comparison on space usage for stops. Space in MiB.

Dataset	Ψ_{16}	Ψ_{64}	Ψ_{256}	gzip	bzip2
Subway	467.07	249.14	193.10	401.72	238.43
	(67.96 %)	(36.25 %)	(28.10 %)	(58.45 %)	(34.69 %)
Bus	499.84	283.12	227.42	957.03	389.74
	(50.36 %)	(28.52 %)	(22.91 %)	(96.42 %)	(39.26 %)

To see the query performance of CTR, we generated 10, 000 random queries of each type, and measured the average time required to solve them.

Table 2 shows the results of spatial queries. Almost any query can be solved in the order of ten μsecs and the heaviest Top-k within *m*secs per query in our experiments. As expected, the query *"ends at X"* performs slightly faster than *"starts at X"*, as the region in Ψ for any stop X is smaller than the region of $, thus needing more time to search a pattern inside the later. It is also expected that the spatial *"uses X"* performs much faster than any other query as it does not operate over Ψ and its samples, using instead the $select_1$ operator over D. For the same reasons, both spatial Top-k algorithms are also independent from the Ψ sample rate parameter. However, it is interesting to point out that even when the binary partitioning algorithm is much faster for small values of k, its sequential counterpart overcomes it for large values of k. This is a reasonable phenomena considering that for large values of k, the number of $select_1$ operations that the

---

[4] 9 bits/stop for subway trips, 13 bits/stop for bus trips.

binary partitioning algorithm needs to perform tends to be the same as in the sequential algorithm, but with the additional cost of maintaining a larger and more complex structure (a priority queue versus a binary heap).

Table 2. Time performance for spatial queries (in $\mu secs$/query).

CTR	Starts at X	Ends at X	Starts at X ends at Y	Uses X	Sequential top 10	Binary top 10	Sequential top 1000	Binary top 1000
Subway Ψ_{16}	6.03	4.53	11.24					
Subway Ψ_{64}	8.22	4.61	16.68	0.3902	50.42	39.36	62.79	75.09
Subway Ψ_{256}	18.78	5.69	38.82					
Bus Ψ_{16}	7.51	6.27	9.24					
Bus Ψ_{64}	9.58	6.52	15.72	0.7944	761.14	588.01	1031.84	1514.07
Bus Ψ_{256}	22.77	11.35	41.31					

Table 3 shows the results of spatio-temporal queries. Looking at the differences between spatial queries and their spatio-temporal counterparts, it can be seen that computing a *count* query over the WM takes roughly around 3 μsec, so its time overhead is relatively small.

Table 3. Time performance for spatio-temporal queries (in $\mu secs$/query).

CTR	Starts at X	Ends at X	Starts at X ends at Y (strong)	Starts at X ends at Y (weak)	Uses X
Subway Ψ_{16}	8.34	7.44	22.42	18.95	
Subway Ψ_{64}	11.21	7.83	28.07	24.32	2.08
Subway Ψ_{256}	21.68	8.58	49.98	46.50	
Bus Ψ_{16}	10.41	9.50	12.25	12.12	
Bus Ψ_{64}	12.95	10.19	18.84	18.75	4.90
Bus Ψ_{256}	26.20	14.87	44.84	44.92	

6 Conclusions and Future Work

As better tracking mechanisms will be installed, the problem of storing and querying trips to support network analysis will gain interest for network management administrations and even end-user applications. For instance, with enough data of vehicle trips from a significant amount of drivers over the network formed by the streets of a city, it would be possible to infer traffic rules by examining turns that nobody takes, their usual driving speed across the network, and other useful information.

We showed that CTR is a powerful structure to represent user trips. Using compact data structures to represent trips over a transportation network allows us not only to keep a much larger amount of data in main memory (compression ratio is around 30 %), but also to efficiently perform spatial and spatio-temporal queries oriented to understand the real usage of the network.

We have presented CTR as a proof of concept development. It is flexible enough to allow new adaptations and functionality improvements we plan to do as future work, such as the analysis of line changes in switching stops (that would require storing the network topology) or providing compression for the time index. Also, future work considers providing new experiments over real data of trips.

References

1. Claude, F., Navarro, G., Ordóñez, A.: The wavelet matrix: an efficient wavelet tree for large alphabets. Inf. Syst. **47**, 15–32 (2015)
2. de Almeida, V.T., Güting, R.H.: Indexing the trajectories of moving objects in networks. GeoInformatica **9**(1), 33–60 (2005)
3. Fariña, A., Brisaboa, N., Navarro, G., Claude, F., Places, A., Rodríguez, E.: Word-based self-indexes for natural language text. ACM TOIS **30**(1), 1 (2012)
4. Frentzos, E.: Indexing objects moving on fixed networks. In: Hadzilacos, T., Manolopoulos, Y., Roddick, J., Theodoridis, Y. (eds.) SSTD 2003. LNCS, vol. 2750, pp. 289–305. Springer, Heidelberg (2003). doi:10.1007/978-3-540-45072-6_17
5. Funke, S., Schirrmeister, R., Skilevic, S., Storandt, S.: Compass-based navigation in street networks. In: Gensel, J., Tomko, M. (eds.) W2GIS 2015. LNCS, vol. 9080, pp. 71–88. Springer, Heidelberg (2015)
6. Grossi, R., Gupta, A., Vitter, J.S.: High-order entropy-compressed text indexes. In: Proceeding 14th SODA, pp. 841–850 (2003)
7. Guttman, A.: R-trees: a dynamic index structure for spatial searching. In: Proceeding SIGMOD, pp. 47–57 (1984)
8. Kellaris, G., Pelekis, N., Theodoridis, Y.: Map-matched trajectory compression. J. Syst. Softw. **86**(6), 1566–1579 (2013)
9. Krogh, B., Pelekis, N., Theodoridis, Y., Torp, K.: Path-based queries on trajectory data. In: Proceedings of the 22nd ACM SIGSPATIAL International Conference on Advances in Geographic Information Systems, pp. 341–350. ACM (2014)
10. Meratnia, N., Park, Y.-Y.: Spatiotemporal compression techniques for moving point objects. In: Bertino, E., Christodoulakis, S., Plexousakis, D., Christophides, V., Koubarakis, M., Böhm, K. (eds.) EDBT 2004. LNCS, vol. 2992, pp. 765–782. Springer, Heidelberg (2004)
11. Munizaga, M.A., Palma, C.: Estimation of a disaggregate multimodal public transport origin-destination matrix from passive smartcard data from santiago, chile. Transp. Res. Part C Emerg. Technol. **24**, 9–18 (2012)
12. Navarro, G., Mäkinen, V.: Compressed full-text indexes. ACM Comput. Surv. **39**(1), 2 (2007)
13. Pelekis, N., Theodoridis, Y.: Mobility Data Management and Exploration. Springer, New York (2014)
14. Pfoser, D., Jensen, C.S., Theodoridis, Y.: Novel approaches in query processing for moving object trajectories. In: Proceeding VLDB, pp. 395–406 (2000)
15. Popa, I.S., Zeitouni, K., Oria, V., Barth, D., Vial, S.: Indexing in-network trajectory flows. VLDB J. **20**(5), 643–669 (2011)
16. Potamias, M., Patroumpas, K., Sellis, T.: Sampling trajectory streams with spatiotemporal criteria. In: Proceeding 18th SSDBM, pp. 275–284 (2006)
17. Richter, K., Schmid, F., Laube, P.: Semantic trajectory compression: representing urban movement in a nutshell. J. Spat. Inf. Sci. **4**(1), 3–30 (2012)

18. Sadakane, K.: New text indexing functionalities of the compressed suffix arrays. J. Algorithms **48**(2), 294–313 (2003)
19. Tao, Y., Papadias, D.: MV3R-Tree: a spatio-temporal access method for timestamp and interval queries. In: Proceeding VLDB, pp. 431–440 (2001)

Longest Common Abelian Factors
and Large Alphabets

Golnaz Badkobeh[1]([✉]), Travis Gagie[2], Szymon Grabowski[3], Yuto Nakashima[4,5],
Simon J. Puglisi[2], and Shiho Sugimoto[4]

[1] Department of Computer Science, University of Warwick, Conventry, UK
g.badkobeh@warwick.ac.uk
[2] Department of Computer Science, Helsinki Institute for Information Technology,
University of Helsinki, Helsinki, Finland
{gagie,puglisi}@cs.helsinki.fi
[3] Institute of Applied Computer Science,
Lodz University of Technology, Łódź, Poland
sgrabow@kis.p.lodz.pl
[4] Department of Informatics, Kyushu University, Kyushu, Japan
{yuto.nakashima,shiho.sugimoto}@inf.kyushu-u.ac.jp
[5] Japan Society for the Promotion of Science, Tokyo, Japan

Abstract. Two strings X and Y are considered Abelian equal if the
letters of X can be permuted to obtain Y (and vice versa). Recently,
Alatabbi et al. (2015) considered the *longest common Abelian factor problem*
in which we are asked to find the length of the longest Abelian-equal
factor present in a given pair of strings. They provided an algorithm
that uses $O(\sigma n^2)$ time and $O(\sigma n)$ space, where n is the length of the
pair of strings and σ is the alphabet size. In this paper we describe an
algorithm that uses $O(n^2 \log^2 n \log^* n)$ time and $O(n \log^2 n)$ space, significantly improving Alatabbi et al.'s result unless the alphabet is small.
Our algorithm makes use of techniques for maintaining a dynamic set of
strings under split, join, and equality testing (Melhorn et al., Algorithmica 17(2), 1997).

1 Introduction

Two strings X and Y are considered to be *Abelian equal* if the letters of X can
be permuted to obtain Y (and vice versa). At the String Masters 2013 meeting,
Thierry Lecroq and Arnaud Lefebvre, posed the *longest common Abelian factor
problem* in which we are asked to find the length of the longest Abelian-equal
factor present in a given pair of strings.

The problem is a variant on the classic longest common factor (LCF) problem, the colorful history of which has been recently chronicled by Apostolico
et al. [3]. The LCF of two strings can be computed in time linear in the length
of the strings via suffix tree construction, and indeed the drive for a linear-time
LCF algorithm was the reason the suffix tree was unearthed when it was.

To our knowledge, the only work on the LCAF problem was presented very
recently by Alatabbi et al. [1]. They describe an algorithm that runs in $O(\sigma n^2)$

© Springer International Publishing AG 2016
S. Inenaga et al. (Eds.): SPIRE 2016, LNCS 9954, pp. 254–259, 2016.
DOI: 10.1007/978-3-319-46049-9_24

worst-case time, using $O(\sigma n)$ working space[1], where n is the length of the strings and $\sigma = |\Sigma|$ is the alphabet size.

Our main result in this paper is an algorithm that uses $O(n^2 \log^2 n \log^* n)$ time and $O(n \log^2 n)$ space, significantly improving Alatabbi et al.'s result unless the alphabet is $o(\log^2 n \log^* n)$. To obtain this result, we make use of techniques for maintaining a set of strings under split, join, and equality testing by Melhorn, Sundar, and Uhrig [8].

We also show how to reduce the space requirements of Alatabbi et al.'s algorithm from $O(\sigma n)$ to $O(n)$, without affecting their running time. Before getting to these new results, however, in Sect. 3 we highlight a link between the LCF and LCAF problems that provides an alternative path to Alatabbi et al.'s upperbound.

2   Preliminaries

Let $S = S[1..n]$ be a string of length n over an alphabet Σ of size $\sigma = |\Sigma|$. For $1 \leq i \leq j \leq n$, $S[i]$ denotes the ith symbol of S, and $S[i..j]$ the contiguous sequence of symbols (or *factor* or *substring*) $S[i]S[i+1]\ldots S[j]$. We will use the same notation for arrays. String $S[i..j]$, where $j - i + 1 = \ell$, will also be called an ℓ-gram from S. Throughout we assume that $\sigma = O(n)$ and $\Sigma = \{1, 2, \ldots, \sigma\}$. If this is not the case, we can first remap the alphabet for both input strings in $O(n \log n)$ time and using $O(n)$ extra space.

The *Parikh vector* for string S, denoted as $P(S)[1 \ldots \sigma]$, is defined as a vector (array) of size σ storing the number of occurrences of each alphabet symbol in S. Formally, $P(S)[c] = k$ iff $|\{i : S[i] = c\}| = k$, for any alphabet symbol c. For two strings S and T of equal length and over a common alphabet, we say that the Parikh vector $P(S)$ is (lexicographically) smaller than the Parikh vector $P(T)$, denoted as $P(S) < P(T)$, iff there exists an alphabet symbol c', $1 \leq c' \leq \sigma$, such that $P(S)[c] = P(T)[c]$ for all $c < c'$ and $P(S)[c'] > P(T)[c']$. The two Parikh vectors are equal, i.e., $P(S) = P(T)$, when $P(S)[c] = P(T)[c]$ for all symbols c.

3   LCAF via LCF

While Alatabbi et al.'s algorithm for computing the LCAF is simple, we note here that the same result can be immediately obtained by a reduction from the LCF problem.

Hui [6] showed that using a generalized suffix tree it is possible to find the LCF for a pair of strings of length n in $O(n)$ time. We use this algorithm n times, for each factor length ℓ, replacing each ℓ-length factor by its Parikh vector followed with a unique terminator (e.g., for the factors taken from A the subsequent terminators can be -1, -2, ..., while for the factors taken from B they can be $-n-1$, $-n-2$, ...). The terminators prevent matches longer than σ. If there exists an LCF of length exactly σ, it must correspond to a pair of

---

[1] We express space usage in words, throughout.

factors, one from A and one from B, of length ℓ. This takes $O(\sigma n)$ time for one value of ℓ, using $O(\sigma n)$ space, hence the total time, for all possible factor lengths, becomes $O(\sigma n^2)$ with $O(\sigma n)$ space (we build and discard the generalized suffix trees one at a time). In this way, we obtain the same time and space bounds as Alatabbi et al.'s solution.

4   Reducing Space Usage in Alatabbi et al.'s Algorithm

Recently, Kociumaka et al. [7] showed that for any tradeoff parameter $1 \leq \tau \leq n$, the LCF problem can be solved in $O(\tau)$ space and $O(n^2/\tau)$ time. Applying this to the LCAF problem, we obtain an algorithm using $O(\tau \sigma n^2)$ time and $O(\sigma n/\tau)$ space, for any $1 \leq \tau \leq \sigma n$.

The specifics of LCAF, however, allow for a better result. We consider each factor length ℓ separately. For a given ℓ, we sort all $n-\ell+1$ factors of A according to their Parikh vectors, using LSD radix sort. Each factor is represented as its start position in A. There are σ passes of the radix sort and the problem seems to be accessing the keys' "digits". However, before each pass of the radix sort we scan A and for each ℓ-sized window collect the count of the corresponding symbol in it. More precisely, just before the ith pass of the radix sort, in which the keys will be distributed according to $P(\cdot)[\sigma - i + 1]$, we compute and store $P(A[j \ldots j+\ell-1])[\sigma-i+1]$ for each factor $A[j \ldots j+\ell-1]$, using $O(n)$ time and $O(n)$ extra space. This allows us to access a digit in the radix sort in constant time. After the ith pass, the $P(\cdot)[\sigma - i + 1]$ statistics are discarded. In this way, sorting of the ℓ-length factors of A takes $O(\sigma n)$ time and requires $O(n)$ working space, including for the output.

We sort the factors of B in the same way. Additionally, for every σth evenly sampled ℓ-length factor of A and B, we store explicitly its Parikh vector using $O(\sigma)$ space. More precisely, we compute and store the Parikh vectors for the factors $A[1 \ldots \ell], A[\sigma + 1 \ldots \sigma + \ell], A[2\sigma + 1 \ldots 2\sigma + \ell], \ldots$, and similarly for $B[1 \ldots \ell], B[\sigma + 1 \ldots \sigma + \ell], B[2\sigma + 1 \ldots 2\sigma + \ell], \ldots$. Because we scan the strings from left to right and compute the successive Parikh vectors incrementally (first making a copy of the previous vector), this phase takes $O(n + (n/\sigma)\sigma) = O(n)$ time and $O(n)$ space.

The computed Parikh vectors serve to speed up factor comparisons during the last phase, which is to intersect the lists of factors from A and B (similar to a two-way merge). By using the sampled Parikh vectors that we have kept at regular intervals of A and B, each factor comparison takes $O(\sigma)$ time, and the intersection therefore takes $O(\sigma n)$ time.

The total cost of the described procedure, over all relevant factor lengths, becomes $O(\sigma n^2)$ and the required space is $O(n)$. This matches the time complexity of Alatabbi et al.'s solution, but reduced space usage by a factor of σ.

5   New Algorithm Based on Dynamic String Sets

To determine if A and B have a common factor of length ℓ, it suffices to be able to count the number of distinct Parikh vectors in a string. To see this, let $D_\ell(A)$,

$D_\ell(B)$, and $D_\ell(A\$B)$ denote the number of distinct Parikh vectors in strings A, B, and $A\$B$, respectively, where $\$$ is a sentinel symbol not occurring in either A or B.

Clearly, if

$$D_\ell(A\$B) < D_\ell(A) + D_\ell(B) \tag{1}$$

then at least one Parikh vector is shared by A and B, and so the LCAF will have length at least ℓ. This gives us a simple algorithm for computing the LCAF: compute $D_\ell(A)$, $D_\ell(B)$, and $D_\ell(A\$B)$ for every $\ell \in [1, n]$; the largest ℓ for which (1) holds is the length of the LCAF.

For the remainder of this section we focus on how to compute $D_\ell(X)$ for a given string X and window length ℓ. Our main tool is a data structure due to Melhorn et al. [8] for maintaining a (dynamic) set of strings under split, join, and equality testing. More precisely, their data structure supports four operations:

(i) make_sequence(s, a_1): creates the sequence s equal to the symbol a_1.
(ii) equal(s_1, s_2): returns true iff the strings s_1 and s_2 are equal.
(iii) join(s_1, s_2, s_3): creates the sequence $s_3 = s_1 s_2$ without destroying s_1 and s_2.
(iv) split(s_1, s_2, s_3, i), which creates two new sequences, $s_2 = a_1 \ldots a_i$ and $s_3 = a_{i+1} \ldots a_n$, without destroying s_1.

All presented operations work with string identifiers (ids). For example, join takes as its parameters the ids of strings s_1 and s_2, and returns the id of the newly created s_3; if some string already in the collection is equal to s_3, their ids will be equal. Importantly, the ids in the collection are positive integers and their maximum value after m operations is m^3.

Two solutions were presented by Melhorn et al.: one deterministic and one randomized—the latter with slightly better expected times for the operations (ii)–(iv)—we only make use here of the deterministic solution. The corresponding four time complexities, for the mth operation in the lifecycle of the structure, are: $O(1)$ for make_sequence, $O(\log m)$ for equal, and $O(\log n(\log m \log^* m + \log n))$ for join and split. After m operations the space used is $O(m \log n(\log^* m + \log n))$.

Our use of Melhorn et al.'s data structure is to store Parikh vectors, which we will simultaneously treat as both strings and arrays of integers. In the context of our application, all we need to be able to do is support increment and decrement of elements of these Parikh vectors. This can be simulated with split and join operations allowed by Melhorn et al.'s string collection data structure, as we now explain.

Consider the successive windows of length ℓ shifted over sequence X. For the first window, we calculate the corresponding Parikh vector in $O(\sigma + \ell) = O(n)$ time and add it to the string collection using a series of make_sequence and join operations. For any following window, starting at some valid position $i + 1$, the respective Parikh vector for $X[i+1 \ldots i+\ell]$ is calculated from the Parikh vector for $X[i \ldots i+\ell-1]$ by incrementing $P(X[i \ldots i+\ell-1])[X[i+\ell]]$ and decrementing $P(X[i \ldots i+\ell-1])[X[i]]$. For presentation clarity, let us implement this operation in two stages: first going from $P(X[i \ldots i+\ell-1])$ to $P(X[i \ldots i+\ell])$ and then going from $P(X[i \ldots i+\ell])$ to $P(X[i+1 \ldots i+\ell])$. Let $s_1 = P(X[i \ldots i+\ell-1])$. Using the dynamic collection of strings, the first transition between the Parikh vectors boils down to the following sequence of steps:

split$(s_1, s_2, s_3, X[i+\ell])$,
split$(s_2, s_4, s_5, X[i+\ell]-1)$,
make_sequence$(s_6, s_1[X[i+\ell]]+1)$,
$s_7 = $ join(s_4, s_6),
$s_8 = $ join(s_7, s_3).

The sequence labeled by s_8 corresponds to $P(X[i \ldots i+\ell])$, hence the first stage is accomplished. The second stage is analogous so we omit it here. This procedure uses a constant number of split, join, and make_sequence operations and so has time complexity $O(\log n(\log m \log^* m + \log n))$.

We observe now that in $O(n \log n(\log m \log^* m + \log n))$ time (where $m = \Theta(n)$) we can obtain the Parikh vectors for all ℓ-grams from X, and the corresponding string ids in passing.

Our goal is to know the number of different string ids produced (which corresponds to the number of distinct Parikh vectors for the ℓ-length factors of X). With this in mind, at each step i we record the id produced in element i of an array of $n-\ell$ elements. Recall that the maximum id is upper-bounded by $\Theta(n^3)$, and so each id can be stored in $O(\log n)$ bits or $O(1)$ words of space. This in turn means the search tree requires $O(n)$ space overall. We then sort this array and then scan it to determine the number of distinct elements.

6　Concluding Remarks

Finding the longest common Abelian factor is a recently posed problem, with a solution given in [1], achieving $O(\sigma n^2)$ worst-case time and needing $O(\sigma n)$ words of space. A significant weakness of that result is its space requirement, which may be unacceptable with a larger alphabet. We have improved this result in two ways.

The algorithm of Sect. 4 keeps the time complexity of Alatabbi et al.'s but reduces its space usage to $O(n)$. This is obtained by very simple means (the key component is LSD radix sort). Our second algorithm removes the dependency on σ and uses $O(n^2 \log^2 n \log^* n)$ time and $O(n \log^2 n)$ space. This algorithm is also simple conceptually, exploiting a reduction of the problem to counting distinct Parikh vectors present in a string for different factor lengths.

We believe better algorithms for the LCAF problem are possible, and the discovery of one is the main open problem we leave; to be specific: is $O(n^2)$ time and $O(n)$ space possible? One obvious line of attack is to use word-level parallelism (in the word-RAM model) for Parikh vector comparisons. The anticipated speed-up factor however is only about $w/\log(n/\sigma)$, where w is the machine word size. Perhaps more interesting would be to attempt to share computations for different factor lengths to obtain faster algorithms. Hardness results, possibly following the 3SUM reduction for other Abelian problems by Amir et al. [2], would also be welcome. In another direction, we may be able to use rounding techniques described by Cicalese et al. [4] to trade off accuracy for time. We are currently working on sampling techniques that we hope can be combined with rounding to yield even faster algorithms.

Finally, we note that achieving $O(n^2)$ time and $O(n)$ space is possible if we are happy with answers that are sometimes incorrect. More precisely, we can use Karp-Rabin hashing in place of Melhorn et al.'s data structure in our algorithm (which is effectively acting as a rolling hash function). This gives a Monte Carlo algorithm that correctly computes the LCAF with high probability; and can be made Las Vegas fairly easily by applying techniques from [5]. We defer the details to the full version of this paper.

References

1. Alatabbi, A., Iliopoulos, C.S., Langiu, A., Rahman, M.S.: Algorithms for longest common abelian factors, arXiv:1503.00049 (2015)
2. Amir, A., Chan, T.M., Lewenstein, M., Lewenstein, N.: On hardness of jumbled indexing. In: Esparza, J., Fraigniaud, P., Husfeldt, T., Koutsoupias, E. (eds.) ICALP 2014. LNCS, vol. 8572, pp. 114–125. Springer, Heidelberg (2014)
3. Apostolico, A., Crochemore, M., Farach-Colton, M., Galil, Z., Muthukrishnan, S.: 40 years of suffix trees. Commun. ACM 59(4), 66–73 (2016)
4. Cicalese, F., Gagie, T., Giaquinta, E., Laber, E.S., Lipták, Z., Rizzi, R., Tomescu, A.I.: Indexes for jumbled pattern matching in strings, trees and graphs. In: Kurland, O., Lewenstein, M., Porat, E. (eds.) SPIRE 2013. LNCS, vol. 8214, pp. 56–63. Springer, Heidelberg (2013)
5. Gagie, T., Gawrychowski, P., Kärkkäinen, J., Nekrich, Y., Puglisi, S.J.: LZ77-based self-indexing with faster pattern matching. In: Pardo, A., Viola, A. (eds.) LATIN 2014. LNCS, vol. 8392, pp. 731–742. Springer, Heidelberg (2014)
6. Hui, L.C.K.: Color set size problem with applications to string matching. In: Apostolico, A., Crochemore, M., Galil, Z., Manber, U. (eds.) CPM 1992. LNCS, vol. 644, pp. 230–243. Springer, Heidelberg (1992). doi:10.1007/3-540-56024-6_19
7. Kociumaka, T., Starikovskaya, T., Vildhøj, H.W.: Sublinear space algorithms for the longest common substring problem. In: Schulz, A.S., Wagner, D. (eds.) ESA 2014. LNCS, vol. 8737, pp. 605–617. Springer, Heidelberg (2014)
8. Mehlhorn, K., Sundar, R., Uhrig, C.: Maintaining dynamic sequences under equality tests in polylogarithmic time. Algorithmica 17(2), 183–198 (1997)

Pattern Matching for Separable Permutations

Both Emerite Neou[1]([✉]), Romeo Rizzi[2], and Stéphane Vialette[1]

[1] Université Paris-Est, LIGM (UMR 8049), CNRS, UPEM, ESIEE Paris, ENPC,
77454 Marne-la-Vallée, France
{neou,vialette}@univ-mlv.fr
[2] Department of Computer Science, Università degli Studi di Verona, Verona, Italy
romeo.rizzi@univr.it

Abstract. Given a permutation π (called the *text*) of size n and another permutation σ (called the *pattern*) of size k, the **NP**-complete permutation pattern matching problem asks whether σ occurs in π as an order-isomorphic subsequence. In this paper, we focus on separable permutations (those permutations that avoid both 2413 and 3142, or, equivalently, that admit a *separating tree*). The main contributions presented in this paper are as follows.

- We simplify the algorithm of Ibarra (*Finding pattern matchings for permutations*, Information Processing Letters 61 (1997), no. 6) to detect an occurrence of a separable permutation in a permutation and show how to reduce the space complexity from $O(n^3 k)$ to $O(n^3 \log k)$.
- In case both the text and the pattern are separable permutations, we give a more practicable $O(n^2 k)$ time and $O(nk)$ space algorithm. Furthermore, we show how to use this approach to decide in $O(nk^3 \ell^2)$ time whether a separable permutation of size n is a disjoint union of two given permutations of size k and ℓ.
- Given a permutation of size n and a separable permutation of size k, we propose an $O(n^6 k)$ time and $O(n^4 \log k)$ space algorithm to compute the largest common separable permutation that occurs in the two input permutations. This improves upon the existing $O(n^8)$ time algorithm by Rossin and Bouvel (*The longest common pattern problem for two permutations*, Pure Mathematics and Applications 17 (2006)).
- Finally, we give a $O(n^6 k)$ time and space algorithm to detect an occurrence of a bivincular separable permutation in a permutation. (Bivincular patterns generalize classical permutations by requiring that positions and values involved in an occurrence may be forced to be adjacent).

1 Introduction

A permutation σ is said to occurs in π, in symbols $\sigma \preceq \pi$, if there exists a subsequence of entries of π that has the same relative order as σ, and in this case σ is said to be a *occurs* in π. Otherwise, π is said to *avoid* the permutation σ. For example, the permutation $\pi = 391867452$ contains the pattern $\sigma = 51342$ as can

S. Inenaga et al. (Eds.): SPIRE 2016, LNCS 9954, pp. 260–272, 2016.
DOI: 10.1007/978-3-319-46049-9_25

be seen in the highlighted subsequence of $\pi = 3\mathbf{9}\mathbf{1}\mathbf{8}67452$ (or $\pi = 391\mathbf{8}\mathbf{6}\mathbf{7}452$ or $\pi = 39186745\mathbf{2}$ or $\pi = \mathbf{3}\mathbf{9}\mathbf{1}\mathbf{8}67452$). However, since the permutation $\pi = 391867452$ contains no increasing subsequence of length four, π avoids 1234. During the last decade, the study of the permutation pattern matching has become a very active area of research and an annual conference (PERMUTATION PATTERN) is devoted to the subject of pattern in permutation and a database[1] of permutation pattern avoidance is maintained by Bridget Tenner.

We consider here the so-called *permutation pattern matching* problem (also sometimes referred to as the *pattern involvement* or *pattern containment* problem): Given two permutations σ and π, this problem is to decide whether $\sigma \preceq \pi$ (the problem is ascribed to Wilf in [5]). The permutation pattern matching problem is **NP**-hard [5]. It is, however, polynomial time solvable by brute-force enumeration if σ has bounded size. Improvements to this algorithm were presented in [1,2], the latter describing a nice $O(|\pi|^{0.47k+o(|\sigma|)})$ time algorithm. Bruner and Lackner [8] gave a fixed-parameter algorithm solving the permutation pattern matching problem with an exponential worst-case runtime of $O(1.79^{\mathrm{run}(\pi)})$, where $\mathrm{run}(\pi)$ denotes the number of alternating runs of π. (This is an improvement upon the $O(k\binom{n}{k})$ runtime required by brute-force search without imposing restrictions on σ and π.) Of particular importance, it has been proved in [11] that the permutation pattern matching problem is fixed-parameter tractable for parameter k.

A few particular cases of the permutation pattern matching problem have been attacked successfully. Of particular interest in our context, the permutation pattern matching problem is solvable in polynomial time for separable patterns. Separable permutations are those permutations where the patterns 2413 and 3142 do not occur. The permutation pattern matching problem is solvable in $O(kn^4)$ time and $O(kn^3)$ space for separable patterns [12] (see also [5]), where k is the size of the pattern and n is the size of the text. Notice that there are numerous characterizations of separable permutations. To mention just a few examples, they are the permutations whose permutation graphs are cographs (*i.e.* P_4-free graphs); equivalently, a separable permutation is a permutation that can be obtained from the trivial permutation 1 by *direct sums* and *skew sums* [16]. While the term separable permutation dates only to the work of Bose, Buss, and Lubiw [5], these permutations first arose in Avis and Newborns work on pop stacks [3].

There exist many generalisations of patterns that are worth considering in the context of algorithmic issues in pattern involvement (see [14] for an up-to-date survey). *Vincular* patterns, also called *generalized* patterns, resemble (classical) patterns, with the constraint that some of the letters in an occurrence must be consecutive. Of particular importance in our context, Bruner and Lackner [8] proved that deciding whether a vincular pattern σ of size k occurs in a permutation π of size n is $W[1]$-complete for parameter k. *Bivincular* patterns generalize classical patterns even further than vincular patterns. Indeed, in bivincular patterns, not only positions but also values of elements involved in a occurrence may be forced to be adjacent

---

[1] http://math.depaul.edu/bridget/patterns.html.

The paper is organized as follows. Section 2 is devoted to presenting the needed material. In Sect. 3, we revisit the polynomial-time algorithm of Ibarra [12] and we propose a simpler dynamic programming approach, and in Sect. 4 we focus on the case where both the pattern and the target permutation are separable. Section 5 is devoted to presenting related problems. Subsection 5.1 is concerned with presenting an algorithm to test whether a separable permutation is the disjoint union of two given (necessarily separable) permutations. In Subsect. 5.2, we revisit the classical problem of computing a longest common separable pattern as introduced by Rossin and Bouvel [15] and propose a slightly faster - yet still not practicable - algorithm. Finally, in Subsect. 5.3, we prove that the pattern matching problem is polynomial-time solvable for vincular separable patterns. To the best of our knowledge, this is the first time the pattern matching problem is proved to be tractable for a generalization of separable patterns. Due to space constraints, most proofs are omitted and deferred to the full version of this paper.

2   Definitions

A *permutation* of size n is a one-to-one function from an n-element set to itself. We write permutations as words $\pi = \pi_1 \pi_2 \ldots \pi_n$, whose letters are distinct and usually consist of the integers $1, 2, \ldots, n$. We designate its i-th element by $\pi[i]$, and for any $i, j \in [n]$ with $i \leq j$, we let $\pi[i:j]$ stand for the sequence $\pi_i \pi_{i+1} \ldots \pi_j$. We let S_n denote the set of all permutations of size n. We shall also represent a permutation π by its *plot* consisting in the set of points at coordinates $(i, \pi[i])$ drawn in the plane. According to this representation, we say that an element $\pi[i]$ is on the *left* (resp. *right*) of another element $\pi[j]$ if $i < j$ (resp. $i > j$). Furthermore, we say that an element $\pi[i]$ is *above* (resp. *below*) another element $\pi[j]$ if $\pi[j] < \pi[i]$ (resp. $\pi[i] < \pi[j]$).

The *reduced form* of a permutation π on a set $\{j_1, j_2, \ldots, j_k\}$ where $j_1 < j_2 < \ldots < j_k$, is the permutation π' obtained by renaming the letters of π so that j_i is renamed i for all $1 \leq i \leq k$. We let $\mathrm{red}(\pi)$ denote the reduced form of π. For example $\mathrm{red}(5826) = 2413$. If $\mathrm{red}(u) = \mathrm{red}(w)$, we say that u and w are *order-isomorphic*.

A permutation $\sigma \in S_k$ is said to *occur* within a permutation $\pi \in S_n$, if there is some k-tuple $1 \leq i_1 \leq i_2 \leq \ldots \leq i_k \leq n$ such that $\mathrm{red}(\pi_{i_1} \pi_{i_2} \ldots \pi_{i_k}) = \sigma$ (*i.e.*, π has a subsequence of size k that is order-isomorphic to σ). The subsequence $\pi_{i_1} \pi_{i_2} \ldots \pi_{i_k}$ is called an *occurrence* of σ in π. If σ does not occur in π, then π is said to *avoid* σ.

For two permutations π_1 of size n_1 and π_2 of size n_2, the *direct sum* of π_1 and π_2 is defined by $\pi_1 \oplus \pi_2 = \pi_1[1]\pi_1[2] \ldots \pi_1[n_1](\pi_2[1] + n_1)(\pi_2[2] + n_1) \ldots (\pi_2[n_2] + n_1)$ [16]. The direct sum operation reduces to putting the elements of π_2 right above the elements of π_1. See Fig. 1 for an example of a direct sum. Similarly, we define the *skew sum* of π_1 and π_2 by $\pi_1 \ominus \pi_2 = (\pi_1[1] + n_2)(\pi_1[2] + n_2) \ldots (\pi_1[n_1] + n_2)\pi_2[1]\pi_2[2] \ldots \pi_2[n_2]$ [16]. The skew sum operation reduces to putting the elements of π_1 left above the elements of π_2. See Fig. 2 for an example of a skew sum.

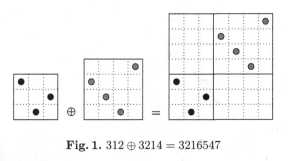

Fig. 1. $312 \oplus 3214 = 3216547$

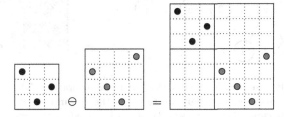

Fig. 2. $312 \ominus 3214 = 7563214$

Separable permutations may be characterized by the forbidden permutation patterns 2413 and 3142. Equivalently, Bose, Buss, and Lubiw [5] define a separable permutation to be a permutation that has a *separating tree* (note that there may be more than one tree for a given permutation): a rooted binary tree in which the elements of the permutation appear (in permutation order) at the leaves of the tree, and in which the descendants of each tree node form a contiguous subset of these elements. Each interior node of the tree is either a *positive* node in which all descendants of the left child are smaller than all descendants of the right node, or a *negative* node in which all descendants of the left node are greater than all descendants of the right node. See Fig. 3 for an illustration. Let $\sigma \in S_k$ be a separable permutation, and T_σ be the corresponding separating tree. For every node v of T_σ, we let $\sigma(v)$ stand for the sequence of elements of σ stored at the leaves of the subtree rooted at v. Also a permutation is said to be separable if and only if it is the permutation with unique element or it can be written as a direct sum or skew sum of two smaller separable permutations. The tree representation and the decomposition with direct sum or skew sum are strongly related: if $\sigma = \sigma_1 \oplus \sigma_2$ (resp. $\sigma = \sigma_1 \ominus \sigma_2$) then there exists a separating tree with a positive (resp. negative) root and the left child of the root is the separating tree of σ_1 and the right child is the separating tree of σ_2.

An occurrence of a *bivincular permutation pattern* $\widetilde{\sigma} = (\sigma, X, Y)$ in π is an occurrence of σ in π such that if $(e_1, e_2) \in X$ then the elements matching e_1 and e_2 must be consecutive in index and if $(e_1, e_2) \in Y$ then the elements matching e_1 and e_2 must be consecutive in value. Moreover if $(e_1, e_2) \in X$ (resp. $(e_1, e_2) \in Y$) and $e_1 \notin \sigma$ then e_2 must matched to the leftmost (resp. bottommost) element of π and if $(e_1, e_2) \in X$ (resp. $(e_1, e_2) \in Y$) and $e_2 \notin \sigma$ then e_1 must matched to the

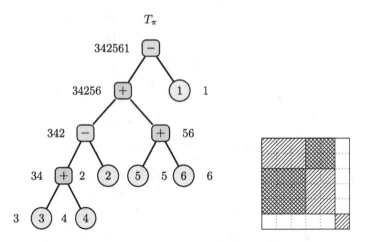

Fig. 3. On the left, a separating tree T_π for the permutation $\pi = 342561$ together with the corresponding $\sigma(v)$ sequences and on the right the decomposition of the root of this tree and of its left child: $342561 = \text{red}(34256) \ominus 1 = (\text{red}(342) \oplus \text{red}(56)) \ominus 1$. (Color figure online)

rightmost (resp. topmost) element of π. Note we only consider "realisable" bivincular permutation pattern which means that $\widetilde{\sigma}$ occurs in σ (by adding elements in X or Y this may not be the case such as $(0,2) \in Y$) and "clean" bivincular permutation pattern which means that there is not redundancy in the elements of X and Y. For example, given $\widetilde{\sigma} = (2143, \{(0,2),(4,3)\}, \{(1,4),(4,5)\})$ **3217845** is an occurrence of $\widetilde{\sigma}$ but **3217845** is not 2 is not the leftmost element. Note that this definition differs from Definition 1.4.1 in [14], but it is more suited for our algorithm.

3   Improved Algorithm to Detect a Separable Pattern

Let $\pi \in S_n$ and $\sigma \in S_k$, and assume that σ is a separable permutation. Ibarra [12] gave a nice $O(kn^4)$ time and $O(kn^3)$ space algorithm to detect an occurrence of σ in π. We revisit the approach of Ibarra and propose a simpler algorithm.

Since σ is a separable permutation, we can assume that we are given in addition a separating tree T_σ for σ (constructing a separating tree of a separable permutation is linear time and space [5]). Let S be a sequence of elements in $[n]$ with no repetitions. A *occurrence* of a node v of T_σ into S is an occurrence of $\text{red}(\sigma(v))$ into $\text{red}(S)$. The *bottom point* $\downarrow(s)$ of an occurrence s of $\sigma(v)$ into S is the minimum value of the sequence s. Similarly, the *upmost point* $\uparrow(s)$ is the maximum value of s. In the following, since all numbers in $[n]$ are positive, we adopt the convention that the maximum value occurring in an empty subset of $[n]$ is 0.

We consider the following family of subproblems that has been first introduced by Ibarra [12]: For every node v of T_σ, every two $i, j \in [n]$ with $i \le j$, and every upper bound ub $\in [n]$, we have the subproblem $\hat{\downarrow}_{v,i,j}[\text{ub}]$, where the semantic is the following:

$$\hat{\downarrow}_{v,i,j}[\text{ub}] \overset{\Delta}{=} \max\{\downarrow(s) : s \text{ is an occurrence of } \sigma(v) \text{ into } \pi[i:j] \text{ with } \uparrow(s) \leq \text{ub}\}.$$

We first observe that this family of problems is already closed under induction (we do not need to introduce the family H as in [12]). These subproblems can be solved by the following equations:

- **Base:** If v is a leaf of T_σ then

$$\hat{\downarrow}_{v,i,j}[\text{ub}] := \max\{\pi[\iota] : \pi[\iota] \leq \text{ub}, i \leq \iota \leq j\}.$$

- **Step:** Let v_L and v_R be the left and right children of v.
  - If v is a positive node of T_σ (*i.e.*, all elements in the interval associated to v_R are larger than all elements in the interval associated to v_L), then

  $$\hat{\downarrow}_{v,i,j}[\text{ub}] := \max\{\hat{\downarrow}_{v_L,i,\iota-1}[\hat{\downarrow}_{v_R,\iota,j}[\text{ub}]] : i < \iota \leq j\}.$$

  - If v is a negative node of T_σ (*i.e.*, all elements in the interval associated to v_R are smaller than all elements in the interval associated to v_L), then

  $$\hat{\downarrow}_{v,i,j}[\text{ub}] := \max\{\hat{\downarrow}_{v_R,\iota,j}[\hat{\downarrow}_{v_L,i,\iota-1}[\text{ub}]] : i < \iota \leq j\}.$$

These relations imply a $O(kn^4)$ time and $O(kn^3)$ space algorithm for detecting an occurrence of a separable permutation of size k in a permutation of size n, as obtained by Ibarra in [12], only simplified.

Proposition 1. *One can reduce the memory consumption of the algorithm above to $O(n^3 \log k)$.*

Proof. Observe first that for computing all the entries $\hat{\downarrow}_{v,\cdot,\cdot}[\cdot]$ for a certain node v with left and right children v_L and v_R, we only need the entries $\hat{\downarrow}_{v_L,\cdot,\cdot}[\cdot]$ and $\hat{\downarrow}_{v_R,\cdot,\cdot}[\cdot]$. The main idea for achieving the memory spearing is the following.

- All problems for a same node v are solved together and their solution is maintained in memory until the problems for the parent of v have also been solved. At that point the memory used for node v is released.
- We use a modified DFS traversal on T_σ: for every node v which has two children, we first process its largest child (in terms of the number of nodes in the subtree rooted at that child), then the other child, and finally v itself.

We claim that the above procedure yields a $O(n^3 \log k)$ space algorithm. We first expand our DFS algorithm to what is known as the White-Gray-Black DFS [9]. First, we color all vertices white. When we call DFS(u), we color u gray. Finally, when DFS(u) returns, we color u black. Thanks to this colour scheme, at each step of the modified DFS, we may partition T_σ into a white-gray subtree (all nodes are either white or gray) and a forest of maximal black subtrees (all nodes are black and the parent of the root - if it exists - is either white or gray). Our space complexity claim is reduced to prove that, at every step of the algorithm, the forest contains at most $O(\log k)$ maximal black subtrees. Let h_σ be the height of T_σ, and consider any partition of T_σ into a white-gray subtree and an non-empty forest T^b of maximal black subtrees. The following property easily follows from the (standard) DFS colour scheme.

Property 1. For every $1 \le i \le h_\sigma$, there exist at most two maximal black subtrees in T^b whose roots are at height i in T_σ. Furthermore, if there are two maximal black subtrees in T^b whose roots are at height i in T_σ (they must have the same parent), then T^b contains no maximal black subtree whose root is at height $j > i$ in T_σ.

According to Property 1 and aiming at maximising $|T^b|$, we may focus on the case where T^b contains one maximal black subtree whose root is at height i, $1 \le i < h_\sigma$, in T_σ (if T^b contains one maximal black subtree whose root is at height 0 in T_σ then $|T^b| = 1$), and T^b contains two maximal black subtrees whose roots are at height h_σ in T_σ (these two maximal black subtrees reduce to size-1 subtrees). The claimed space complexity for the dynamic programming algorithm (*i.e.*, $|T^b| = \log(k)$) now follows from the fact that we are using a modified DFS algorithm where we branch of the largest subtree first after having marked a vertex gray. Indeed, the maximal black subtree whose root is at height 1 in T_σ contains at least half of the nodes of T_σ, and the same argument applies for subsequent maximal black subtrees in the forest T^b.   □

4   Both π and σ Are Separable Permutations

When both π and σ are separable permutations we can strive for more efficient solutions since we can construct in linear time the two separating trees T_π and T_σ. It turns out, however, that the standard (*i.e.* binary) separating trees are not well-suited to handle this task. We use here the notion of *compact separating tree* (also known as *decomposition tree* [16]). Informally, in compact separating tree, we strive for every node to have as many children as possible (so that the compact separating tree of the identity permutation has only the root as its - positive - internal node). A simple linear time post-processing can be used to produce the decomposition tree out of the binary separating tree. We will adopt the convention that a compact separating tree of a separating tree T_π is denoted $\tilde{T}_\pi$. The compact tree can be understood with direct/skew sums as the largest decomposition in direct/skew sums: if $\pi = \pi_1 \oplus \ldots \oplus \pi_\ell$ then the (unique) compact separating tree of π is the tree with a positive root and with the compact separating tree of π_1 as first child, the compact separating tree of π_i as i^{th} child and the compact separating tree of π_ℓ as ℓ^{th} child. See Figs. 4 and 5 for examples. Note that when π is decomposed into direct (resp. skew) sums it forms a stair up (resp. down) of rectangles.

Now, recall that the *tree inclusion* problem for ordered and labeled trees is defined as follows: Given two ordered and labeled trees T and T', can T be

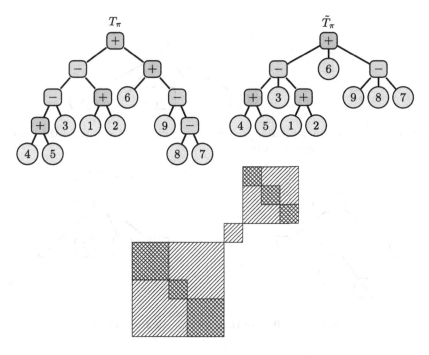

Fig. 4. A separating tree T_π for the permutation $\pi = 453126987$, the corresponding separating tree $\tilde{T}_\pi$ and the decomposition $453126987 = \text{red}(45312) \oplus \text{red}(6) \oplus \text{red}(987) = (\text{red}(45) \ominus \text{red}(3) \ominus \text{red}(12)) \oplus \text{red}(6) \oplus (\text{red}(9) \ominus \text{red}(8) \ominus \text{red}(7))$ (Color figure online)

obtained from T' by deleting nodes? (Deleting a node v entails removing all edges incident to v and, if v has a parent u, replacing the edge from u to v by edges from u to the children of v; see Fig. 6.) This problem has been recognized as an important query primitive in XML databases. The rationale for considering compact separating trees stems from the following property.

Property 2. Let π and σ be two separable permutations. We have $\sigma \preceq \pi$ if and only if the compact separating tree $\tilde{T}_\sigma$ is included into the compact separating tree $\tilde{T}_\pi$.

Kilpeläinen and Manilla [13] presented the first polynomial time algorithm using quadratic time and space for the tree inclusion problem. Since then, several improved results have been obtained for special cases when T and T' have a small number of leaves or small depth. However, in the worst case these algorithms still use quadratic time and space. The best algorithm is by Bille and Gørtz [4] who gave an $O(n_T)$ space and

$$O\left(\min\left\{\begin{array}{l} l_{T'}\, n_T \\ l_{T'}\, l_T \,\log\log n_T + n_T \\ \frac{n_T\, n_{T'}}{\log n_T} + n_T \log n_T \end{array}\right\}\right)$$

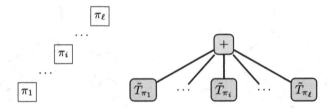

Fig. 5. On the left the permutation $\pi = \pi_1 \oplus \ldots \oplus \pi_i \oplus \ldots \oplus \pi_\ell$ and on the right its corresponding compact separating tree.

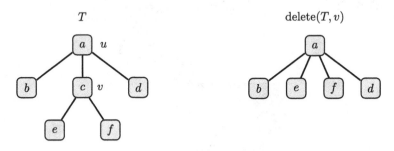

Fig. 6. The effect of removing a node from a tree.

time algorithm, where n_T (resp. $n_{T'}$) denotes the number of node of T (resp. T') and l_T (resp. $l_{T'}$) denotes the number of leaves of T (resp. T').

However, all efficient solutions developed so far for the tree inclusion problem result in very complicated and hard-to-implement algorithms. For example, the main idea in the efficient algorithm presented in [4] is to construct a data structure on T supporting a small number of procedures, called the set procedures, on subsets of nodes of T. We propose a dynamic programming based approach for solving this problem.

Proposition 2. *There exits an $O(n^2 k)$ time and $O(nk)$ space algorithm to find an occurrence of a separable pattern of size k in a separable permutation of size n.*

5 Related Problems

Some related problems (deciding the union of a separable permutation, finding a maximum size separable subpermutation and pattern matching issues for bivincular separable patterns) are gathered in this section. All algorithms rely on dynamic programming.

5.1 Deciding the Union of a Separable Permutations

This subsection is devoted to shuffling permutations. Given three permutations π, σ and τ, the problem is to decide whether π is the disjoint union of two patterns that are order-isomorphic to σ and τ, respectively. For example 937654812 is the disjoint union of two subsequences that are order-isomorphic to 2431 and 53241, as can be seen in the highlighted form **937654812**. This problem is of interest since it is strongly related to two others combinatorial problems that naturally arise in the context of pattern in permutations. The first one is to decide whether the permutation pattern matching problem for parameter $n - k$ is fixed-parameter tractable (FPT). (Recall that the permutation pattern matching problem for parameter k is fixed-parameter tractable [11].) The second one is to decide whether a permutation is a square: Given a permutation π, does there exists a permutation σ such that π is the disjoint union of two subsequences that are both order-isomorphic to σ? This problem has recently been proved to be **NP**-complete [10] for general permutations.

Proposition 3. *Given three separable permutations π of size n, σ of size k and τ of size ℓ, there exists an $O(nk^3\ell^2)$ time and $O(nk^2\ell^2)$ space algorithm to decide whether π is the disjoint union of two patterns that are order-isomorphic to σ and τ, respectively.*

Note that the complexity of the problem is still open if we do not restrict the input permutations to be separable [10].

5.2 Finding a Maximum Size Separable Subpermutation

The longest common pattern problem for permutations is, given a set of permutation, to find the largest permutation that occurs in each input permutation. The problem is intended to be the natural counterpart to the classical longest common subsequence problem. Rossin and Bouvel [15] gave an $O(n^8)$ time algorithm for computing the largest common separable permutation that occurs in two permutations of size (at most) n, one of these two permutations being separable. This problem was further generalised in [6] where it is shown that that the problem of computing the largest separable permutation that occurs in k permutations of size (at most) n is solvable in $O(n^{6k+1})$ time and $O(n^{4k+1})$ space. Notice that this later problem is **NP**-complete for unbounded k, even if all input permutations are actually separable. The following proposition improves upon the algorithm of Rossin and Bouvel [15].

Proposition 4. *Given a permutation of size n and a separable permutation of size k, one can compute in $O(n^6k)$ time and $O(n^4 \log k)$ space the largest common separable permutation that occurs in the two input permutations.*

5.3 Vincular and Bivincular Separable Patterns

We prove here that detecting a vincular or a bivincular separable pattern in a permutation is polynomial time solvable. Since a vincular pattern is a special case of bivincular pattern (when $Y = \{\emptyset\}$), we focus on bivincular patterns. Note that the algorithm of Sect. 3 cannot be used to find an occurrence of a bivincular pattern as we do not have any control on the positions and on the values of the matched elements.

Let $\widetilde{\sigma}$ be a separable bivincular pattern (this is a shortcut for σ being separable) of size k and π is a permutation of size n. We can represent bivincular patterns (as well as occurrences of bivincular patterns in permutations) by theirs plots. Such plot consists in the set of points at coordinates $(i, \sigma[i])$ drawn in the plane together with forbidden regions denoting adjacency constraints (similarly to what is done with mesh patterns, see [7]). A vertical forbidden region between two points denotes the fact that the occurrence of these two points must be consecutive in positions. Similarly, a horizontal forbidden region between two points denotes the fact that the occurrence of these two points must be consecutive in value. Now, given a permutation π and a pattern $\widetilde{\sigma}$, the bivincular pattern $\widetilde{\sigma}$ occurs in π if there exists a set of points in the plot of π that is order-isomorphic to σ and if the forbidden regions do not contain any point (see Fig. 7).

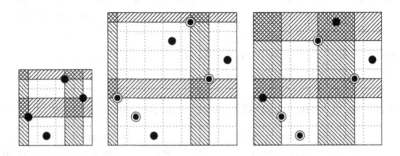

Fig. 7. From left to right, the bivincular pattern $\widetilde{\sigma}$ = $(2143, \{(0, 2), (4, 3)\}$, $\{(1, 4), (4, 5)\})$, an occurrence of $\widetilde{\sigma}$ in 3216745, an occurrence of σ in 3216745 but not an occurrence of $\widetilde{\sigma}$ in 3216745 because the point $(1, 3)$ and $(5, 7)$ are in the forbidden areas.

Proposition 5. *Given a permutation π of size n and a bivincular separable pattern $\widetilde{\sigma}$ of size k, there exists a $O(n^6 k)$ time and space algorithm to decide whether $\widetilde{\sigma}$ occurs in π.*

Before explaining the main idea of the algorithm, we need the notion of rectangle in a permutation. Given a permutation π, a rectangle R with bottom left corner (i, lb) and top right corner (j, ub) is the pattern $\pi[i : j]$ in which all entries greater than ub and smaller than lb are removed. We say that a rectangle R contains an occurrence of σ if and only if there exists a subsequence of Re which is order-isomorphic to σ. The following lemma is the key element for proving Proposition 5:

Lemma 1. *Let $\sigma = \sigma_L \oplus \sigma_R$ (resp. $\sigma = \sigma_L \ominus \sigma_R$). σ occurs in π if and only if there exist rectangles R_L and R_R in π, such that R_L is left below R_R (resp. R_L is left above R_R), R_L contains an occurrence of σ_L and R_R contains an occurrence of σ_R.*

Given a positive (resp. negative) node v of σ with left child v_L and right child v_R, and a rectangle R of π, deciding whether $\sigma(v)$ occurs in R reduced to deciding whether there exists a split of the rectangle R into two rectangles R_L and R_R such that R_L is left below R_R (resp. R_L is left above R_R), R_L contains an occurrence of $\sigma(v_L)$ and R_R contains an occurrence of $\sigma(v_R)$. This recursive algorithm solves the permutation pattern matching, but not for the bivincular case as we have no control over the values and the positions of the elements in the occurrence. Notice now that, given two rectangles that are consecutive horizontally, say $R_1 = ((*,*),(j,*)$ and $R_2 = ((j+1,*)(*,*))$, if the rightmost element of the occurrence in R_1 is on the right edge of R_1 and the leftmost element of the occurrence in R_2 is on the left edge of R_2 then those two elements are consecutive in position. In the same way, given two rectangles that are consecutive vertically, say $R_1 = ((*,*),(*,\mathrm{ub}))$ and $R_2 = ((*,\mathrm{ub}+1)(*,*))$, if the topmost element of the occurrence in R_1 is on the top edge of R_1 and the bottommost element of the occurrence in R_2 is on the bottom edge of R_2 then those two elements are consecutive in value.

The proposed algorithm implements the above idea to ensure that two elements are consecutive in position or in value in the sought occurrence: The algorithm splits the rectangle R into R_L and R_R such that R_L and R_R are always consecutive horizontally and vertically: if v is a positive node then R is splitted into $R_L = ((*,*),(j,\mathrm{ub}))$ and $R_R = ((j+1,\mathrm{ub}+1),(*,*))$, and otherwise (if v is a negative node) then R is splitted into $R_L = ((*,\mathrm{lb}),(j,*))$ and $R_R = ((j+1,*),(*,\mathrm{lb}-1))$.

Acknowledgments. We thank the anonymous reviewers whose comments and suggestions helped improve and clarify this manuscript.

References

1. Ahal, S., Rabinovich, Y.: On complexity of the subpattern problem. SIAM J. Discrete Math. **22**(2), 629–649 (2008)
2. Albert, M.H., Aldred, R.E.L., Atkinson, M.D., Holton, D.A.: Algorithms for pattern involvement in permutations. In: Eades, P., Takaoka, T. (eds.) ISAAC 2001. LNCS, vol. 2223, pp. 355–366. Springer, Heidelberg (2001)
3. Avis, D., Newborn, M.: On pop-stacks in series. Utilitas Math. **19**, 129–140 (1981)
4. Bille, P., Gørtz, I.L.: The tree inclusion problem: in linear space and faster. ACM Trans. Algorithms **7**(3), 38 (2011)
5. Bose, P., Buss, J.F., Lubiw, A.: Pattern matching for permutations. Inf. Process. Lett. **65**(5), 277–283 (1998)
6. Bouvel, M., Rossin, D., Vialette, S.: Longest common separable pattern among permutations. In: Ma, B., Zhang, K. (eds.) CPM 2007. LNCS, vol. 4580, pp. 316–327. Springer, Heidelberg (2007)
7. Brändén, P., Claesson, A.: Mesh patterns and the expansion of permutation statistics as sums of permutation patterns, ArXiv e-prints (2011)
8. Bruner, M.-L., Lackner, M.: A fast algorithm for permutation pattern matching based on alternating runs. In: Fomin, F.V., Kaski, P. (eds.) SWAT 2012. LNCS, vol. 7357, pp. 261–270. Springer, Heidelberg (2012)

9. Cormen, T.H., Leiserson, C.E., Rivest, R.L., Stein, C.: Introduction to Algorithms, 3rd edn. MIT Press, Cambridge (2009)
10. Giraudo, S., Vialette, S.: Unshuffling permutations. In: Kranakis, E., et al. (eds.) LATIN 2016. LNCS, vol. 9644, pp. 509–521. Springer, Heidelberg (2016). doi:10.1007/978-3-662-49529-2_38
11. Guillemot, S., Marx, D.: Finding small patterns in permutations in linear time. In: Chekuri, C. (ed.) Proceedings of the Twenty-Fifth Annual ACM-SIAM Symposium on Discrete Algorithms (SODA), SIAM 2014, Portland, Oregon, USA, pp. 82–101 (2014)
12. Ibarra, L.: Finding pattern matchings for permutations. Inf. Process. Lett. **61**(6), 293–295 (1997)
13. Kilpeläinen, P., Manilla, H.: Ordered and unordered tree inclusion. SIAM J. Comput. **24**(2), 340–356 (1995)
14. Kitaev, S.: Patterns in Permutations and Words. Springer, Heidelberg (2013)
15. Rossin, D., Bouvel, M.: The longest common pattern problem for two permutations. Pure Math. Appl. **17**, 55–69 (2006)
16. Vatter, V.: Permutation classes. In: Bóna, M. (ed.) Handbook of Enumerative Combinatorics, pp. 753–818. Chapman and Hall/CRC (2015)

Author Index

Printed in the United States
By Bookmasters